Analog Circuits and Signal Processing

The *Analog Circuits and Signal Processing* book series, formerly known as the *Kluwer International Series in Engineering and Computer Science*, is a high level academic and professional series publishing research on the design and applications of analog integrated circuits and signal processing circuits and systems. Typically per year we publish between 5-15 research monographs, professional books, handbooks, and edited volumes with worldwide distribution to engineers, researchers, educators, and libraries. The book series promotes and expedites the dissemination of new research results and tutorial views in the analog field. There is an exciting and large volume of research activity in the field worldwide. Researchers are striving to bridge the gap between classical analog work and recent advances in very large scale integration (VLSI) technologies with improved analog capabilities. Analog VLSI has been recognized as a major technology for future information processing. Analog work is showing signs of dramatic changes with emphasis on interdisciplinary research efforts combining device/circuit/technology issues. Consequently, new design concepts, strategies and design tools are being unveiled. Topics of interest include: Analog Interface Circuits and Systems; Data converters; Active-RC, switched-capacitor and continuous-time integrated filters; Mixed analog/digital VLSI; Simulation and modeling, mixed-mode simulation; Analog nonlinear and computational circuits and signal processing; Analog Artificial Neural Networks/Artificial Intelligence; Current-mode Signal Processing; Computer-Aided Design (CAD) tools; Analog Design in emerging technologies (Scalable CMOS, BiCMOS, GaAs, heterojunction and floating gate technologies, etc.); Analog Design for Test; Integrated sensors and actuators; Analog Design Automation/Knowledge-based Systems; Analog VLSI cell libraries; Analog product development; RF Front ends, Wireless communications and Microwave Circuits; Analog behavioral modeling, Analog HDL.

Rupam Das • Hadi Heidari
Editors

Wireless Power Technologies for Biomedical Devices

Springer

Editors
Rupam Das
Electronic Engineering
University of Exeter
Exeter, UK

Hadi Heidari
University of Glasgow
Glasgow, UK

ISSN 1872-082X ISSN 2197-1854 (electronic)
Analog Circuits and Signal Processing
ISBN 978-3-031-52836-1 ISBN 978-3-031-52834-7 (eBook)
https://doi.org/10.1007/978-3-031-52834-7

This Springer imprint is published by the registered company Springer Nature Switzerland AG
The registered company address is: Gewerbestrasse 11, 6330 Cham, Switzerland

Contents

Chapter 1
Wireless Power Transfer for Self-Reporting Cardiovascular Implantable Medical Devices

Jungang Zhang

1.1 Wireless Power Transfer Technology in cIMDs

While recent advancements in bio-energy harvesting have proven effective in powering low-power cIMDs, such as leadless pacemakers, the power output obtained from the human body is not capable of providing sufficient power cIMDs with higher energy demands. Table 1.1 provides a comprehensive overview of various characteristics and power requirements of different cardiovascular implantable medical devices (cIMDs) [1–14] For example, during defibrillation, an ICD requires the delivery of high-voltage electric shocks, providing a minimum pulse power of 5 W for 10 s. As a result, it demands a higher power capability compared to a pacemaker. ICDs commonly rely on lithium/silver vanadium oxide (Li/SVO) batteries for power, but they encounter a problem of delayed shock delivery caused by increasing resistance during discharging. Electromagnetic wireless power transmission (WPT) offers dynamic adjustment of the input power to match the load demands, providing reliable and sufficient support. WPT technology holds promise for achieving the goal of battery-free and wireless-powered cIMDs. This section explores the fundamental principles of different WPT links and their state-of-the-art applications in powering cIMDs, along with the safety standard concerning human body exposure to electromagnetic fields.

J. Zhang (✉)
University of Glasgow, Glasgow, UK
e-mail: Jungang.Zhang@glasgow.ac.uk

R. Das, H. Heidari (eds.), *Wireless Power Technologies for Biomedical Devices*, Analog Circuits and Signal Processing, https://doi.org/10.1007/978-3-031-52834-7_1

Table 1.1 Specifications of cIMDs and their power requirements

cIMDs	Pacemaker	ICD	VAD	Smart stent
Weight/size	2 g[a] [1] 20–50 g[b] [2]	60 g[c] [3] 130 g[d]	145 g[e] [4] 270 g[f] [5]	8–38 mm[g], 2.5–4 mm[h] [6]
Lifespan	7.2 years [7]	4.9 ± 1.6 years [8]	29.7 ± 14.9 days [9]	–
Power level	1–10 μW [10]	10–100 μW [11, 12]	5-25 W [13]	mW level [14]
Power load Requirement	High energy density Low-rate capability	High energy density High-rate capability (30–50 J)	High range of power High-rate capability	High energy density Medium-rate capability

[a]Leadless pacemaker, [b]transvenous pacemaker
[c]Subcutaneous ICD, [d]transvenous ICD
[e]50 cc blood displacement volume, [f]90 cc displacement volume
[g]in length, [h]in diameter

1.1.1 Antenna Design & Miniaturization

An antenna is a critical component in WPT used for transmitting and receiving signals. Designing an antenna is crucial and indispensable to ensure efficient power transmission. Here, our focus is on the compact antennas used in cIMDs, where we will review their types, geometries, and performance. In particular, planar antennas, such as the rectangular microstrip antenna [15], circular patch antenna [16, 17], and PIFA antenna [18], are preferred due to their low profile, easy fabrication, and ease of integration with circuits in WPT systems. In order to meet the size requirement of the IMDs, various miniaturization technologies can be used to reduce antenna size:

(a) Geometry configuration: The use of a folded configuration, such as meander-shape [19, 20], spiral-shape [21, 22], or fractal structure [23], can increase the current flowing path and effectively reduce the size of the antenna. By employing a meander or spiral configuration, the effective dimensions of an antenna can be increased, thereby providing a longer physical path for surface current flow. This design approach results in a lower resonant frequency compared to conventional antennas [24]. Additionally, the reduction factor β, which determines the size reduction of the antenna, is closely related to the number of meander elements per wavelength and the width of the rectangular loops. This relationship is defined by Eq. (1.1) [25, 26].

$$\beta = l/L \tag{1.1}$$

where, L is the length of the conventional monopole and l is the length of the meander antenna with the same resonant frequency.

(b) High dielectric substrate materials: The dielectric properties and frequencies play a significant role in the propagation of electromagnetic waves. By using a high dielectric constant substrate, the guided wavelength and the length of a patch can be reduced. This relationship can be described by the following equation [27]:

$$L = \frac{\lambda_g}{2} = \frac{C}{2f\sqrt{\varepsilon_{eff}}} \tag{1.2}$$

where, L represents the length of the patch antenna, λ_g is the guided wavelength, C denotes the speed of light, and ε_{eff} is the effective dielectric constant. However, the drawback of this approach is that it narrows the bandwidth of the antenna and may result in surface-wave losses [28].

(c) Introducing a superstrate also favours antenna miniaturization. In addition, it not only helps in reducing the size of the antenna but also enhances its gain and bandwidth. This is achieved by creating a larger in-phase electric field on the surface of the superstrate [29].

(d) Shorting pins that connect the patch and ground plane of a microstrip patch antenna can be used to reduce the antenna's size [24]. The shorting pin can be represented as an equivalent inductance and connected in parallel to the resonant LC circuit of the patch [30], the resonant frequency of the patch antenna can be lowered.

Table 1.2 presents a comprehensive overview of different antenna types, including their geometry modifications, antenna characteristics, and specific absorption rate (SAR) values in cardiac tissue [21, 31–36] .

In addition to miniaturized planar-type antennas, substantial research has focused on designing 3D conformal antennas that can be integrated into metal stents, eliminating the need for additional antenna volume in coronary stents. This approach allows for efficient use of space while enabling wireless power transfer in implantable devices. Figure 1.1 illustrates the two types of antennas applied for cardiovascular implants: planar antenna design and stent-based 3D conformal antenna design. For example, an innovative concept called the 'Stentenna' was proposed by K. Takahata et al. [15], where a stent is utilized as the antenna. The Stentenna design incorporates a capacitive pressure sensor to create an inductive-capacitor (LC) tank, enabling inductive coupling with an external antenna. This unique integration allows for a highly compact system that combines antennas, stents, and biosensors seamlessly. A further investigation of the 'Stentenna' [16], conducted by the same research team, involved the design of a zigzag pattern within a stainless-steel tube, eliminating bridge struts. When the tube expands, it forms a helical structure, resulting in a significant increase in stent inductance (multiplied by a factor of 3.2). To enhance conductivity, a copper layer was applied to the stent, achieving optimal resonance of the L-C circuit with a higher Q factor (35 at 150 MHz). To ensure biocompatibility and electrical insulation, the stent antenna was coated with a gold layer and a conformal Parylene-C layer. The introduction of

Table 1.2 Antenna types and geometry modifications applied in cardiac implants

Antenna type	Geometry modification	Operating frequency	Dimension (mm^3)	Antenna Parameters			SAR (W/Kg)	Bandwidth	Gain (dBi)	Test scenarios
				Substrate	Superstrate	Feed				
Microstrip antenna [31]	Meandered-shaped	2.46 GHz[a]	40 × 30 × 3.3	Rogers R03010 (ε_r=10.2)	Rogers R03010	Waveguide	–	1.637GHz	–	2/3 muscle equivalent phantom ($\varepsilon_r = 38.1$, $\sigma = 0.53$ S/m)
Microstrip patch [32]	Meander-line	2.4 GHz[b]	3 × 4 × 0.5	Rogers RT/duroid6010 (ε_r=10.2)	Rogers RT/duroid6010	Coaxial probe	31.04[ii] 270.28[i]	21.88%	−25.95	Heart phantom box & Minced pork muscle
Microstrip patch [21]	Spiral-shaped	403 MHz[c]	11.9 × 18.2 × 3.8	Rogers RT/duroid6010	Rogers RT/duroid6010	Coaxial probe	2.74[i]	–	–	2/3 human muscle box ($\sigma = 0.6463$S/m $\varepsilon_r = 42.807$)
Microstrip patch [33]	Spiral-shaped[α]	402 MHz[c], 2.45 GHz[d], 1.6 GHz[e]	17.15 × 6.5 × 0.377	Rogers RT/duroid 6010	Rogers RT/duroid 6010	Coaxial probe	588[i] 441[i] 305[i]	148 MHz 219 MHz 171 MHz	−30.5, −18.2 −22.6	Homogeneous body model
Circular patch [34]	Spiral-shaped	403 MHz[c]	Radius = 5 mm, Thickness = 0.89 mm	Rogers RO 3006 (ε_r=10.2)	Rogers RO 3210 (ε_r=6.4)	Coaxial probe	–	–	–	Homogeneous & Inhomogeneous heart muscle

Microstrip patch [35]	Cut off two diagonals	2.45GHz ISM band	40 × 40 × 0.635	Rogers RO 6010	–	Coaxial probe	40.6[ii]	37.74%	−15.87	Three-layered chest model & minced pork
Microstrip patch [36]	H-shaped cavity slot	1.4 GHz 2.45 GHz	8 × 8 × 0.635	Rogers RT/duroid6010	–	Coaxial probe	81.7[ii] 84.6[ii]	300 MHz 380 MHz	−27.68 −27.10	Realistic human model & pork mince

[a]ISM band (2.4-2.48 GHz); [b]ISM band; [c]MICS band (402–405 MHz), [d]ISM band (433.1–434.8 MHz and 2.4–2.48 GHz); [e]mid-field band (1520–1693 MHz)
[i]SAR_{1g}, [ii]SAR_{10g}
[α]With an open-end slot in the ground

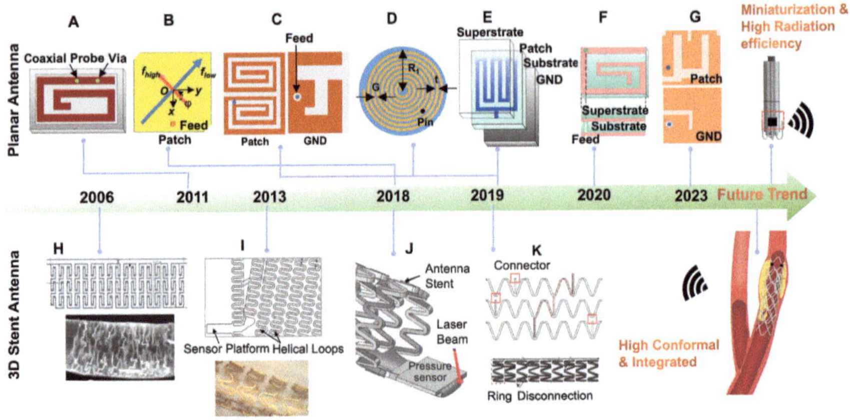

Fig. 1.1 The timeline and categories of planar and 3D antenna design for cardiovascular implants: (**a**) spiral-shaped microstrip patch antenna, (**b**) microstrip patch antenna with two cutting diagonals, (**c**) spiral-shaped antenna, (**d**) spiral-shaped circular patch antenna, (**e**) meander line patch antenna, and (**f**) spiral-shaped patch antenna, (**g**) H-shaped cavity slot patch antenna. (**h**) a stent antenna consists of series of crossband, and involute contours attached to stringers, (**i**) the stent antenna arranged in angle bands in a zigzag pattern, showing a helical shape in 3D, (**j**) helical structure stent antenna, (**k**) single-connector 3D stent antenna

the 'Stentenna' represents a significant innovation, transforming traditional passive metal stents into compact and intelligent stents. Notably, Chen et al. [37] conducted an in-vivo test in a pig by deploying a helical-structure stent using a balloon catheter. This innovative stent design incorporated both an inductive stent and a capacitive pressure sensor. The results showed an impressive wireless monitoring of blood pressure within the vessel, with a resolution of 12.4 mmHg, while also adhering to the PCI standard procedure. This remarkable achievement marks a significant milestone in the development of smart stents. In 2019, C.-H. Liu et al. [38] introduced a stent antenna that exhibited excellent radiation performance. Unlike conventional multi-connector stent antennas, this design featured multiple rings connected by a single connector. The single-connector configuration resulted in a more favorable resonant frequency compared to the multi-connector design, owing to the random current distribution observed in the latter. Consequently, at a resonant frequency of 2.07 GHz, this single-connector antenna demonstrated impressive performance metrics, including an S_{11} of -17 dB, a gain value of 1.38 dBi, and a radiation efficiency of 74.5%.

Besides electromagnetic (EM)-based antennas, another notable development in recent years has been the emergence of magnetoelectric (ME) antennas utilizing acoustic waves. Acoustic-based wireless power transfer (WPT) has gained significant attention as another alternative approach for powering compact IMDs. A notable advantage of acoustic-based wireless power transfer (WPT) is its ability to facilitate the development of magnetoelectric (ME) antennas at the nanoscale. In contrast to electromagnetic (EM)-based wireless power transfer

(WPT), which relies on electromagnetic induction or radiation, acoustic power transfer utilizes ultrasound waves or vibrations to transmit energy. Specifically, nanomechanical magnetoelectric (ME) antennas driven by acoustic waves utilize ferromagnetic/piezoelectric thin-film materials [39]. These antennas transmit energy by harnessing the direct and converse piezoelectric effects as well as the magnetostriction effect at the acoustic resonance frequency. Due to the shorter wavelength of acoustic waves compared to electromagnetic (EM) waves at the same frequency, these magnetoelectric (ME) antennas allow for a significant reduction in size, typically by 1–2 orders of magnitude, compared to conventional laminate antennas [40, 41]. This substantial miniaturization potential greatly promotes antenna miniaturization efforts. However, the adoption of ME antennas for acoustic-based WPT is limited due to stringent material requirements, such as a high piezomagnetic coefficient, as well as fabrication challenges. Moreover, ME antennas face challenges in terms of low output power and poorer S-parameter performance compared to EM-based antennas. These limitations hinder the widespread adoption of ME antennas in the field of wireless power transfer [42, 43].

1.1.2 Specific Absorption Rate and Dielectric Properties in Human Tissue

The electromagnetic (EM) wireless power transfer (WPT) system relies on the propagation of EM fields to transmit energy from an external source to an internal implant. However, the absorption of EM waves by the surrounding tissue can result in significant power loss and potential thermal effects. As a result, ensuring the safety of human exposure to EM radiation becomes the primary concern in wireless power and data transmission for implantable devices. In order to ensure the safety of WPT systems, it is necessary to comply with safety regulations regarding the power dissipation, specifically in terms of the Specific Absorption Rate (SAR) value measured in watts per kilogram (W/kg). SAR is a measure of the electromagnetic energy absorbed by the tissue per unit mass of tissue when exposed to EM radiation, leading to an increase in tissue temperature [44], as depicted in Eq. (1.3)

$$SAR = \frac{\sigma E^2}{\rho} \tag{1.3}$$

where E represents the induced electric field, σ and ρ denote the conductivity (kg/m^3) and the mass density (S/m) of the tissue. The Federal Communications Commission (FCC) has established a SAR limit of 1.6 W/kg (SAR-1 g) [45], while the IEEE SAR safety standard specifies a SAR limit of 2 W/kg (SAR-10 g) [46]. These standards serve as benchmarks to ensure the safety of wireless power and data transmission in implantable devices, minimizing the risk of adverse health effects caused by excessive tissue heating. In addition, the dielectric properties of

Table 1.3 Dielectric properties of tissue layers at different operating frequencies

Frequency	Tissue layer	Thickness(mm)	Conductivity (S/m)	Relative permittivity
40.68 MHz [48]	SkinMuscleFat	—	0.38 0.67 0.034	124.382.67.3
403.5 MHz [49]	Skin	3	0.68 (dry) 0.67 (wet)	46.71 (dry)49.84 (wet)
	Fat	10	0.041	5.58
	Muscle	20	0.79	57.1
915 MHz [50]	Skin	–	0.74 (dry) 0.77 (wet)	46.25 (dry)44.03 (wet)
1.427GHz [51]	Fat	–	0.14	8.84
	Muscle	–	0.70	53.83
	Skin	–	1.04	39.6
	Muscle	–	1.15	54.1
2.4GHz [52]	Skin	5	1.46	38
	Fat	10	0.11	5.3
	Muscle	10	1.77	52.7
3.5GHz [53]	Skin	2.5	2.35	41.41
	Fat	5	0.42	10.5
	Muscle	35	2.72	52.12
10GHz [54]	Skin	–	8.01	31.29
	Fat	–	0.58	4.62
	Muscle	–	10.62	42.76

biological tissue, including permittivity and conductivity, are significantly different from those of vacuum, resulting in a large mismatch between the wavelength λ in free-space and the wavelength $\lambda \big/ \sqrt{\epsilon_r}$ in tissue, where ϵ_r represents the relative permittivity of the tissue. Furthermore, the Specific Absorption Rate (SAR) value is highly dependent on the operational frequency [47]. Therefore, selecting an appropriate operational frequency is crucial to minimize power loss resulting from the interaction between the antenna and the tissue, while also ensuring compliance with SAR regulations. It is essential to investigate the effect of electromagnetic radiation on the human body at the appropriate frequency. Extensive research has been conducted to estimate the dielectric properties of human phantom models at various operating frequencies. Table 1.3 provides a comprehensive overview of the electrical properties of skin, fat, and muscle layers with respect to frequency variations [48–54]. The data presented in Table 1.2 reveals consistent trends of decreasing relative permittivity and increasing conductivity as the operating frequency increases. Based on these trends, it is widely acknowledged that selecting a lower operational frequency is preferable to ensure safe tissue absorption and minimize power loss.

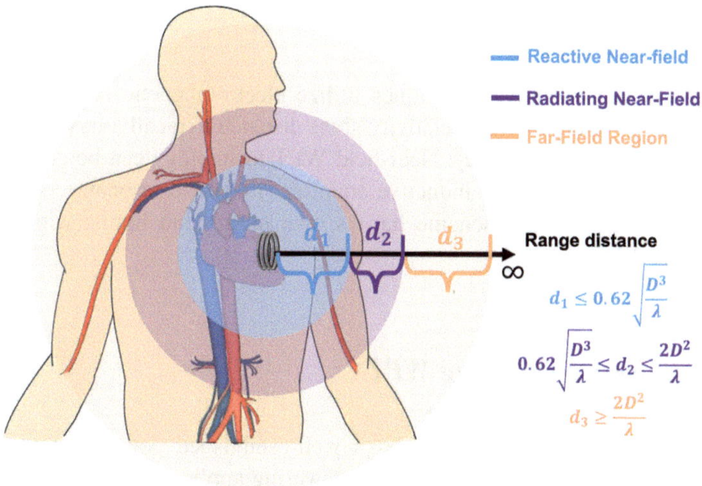

Fig. 1.2 Field regions of an antenna in wireless power transmission

In general, WPT technology can be classified into near-field and far-field WPT, based on the transmission distance and the propagation mode, including reactive and radiative modes. Figure 1.2 depicts the field regions of an antenna. In addition, the near-field region can be subdivided into reactive near-field and radiative near-field (Fresnel region). The transfer distance ranges of three fields are defined as follows [55]:

$$\text{Reactive near field } d_1 \leq 0.62\sqrt{\frac{D^3}{\lambda}} \tag{1.4}$$

$$\text{Radiative near field } 0.62\sqrt{\frac{D^3}{\lambda}} \leq d_2 \leq \frac{2D^2}{\lambda} \tag{1.5}$$

$$\text{Far Field } d_3 \geq \frac{2D^2}{\lambda} \tag{1.6}$$

where d represents radius of the transmission region, D denotes the largest dimension of the antenna, and λ refers to the wavelength.

1.2 Near-Field WPT

Near-field or non-radiative techniques utilize electromagnetic coupling between coils to transfer energy over a relativity short distance, typically varying from kHz to MHz in operating frequency. Near-field WPT technology can be classified into several categories, including inductive coupling, magnetic resonant coupling, and capacitive coupling. Each scheme has its own merits and limits, which will be discussed in detail.

1.2.1 Inductive Coupling WPT

Inductive coupling WPT (ICWPT) is a well-established power transfer method that has been widely applied in different powering applications. A typical ICWPT consists of a transmitter coil (Tx) and a receiver coil (Rx). When the Tx is powered by an alternating current (AC) source, a time-varying magnetic field generated between the two coils, which in turn induces a voltage in the Rx. The ICWPT shows remarkably high efficiency when the lateral and angle alignment between Tx and Rx is perfect. Specifically, the power transmission efficiency (PTE) is mainly determined by the coupling coefficient "k" and the quality factor "Q," as presented in the following Eqs. (1.7), (1.8) and (1.9) [56]:

$$k = \frac{M}{\sqrt{L_1 L_2}} = \frac{d_1^2 d_2^2}{\sqrt{d_1 d_2}\left(\sqrt{d_1^2 + D^2}\right)^3} \tag{1.7}$$

$$Q = \frac{2\pi f L}{R_{effective}} \tag{1.8}$$

$$\eta = \frac{k^2 Q_1 Q_{2L}}{1 + k^2 Q_1 Q_{2L}} \cdot \frac{Q_{2L}}{Q_L} \tag{1.9}$$

where, M is mutual inductance, L_1, L_2 are the self-inductance of a pair of coils separately. In the derived equation, d_1,d_2 are the diameter of the transmitter and receiver antenna, and D is the separation distance. In Eq. (1.8) and (1.9), L is the inductance and $R_{effective}$ is the self-resistance of antenna, Q_1, Q_2 are the quality factor of two coils, $Q_{2L} = Q_2 Q_L / Q_2 + Q_L$ and $Q_L = R_{LOAD}/2\pi f L_2$. Although ICWPT can achieve high PTE, it is vulnerable to the distance and the alignment between Tx and Rx under the non-resonant state, also known as weakly coupled or loosely coupling. Particularly, the strength of the magnetic field is inversely proportional to the distance ($1/D^3$). Therefore, the PTE of inductive coupling attenuates significantly with the separation increase.

1.2.2 Magnetic Resonant Coupling WPT

IC-WPT shows relatively high PTE but it is strongly affected by the short transmission distance and the precise alignment between Tx and Rx. Hence, to overcome the aforementioned limitations of IC-WPT, a strongly coupled WPT scheme called magnetic resonant coupling (MRC) was initially investigated by A. Kurs et al. [57] in 2007 to extend the transmission distance up to 2 meters by introducing a four-coil WPT system. The typical resonant coupled four-coil topology includes a source coil, a load coil, and two intermediate coils (Tx and Rx) loaded with capacitors. A strong magnetic coupling is generated between the Tx and Rx coils when they are tuned to the same resonant frequency. This allows adjacent coils to be coupled to each other through magnetic resonance coupling, which compensates for magnetic flux leakage, resulting in a higher PTE under a longer transmission distance compared to the IC-WPT.

Resonance frequency is a crucial factor for enhancing the PTE of MRC link, as depicted in Eq. (1.10) [57]:

$$f_r = \frac{1}{2\pi\sqrt{LC}} \tag{1.10}$$

where, L and C are the self-inductance and parasitic capacitance of the resonant transmitter or receiver coils. In addition, the coils can be modelled by a series RLC circuit and the quality factor (Q) is given by Eq. (1.11) [58]:

$$Q = \frac{2\pi f_r}{R} \tag{1.11}$$

where, R is the parasitic resistance of the coils. Additionally, the PTE is calculated based on the equivalent circuit, which is illustrated in the paper [59]. The MRC-WPT system can be considered as a two-port network and the PTE of the wires power transfer system can be analyzed by scattering parameters S_{11} and S_{21}, as presented in Eqs. (1.12) and (1.13) [59, 60]

$$\eta = \frac{|S_{21}|^2/Z_0}{\mathrm{Re}\left[\ |S_{11}+1|^2/Z_{in}\right]} \tag{1.12}$$

where, S_{11} is input reflection coefficient, $Z_0 = 50\ \Omega$ is the characteristic impendence, Z_{in} is the input impedance, and S_{21} is the forward transmission coefficient can be expressed as follows:

$$S_{21} = \frac{2k_{12}k_{23}k_{34}Q_2Q_3\sqrt{Q_1Q_4}}{1+{k_{12}}^2Q_1Q_2+{k_{23}}^2Q_2Q_3+{k_{34}}^2Q_3Q_4+{k_{12}}^2{k_{34}}^2Q_1Q_2Q_3Q_4} \tag{1.13}$$

where, k_{12} (source to Tx), k_{23} (Tx to Rx), k_{34} (Rx to load) are the coupling coefficient of adjacent coils, which can be found in the Eq. (1.7).

Various configurations of MRC-WPT, such two-coil, three-coil, and four-coil of MRC-WPT, show different characteristics, and their schematics and circuit models are discussed and compared in reference [61]. With the purpose of powering pacemakers, G. Monti et al. [62] adopted a two-coil MRC-WPT scheme to deliver up to 1 mW power at the resonance frequency of 403 MHz. Particularly, they designed a planar spiral-shaped resonator and a planar spiral-shaped resonator on Arlon 880 substrate, which served as the Tx and Rx, respectively. With regard to the tolerance of distance and alignment, sensitivity experiments were performed on the MCR-WPT system, and the results showed that the PTE drops sharply when the transmission distance exceeds 5 mm, or the rotation angle is greater than 30 degrees. To further increase transmission distance and mitigate the misalignment influence on the PTE, a 4-coil MRC-WPT was implemented in [63] for powering deep biomedical implants. Two circular external coils (source and Tx) and two spherical shaped implantable coils (Rx and load) were designed for maximining PTE under the misalignment conditions. This system has demonstrated a PTE of 78.5% over a transmission distance of 10 mm and remains effective with a range of 15 mm and up to 50 ° of angular misalignment. The 4-coil structure MRC-WPT schemes enhanced the coupling link by adding extra two resonant coils compared to the two-coil scheme, resulting in a higher PTE over a longer distance. However, the misalignment issue remains and tuning the coils to the same frequency can cause low power output due to the frequency splitting [64].

1.2.3 Capacitive Coupling WPT

Capacitive coupling WPT (CC-WPT) delivery power by using the induced displacement current generated from the electric field between conductive elements, which differs from IC-WPT and MRC-WPT that transfer power through magnetic fields [65, 66]. Thus, CC-WPT benefits from a lower EM interference level and better misalignment tolerance compared to the magnetic coupling WPT methods. Generally, CC-WPT system consists of two metal plates serve as Tx and another two identical plates as Rx, separated by a dielectric medium. This configuration forms receive and return paths of the displacement current, thus creating a power flow loop. The coupling capacitor, similar to the 'Q' factor in inductive coupling, is a crucial element in determining the amount of power transferred in CC-WPT. The equations for coupling capacitance and displacement current are denoted as Eqs. (1.14) and (1.15) [67, 68]:

$$C_T = \varepsilon_0 \varepsilon_r (\omega) \frac{A}{D} \tag{1.14}$$

$$I_{disp} = \varepsilon_0 \varepsilon_r(\omega) A \frac{\partial \vec{E}}{\partial t} \quad (1.15)$$

where, ε_0, $\varepsilon_r(\omega)$ are the free space permittivity and frequency variant permittivity, A is the effective area, D is the separation between the plates, and $\frac{\partial \vec{E}}{\partial t}$ is the rate of the electric field delay. There are various losses consider in CC-WPT system for powering the IMDs, including the tissue loss, conduction loss, source loss. However, when operating in the MHz to GHz range, the conduction and source losses are negligible compared to tissue loss. In particular, the tissue loss can be represented as an equivalent resistance, and the PTE of a capacitive coupling link can be expressed using the Eqs. (1.16) and (1.17) [69]:

$$R_T = Real\left(\frac{-jD}{\omega \varepsilon_0 \varepsilon_r(\omega) A}\right) \quad (1.16)$$

$$\eta = \frac{R_L}{R_L + R_T}\left(1 - |\Gamma|^2\right) \quad (1.17)$$

where, R_L is the load resistor, Γ is the reflection coefficient.

From the Eq. (1.16), the coupling capacitance is determined by the size of conductors and constrained by the small ε_0 value ($\varepsilon_0 = 8.85 \times 10^{-12}$ F/m). This results in a drawback of low capacitance and short transmission range (in the millimeter range). In order to optimize the PTE, a resonant compensation network (including capacitors and inductors) is commonly used to resonate with the coupler and boost power transmission. Additionally, there is a trade-off when designing the CC-WPT link between the conductor size, transmission distance, and the operational frequency. Although CC-WPT has several advantages over inductive coupling, such as low EM interference and high immunity to misalignment, its limited conditions and performance have not been widely explored in biomedical implant applications. Early research on capacitive coupling in IMDs dates back to 2009 [66] .This research presented a design for a power and data telemetry link for biomedical implants, highlighting the benefits of capacitive coupling. An in-vitro study conducted by A. Aldaoud et al. [14] compared the performance of inductive and capacitive power links for powering cardiovascular stents. The results demonstrated that the overall PTE of the inductive coupling was higher than that of the capacitive coupling.

1.3 Midfield WPT

Near-field wireless power transfer (WPT) schemes typically operate at low-frequency levels (<10 MHz) to mitigate tissue absorption when electromagnetic fields penetrate the body. However, this low-frequency carrier results in a large size receiver. In addition, when the receiver is considerably smaller than the distance between it and the source, it often leads to weak coupling between the coils. As previously mentioned, there is a general belief that lower frequencies exhibit higher transfer efficiency, as tissue absorption tends to increase with frequency. However, this theory does not suitable for high frequencies, as demonstrated by A.Poon et al. [70], higher frequency carriers facilitate receiver miniaturization and longer transmission distances without increasing tissue absorption. Midfield WPT utilizes an optimal source design with a patterned metal plate instead of inductive coils. The implementation of the patterned metal plate excites the propagation modes within the deep tissue, enabling concentrated and adaptable energy transfer. This advancement allows for the reduction in the size of the receiver by several orders of magnitude while achieving significant power transfer over the near-field inductive coupling. Besides, the midfield theory was proved by utilizing a full-wave analysis of wireless power transfer to implant device. This theory demonstrated the existence of an optimal frequency, typically in the low gigahertz range, that enhances the power transfer efficiency for small-scale receivers, as presented in Eq. (1.18) [70]:

$$f_{opt} \approx \frac{1}{2\pi}\sqrt{\frac{c\sqrt{\epsilon_0}}{\tau d\left(\epsilon_0 - \epsilon_\infty\right)}} \tag{1.18}$$

where, c is speed of light, ϵ_0, ϵ_∞, τ are the dielectric properties of tissue in the Debye relaxation model, d is the distance between Tx and Rx.

Mid-field wireless power transfer (WPT) has garnered considerable attention as a promising method for powering implantable medical devices (IMDs), especially in the field of cardiovascular implants. The theory of midfield wireless power transfer (WPT) was successfully demonstrated by achieving power transfer of 195 μW to a 2-mm diameter receiver (Rx) coil located in the heart, 5 cm distance from the external patterned metal plate (Tx) [71], as illustrated in Fig. 1.3B(a–c). Furthermore, in their study, Ho et al. calculated the approximate optimal frequencies for different types of human tissue at a distance of 5 cm. For instance, they found that the optimal frequency for blood was approximately 1.58 GHz, while for muscle tissue it was around 1.76 GHz. Subsequently, in a study by Das et al. [72], the mid-field WPT technique was employed to power a leadless pacemaker at the optimal frequency of 1.5 GHz, determined using the Eq. (1.18) with heart tissue parameters. The mid-field transmitter antenna was designed as a slot array-like metal plate on a Teflon substrate, while the receiver (Rx) consisted of a spiral-shaped PIFA antenna fabricated on a flexible polyamide substrate, as depicted in Fig. 1.3C(a–c). The mid-field WPT system successfully delivered 500 mW power at the transmitter (Tx) and received 2.9 mW power at the receiver (Rx), which proved to be sufficient for powering various cardiovascular implants. In another study by Basir and Yoo [73], an efficient WPT system utilizing the mid-field technique was proposed for deep-body implants. The proposed configurations of the transmitter (Tx) and receiver (Rx) are illustrated in Fig. 1.3D(a–c). The mid-field transmitter was designed

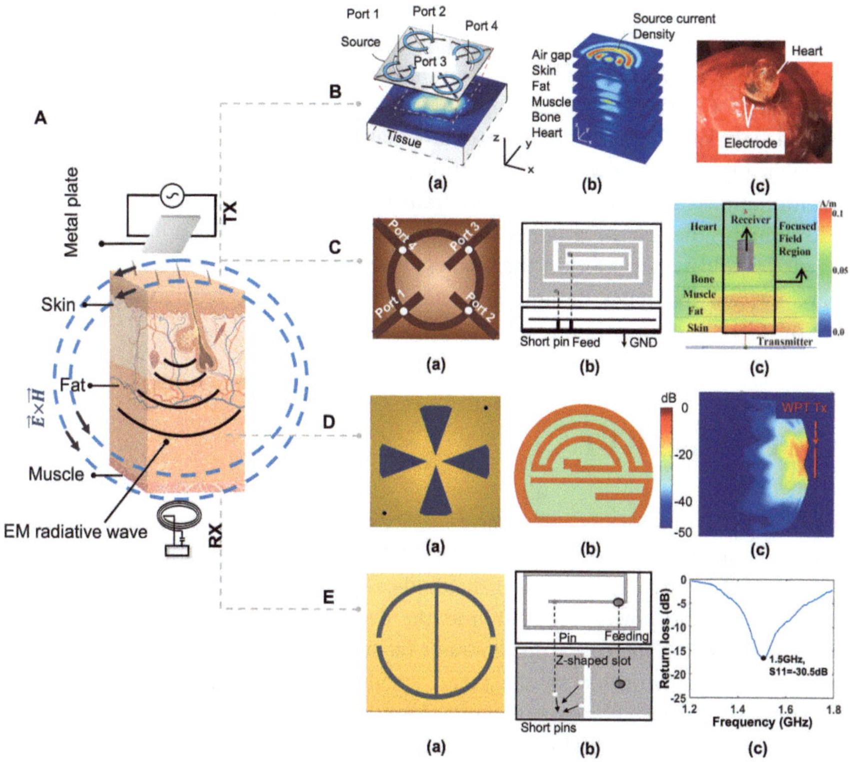

Fig. 1.3 Work principle of midfield WPT and its applications in cardiovascular implants

with a slotted structure and four ports on a Teflon substrate, enabling efficient power concentration in the tissue through precise phase control. Additionally, the authors developed a compact implantable antenna featuring a spiral-shaped structure capable of quad-band operation. Measurement results demonstrated that the system successfully delivered 6.7 mW of power to deep tissue implants over a distance of 5 cm, highlighting its significant potential for powering cardiovascular implants. Nguyen et al. [74] further contributed to the midfield WPT by introducing a compact mid-field transmitter (Tx) configuration designed specifically for powering miniature implants, as depicted in Fig. 1.3E(a–c). The receiver (Rx) utilized a planar PIFA (Planar Inverted-F Antenna) with a Z-shaped slot on the ground layer and three short pins to maximize the gain value. Operating at an optimal frequency of 1.5 GHz, this mid-field WPT system demonstrated a highly concentrated magnetic field within the tissue, as observed through current density and magnetic field analysis. The Rx was positioned 55 mm away from the Tx within a heart tissue layer, and the measured output power at the Rx reached 5.6 mW from a 1 W power coupling at the Tx. Furthermore, the researchers investigated the performance of the mid-field WPT system under various misalignment conditions, including lateral and angular misalignments. The results showed that interference caused by misalignment was effectively mitigated in the mid-field WPT system, as evidenced by stable S_{21} values and PTE across all misalignment cases. This finding is highly promising for the application of implantable devices.

Mid-field WPT offers a notable solution that is better suited for powering cardiovascular implants compared to conventional near-field WPT methods. It overcomes the limitations associated with short transmission distances and misalignments. Additionally, the use of a controllable metal plate allows for adjustment of the magnetic field pattern within the tissues, thereby enhancing the PTE. Moreover, this approach facilitates the miniaturization of implantable receivers operating in the low GHz frequency range.

1.4 Far-Field WPT

Far-field WPT employs beams, such as microwaves or laser beams, in the radiated electromagnetic fields to transfer power, instead of relying on magnetic coupling as reviewed in inductive coupling methods. As aforementioned, near-field WPT experiences rapid power attenuation as the distance between the transmitter (Tx) and receiver (Rx) increases. In contrast, far-field scheme allows wave energy to spread out over a considerable distance, spanning meters. However, the extended transmission distance comes with a drawback-significant path loss leading to lower power transfer efficiency. This limitation arises from the fact that energy transmitted through electromagnetic (EM) waves radiates and dissipates in all directions (omnidirectional nature), resulting in reduced power concentration at the receiver [75]. Figure 1.4a illustrates the far-field power link, encompassing various energy-conversion sub-stages: direct current (DC) to radio frequency (RF) conversion (P_{Tx}), effective radiative power from the transmitter (P_{in}), free-space transmission (P_{inc}), incident power at the receiver (P_{Rx}), and RF to DC conversion (P_{out}). The overall power transfer efficiency can be calculated using Eq. (1.19) [76].

$$\eta = \frac{P_{out}}{P_{in}} = \frac{P_{TX}}{P_{in}} \frac{P_{erp}}{P_{TX}} \frac{P_{inc}}{P_{erp}} \frac{P_{RX}}{P_{inc}} \frac{P_{out}}{P_{RX}} = \eta_1 \eta_2 \eta_3 \eta_4 \eta_5 \tag{1.19}$$

Multi-path loss poses a significant concern to the PTE of far-field WPT. As depicted from above equation, the transmit efficiency ($\eta_1 \times \eta_2$) and the receive efficiency ($\eta_4 \times \eta_5$) play a crucial role in optimizing the overall PTE in far-field WPT systems. To improve the transmit efficiency in far-field WPT, one effective approach is to utilize directional antennas or antenna arrays. These antennas can help mitigate path loss by precisely focusing the electromagnetic radiation towards the intended receiver [77]. By employing antenna arrays with beam steering capabilities, it becomes possible to manipulate the beam pattern, achieving a

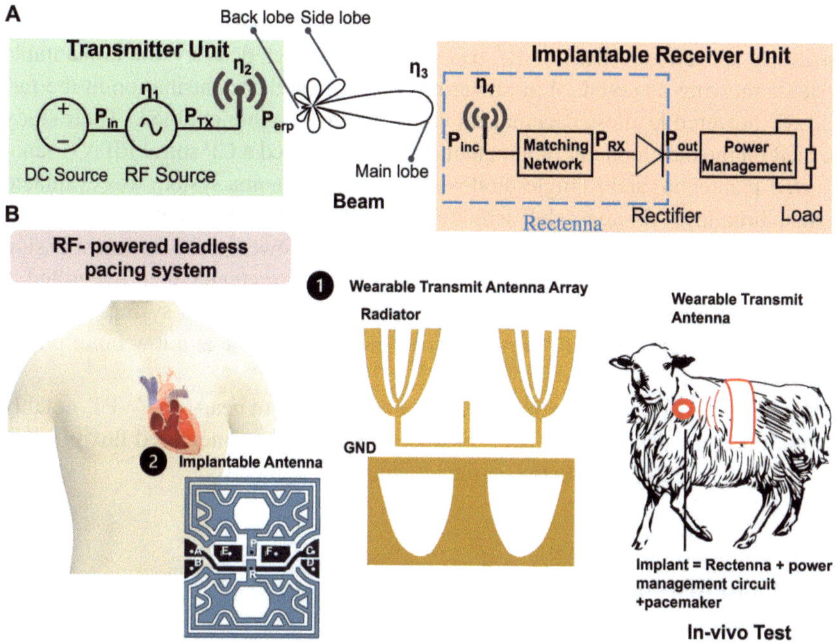

Fig. 1.4 Work principle of midfield WPT and its applications in cardiovascular implants

directional main lobe with minimal side lobes. This enhances the directivity of the transmission, leading to improved transmit efficiency [78].

On the other hand, the receive efficiency is greatly influenced by the performance of the rectifier circuit, which converts the received RF power into DC power for the load. An efficient rectenna, which combines the antenna and rectifier circuit, plays a critical role in maximizing the receive efficiency, particularly in the conversion of RF to DC [79]. Designing and optimizing the rectenna is essential for enhancing the overall power conversion efficiency of the system [20].

In recent years, RF-based far-field power strategies have gained significant interest in the field of wearable and implantable devices, particularly in the context of IoT medical systems. A notable example is presented in the paper [80], which introduces an RF-powered leadless pacing system. The system comprises a wearable transmit-antenna array and an implantable rectenna (Rx), as depicted in Fig. 1.4b. For the receiver component, a planar dipole antenna with a fractal geometry was chosen and fabricated on a Rogers TMM10i substrate (with $\tan\delta = 0.002$ and $\varepsilon r = 9.8$). The receiver was then matched with a modified-Greinacher rectifier circuit to enhance the power conversion efficiency. On the other hand, the transmit array was designed using the RT/Duroid 6010 material (with a thickness of 0.254 mm). It featured two semi-oval shaped truncations in the ground plane to enable directional beam formation of the antenna. The overall wireless pacing system was successfully validated in an *in-vivo* ovine model, achieving a power transfer efficiency (PTE) of 65% at an operating frequency of 954 MHz. This demonstration exemplifies the feasibility and effectiveness of RF-based far-field power delivery for implantable devices, showing its potential in powering cIMDs. Another contribution in the far-field RF-powered leadless pacemakers was made by A. Abdi et al [81] Their study focused on the development of a rectenna, which included a CP spiral PIFA (Planar inverted-F antenna) and a single-diode detector. The rectenna system was evaluated using a three-layer tissue model, achieving 40% RF to DC efficiency. Subsequently, Ipbal et al. [82] developed a simultaneous wireless power and data transmission system for biomedical implants. They introduced a rectenna that integrated a rectifier with a self-diplexing implantable antenna. Notably, this system achieved an impressive peak RF to DC conversion efficiency of 76.1% at a low input power of 2 dBm.

Far-field WPT presents distinct advantages compared to near-field WPT, notably enabling power transfer over larger distances and providing increased flexibility in dealing with misalignment between the transmitter and receiver.

1.5 Conclusion

The long-term implantation of certain cardiac implantable medical devices (cIMDs) is vital for effectively managing chronic cardiovascular conditions and restoring normal functions of the cardiovascular system. To address this need, wireless and battery-free power technologies have emerged as valuable solutions, enabling the development of non-invasive diagnostic and therapeutic IMDs. These advancements eliminate the requirement for frequent surgical interventions, allowing for more convenient and efficient treatment options. Table 1.4 provides a comprehensive comparison of various battery-less power strategies for IMDs, offering insights into the different approaches and their respective advantages and limitations, aiding in the selection of the most suitable power strategy for specific IMD applications.

Table 1.4 Comparison of battery-less power strategies for cIMDs

Energy Harvesting Type	Technique	Elements	Materials	Generated Power/Energy level	Merit	Demerit
In-body energy harvest	Biofuel cells	Electrode	Ag/Pt/Au/CNTs	Output voltage <1 V	Biocompatible Abundant fuels in the body	Low output voltage Low conversion efficiency
	Piezoelectric transducer	Cantilever beam Piezoelectric film	Ceramic materials/Polymers/Composite materials/nanomaterials	Power density in the range of $\mu W/cm^3$ and mW/cm^3	High energy density	Complex structure and circuitry Affected by the vibration frequency
	Triboelectric transducer	Triboelectric material pair	Two materials with strong electron affinity	Power density $\mu W/cm^3$ level	Simple construction Easy fabrication	Low energy densities
	Thermoelectric transducer	Thermocouple	Semiconductors/conductive polymer/hybrid materials	Power density $\mu W/cm^3$ level	No moving component Long-term energy supply	Low conversion efficiency Time-variant temperature difference required in the body Limited implantable locations

(continued)

Table 1.4 (continued)

Energy Harvesting Type	Technique	Elements	Materials	Generated Power/Energy level	Merit	Demerit
External wireless power transfer	Inductive coupling	Coil	Aluminum/copper/iron	Output power consider the SAR limitation, size, distance: in the range of μW to mW	High transfer efficiency	Short transmission distance (cm) Accurate alignment requirement
	Capacitive coupling	Conductive plate	Conductor materials		Non-radiative	Very short distance (mm)
	Magnetic resonant coupling	Resonator	Aluminum/copper/iron		Medium transmission distance (m)	Multiple coils occupied space
	Mid-field	Antenna	Patterned metal plate (Tx)		Efficiently concentrate the power in the tissue, high tolerance with alignment	–
	Far-field	Antenna	Depends on the substrate		Long transmission distance (m)	Low efficiency

References

1. N. Bhatia, M. El-Chami, J. Geriatr. Cardiol. **15**, 249 (2018)
2. M. Kotsakou, I. Kioumis, G. Lazaridis, G. Pitsiou, S. Lampaki, A. Papaiwannou, A. Karavergou, K. Tsakiridis, N. Katsikogiannis, I. Karapantzos, Ann. Transl. Med. **3**, 42 (2015)
3. P. Magnusson, J.V. Pergolizzi, J.A. LeQuang, *Cardiac Pacing and Monitoring-New Methods, Modern Devices* (IntechOpen, 2018)
4. M. Hosseinipour, R. Gupta, M. Bonnell, M. Elahinia, J. Rehabil. Assistive Technol. Eng. **4**, 2055668317725994 (2017)
5. J.H. Karimov, K. Fukamachi, R.C. Starling, *Mechanical Support for Heart Failure: Current Solutions and New Technologies* (Springer, 2020)
6. J. Al Suwaidi, P.B. Berger, D.R. Holmes Jr., JAMA **284**, 1828 (2000)
7. D. Katz, T. Akiyama, Ann. Noninvasive Electrocardiol. **12**, 223 (2007)
8. A.S. Manolis, T. Maounis, S. Koulouris, V. Vassilikos, Clin. Cardiol. **40**, 759 (2017)
9. Z. Taimeh, R.J. Koene, J. Furne, A. Singal, P.M. Eckman, M.D. Levitt, M.R. Pritzker, J. Heart Lung Transplant. **36**, 625 (2017)
10. A. Kumar, R. Kiran, V.S. Chauhan, R. Kumar, R. Vaish, Mater. Res. Express **5**, 075701 (2018)
11. G. Van Welsenes, C. Borleffs, J. Van Rees, J. Atary, J. Thijssen, E. Van der Wall, M. Schalij, Neth. Hear. J. **19**, 24 (2011)
12. K.W. Beard, *Linden's Handbook of Batteries* (McGraw-Hill Education, 2019)
13. B.H. Waters, J. Park, J.C. Bouwmeester, J. Valdovinos, A. Geirsson, A.P. Sample, J.R. Smith, P. Bonde, J. Heart Lung Transplant. **37**, 1467 (2018)
14. A. Aldaoud, J.-M. Redoute, K. Ganesan, G.S. Rind, S.E. John, S.M. Ronayne, N.L. Opie, D.J. Garrett, S. Prawer, IEEE J. Electromagn. RF Microwaves Med. Biol. **2**, 193 (2018)
15. P. Soontornpipit, C.M. Furse, Y.C. Chung, IEEE Trans. Microwave Theory Tech. **2004**, 52 (1944)
16. K. Zhang, C. Liu, X. Liu, H. Guo, X. Yang, Int. J. Antennas Propag. (2017)
17. S. Ding, S. Koulouridis, L. Pichon, A dual-band miniaturized circular antenna for deep in body biomedical wireless applications. Presented at 2019 13th European Conference on Antennas and Propagation (EuCAP) (2019)
18. S. Salama, D. Zyoud, R. Daghlas, A. Abuelhaija, Design of a planar inverted F-antenna for medical implant communications services band. Presented at Journal of Physics: Conference Series (2020)
19. M.S. Islam, M.T. Islam, M.A. Ullah, G.K. Beng, N. Amin, N. Misran, IEEE Access **7**, 127850 (2019)
20. J. Zhang, M. Wagih, D. Hoare, N. Mirzai, J. Mercer, R. Das, H. Heidari (2023)
21. W. Huang, A.A. Kishk, Int. J. Antennas Propag. **2011**, Article ID: 919821 (2011)
22. A. Ahmad, F. Arshad, S.I. Naqvi, Y. Amin, H. Tenhunen, J. Loo, IETE J. Res. **66**, 22 (2020)
23. W.A. Awan, N. Hussain, T.T. Le, AEU Int. J. Electron. Commun. **110**, 152851 (2019)
24. M. Zada, H. Yoo, IEEE Trans. Antennas Propag. **66**, 7378 (2018)
25. T.J. Warnagiris, T.J. Minardo, IEEE Trans. Antennas Propag. **46**, 1797 (1998)
26. J. Rashed, C.-T. Tai, IEEE Trans. Antennas Propag. **39**, 1428 (1991)
27. C. A. Balanis, Wiley (2016)
28. Z. Bendahmane, S. Ferouani, C. Sayah, Prog. Electromagn. Res. **102**, 163 (2020)
29. J.H. Kim, C.-H. Ahn, J.-K. Bang, IEEE Trans. Antennas Propag. **64**, 1164 (2016)
30. R. Porath, IEEE Trans. Antennas Propag. **48**, 41 (2000)
31. R. R. Hasan, M. A. Rahman, S. Sinha, M. N. Uddin, T.-S. R. Niloy, In body antenna for monitoring pacemaker. Presented at 2019 International Conference on Automation, Computational and Technology Management (ICACTM) (2019)
32. M. Zada, I.A. Shah, A. Basir, H. Yoo, IEEE Trans. Antennas Propag. **69**, 1152 (2020)
33. I.A. Shah, M. Zada, H. Yoo, IEEE Trans. Antennas Propag. **67**, 4230 (2019)
34. M. Ramzan, X. Fang, Q. Wang, N. Neumann, D. Plettemeier, Miniaturized planar implanted spiral antenna inside the heart muscle at MICS band for future leadless pacemakers. Presented

at 2019 13th International Symposium on Medical Information and Communication Technology (ISMICT) (2019)
35. Z.-J. Yang, L. Zhu, S. Xiao, IEEE Trans. Antennas Propag. **66**, 5180 (2018)
36. J. Zhang, R. Das, D. Hoare, H. Wang, A. Ofiare, N. Mirzai, J. Mercer, H. Heidari, IEEE Trans. Antennas Propag. **71**(6), 4759–4771 (2023)
37. X. Chen, B. Assadsangabi, Y. Hsiang, K. Takahata, Adv. Sci. **5**, 1700560 (2018)
38. C.-H. Liu, S.-C. Chen, H.-M. Hsiao, Sensors **19**, 4616 (2019)
39. T. Nan, H. Lin, Y. Gao, A. Matyushov, G. Yu, H. Chen, N. Sun, S. Wei, Z. Wang, M. Li, Nat. Commun. **8**(1), 296 (2017)
40. H. Lin, M. Zaeimbashi, N. Sun, X. Liang, H. Chen, C. Dong, A. Matyushov, X. Wang, Y. Guo, Y. Gao, NEMS magnetoelectric antennas for biomedical application. Presented at 2018 IEEE International Microwave Biomedical Conference (IMBioC) (2018)
41. M. Zaeimbashi, M. Nasrollahpour, A. Khalifa, A. Romano, X. Liang, H. Chen, N. Sun, A. Matyushov, H. Lin, C. Dong, Nat. Commun. **12**, 1 (2021)
42. S. Shevlin, Nat. Mater. **18**, 191 (2019)
43. Y. Cheng, B. Peng, Z. Hu, Z. Zhou, M. Liu, Phys. Lett. A **382**, 3018 (2018)
44. M.A. Jensen, Y. Rahmat-Samii, Proc. IEEE **83**, 7 (1995)
45. Federal Communications Commission, Report and order 1996
46. IEEE Standards Coordinating Committee, IEEE **C95**, 1–1991 (1992)
47. X. Liu, H.-J. Chen, Y. Alfadhl, X. Chen, C. Parini, D. Wen, J. Appl. Phys. **113**, 074902 (2013)
48. M. Grzeskowiak, F. El Hatmi, A. Diet, M. Benamara, D. Delcroix, T. Alves, S. Protat, S. Mostarshedi, O. Picon, Y. Le Bihan, C. R. Phys. **16**, 819 (2015)
49. M.M. Rahman, G. Rather, J. Sci.: Adv. Mater. Devices **5**, 134 (2020)
50. K.N. Bocan, M.H. Mickle, E. Sejdić, IEEE Trans. Microwave Theory Tech. **66**, 3547 (2018)
51. C.-K. Wu, T.-F. Chien, C.-L. Yang, C.-H. Luo, Int. J. Antennas Propag. **10**, 1–12 (2012)
52. M. Vallejo, J. Recas, P.G. Del Valle, J.L. Ayala, Sensors **13**, 7546 (2013)
53. A.A.Y. Ibraheem, M. Manteghi, Prog. Electromagn. Res. **40**, 57 (2014)
54. M.A.I. Oni, S.H. Shehab, S.R. Khurshid, M. Badruzzaman, S. Dey, Am. Acad. Scholarly Res. J. **7**, 96 (2015)
55. K. Chang, *RF and Microwave Wireless Systems*, vol 161 (Wiley, 2004)
56. M. Adeeb, A. Islam, M. Haider, F. Tulip, M. Ericson, S. Islam, Act. Passive Electron. Compon. **2011**, 11 (2012)
57. A. Kurs, A. Karalis, R. Moffatt, J.D. Joannopoulos, P. Fisher, M. Soljačić, Science **317**, 83 (2007)
58. A.P. Sample, D.T. Meyer, J.R. Smith, IEEE Trans Ind Electron **58**, 544 (2010)
59. V. Vijayakumaran Nair, J.R. Choi, Energies **9**, 156 (2016)
60. Y. Zhang, Z. Huo, X. Wang, X. Han, W. Wu, B. Wan, H. Wang, J. Zhai, J. Tao, C. Pan, Nat. Commun. **11**, 1 (2020)
61. X. Mou, D.T. Gladwin, R. Zhao, H. Sun, IET Power Electron. **12**, 3005 (2019)
62. G. Monti, P. Arcuti, L. Tarricone, IEEE Trans. Microwave Theory Tech. **63**, 3814 (2015)
63. P. Yellappa, Y.R. Lee, J.R. Choi, Analysis of coil misalignment issue in resonance-based wireless power transmission system for implantable biomedical applications. Presented at 2018 IEEE International Symposium on Circuits and Systems (ISCAS) (2018)
64. R. Huang, B. Zhang, D. Qiu, Y. Zhang, IEEE Trans. Magn. **50**, 1 (2014)
65. W.Y. Po, S. Warisawa, Coil shape antenna design to enhance the performance of 1 pair capacitive coupling for implant medical device application. Presented at 2020 IEEE Wireless Power Transfer Conference (WPTC) (2020)
66. A.M. Sodagar, P. Amiri, Capacitive coupling for power and data telemetry to implantable biomedical microsystems. Presented at 2009 4th International IEEE/EMBS Conference on Neural Engineering (2009)
67. R. Jegadeesan, K. Agarwal, Y.-X. Guo, S.-C. Yen, N.V. Thakor, IEEE Trans. Microwave Theory Tech. **65**, 280 (2016)
68. R. Erfani, F. Marefat, A.M. Sodagar, P. Mohseni, IEEE Trans Circuits Syst II Express Briefs **65**, 923 (2017)

69. M. Kline, I. Izyumin, B. Boser, S. Sanders, Capacitive power transfer for contactless charging. Presented at 2011 Twenty-Sixth Annual IEEE Applied Power Electronics Conference and Exposition (APEC) (2011)
70. A.S. Poon, S. O'Driscoll, T.H. Meng, Optimal operating frequency in wireless power transmission for implantable devices. Presented at 2007 29th Annual International Conference of the IEEE Engineering in Medicine and Biology Society (2007)
71. J.S. Ho, S. Kim, A.S. Poon, Proc. IEEE **101**, 1369 (2013)
72. R. Das, Y. Cho, H. Yoo, High efficiency unidirectional wireless power transfer by a triple band deep-tissue implantable antenna. Presented at 2016 IEEE MTT-S International Microwave Symposium (IMS) (2016)
73. A. Basir, H. Yoo, IEEE Trans. Microwave Theory Tech. **68**, 1943 (2020)
74. N. Nguyen, N. Ha-Van, C. Seo, IEEE Antennas Wirel. Propag. Lett. **19**, 2270 (2020)
75. R. Gonçalves, N.B. Carvalho, P. Pinho, C. R. Phys. **18**, 78 (2017)
76. M. Xia, S. Aissa, IEEE Trans. Signal Process. **63**, 2835 (2015)
77. J. Huang, Y. Zhou, Z. Ning, H. Gharavi, IEEE Wirel Commun **26**, 163 (2019)
78. G. Sun, Y. Liu, H. Li, S. Liang, A. Wang, B. Li, Int. J. Antennas Propag. (2018)
79. J. Zhang, D. Hoare, R. Das, M. Holsgrove, J. Czyzewski, N. Mirzai, J. Mercer, H. Heidari, Wireless impedance platform for autonomous vascular implantable devices. Presented at 2022 29th IEEE International Conference on Electronics, Circuits and Systems (ICECS) (2022)
80. S.M. Asif, A. Iftikhar, J.W. Hansen, M.S. Khan, D.L. Ewert, B.D. Braaten, IEEE Access **7**, 1139 (2018)
81. A. Abdi, H. Aliakbarian, IEEE J. Transl. Eng. Health Med. **7**, 1 (2019)
82. A. Iqbal, M. Al-Hasan, I.B. Mabrouk, A. Basir, M. Nedil, H. Yoo, IEEE Trans. Microwave Theory Tech. **69**, 3438 (2021)

Chapter 2
Advancements in Energy Harvesting for Implantable Cardiovascular Devices

Bhavani Prasad Yalagala, Jungang Zhang, Rupam Das, and Hadi Heidari

Energy harvesting systems can convert various forms of energy present in the human body, such as heat, chemical energy from molecules like glucose, and physical energy from actions like breathing and limb movement. These systems play a vital role in generating electrical energy, which can be used to recharge implanted batteries or even serve as their alternative power source [1]. Recently, the need for wearable and implantable devices are increasing day by day owing to the development of the various scientific technological advancements especially in medical electronics. The major key limiting factor for the portable, implantable, and wearable electronics is the miniaturised, stable, and rechargeable power source to drive the electronic devices [2]. Currently, most of them were powered using the miniaturised batteries or some of the energy storage cells but still needs to replace after regular interval of time or need to be dispose the device after certain period [3]. Hence, the energy harvesting system harvesting energy from the internal organ movement is very much necessary for the long-term sustainability of the wearable and implantable medical electronic devices (Figs. 2.1 and 2.2).

On the other hand, human body has abundance of energy in the form of the mechanical vibrations such as the walking, moving, jumping, and running. Harvesting energy from the mechanical vibrations of the body especially from the internal human organs such as contraction and relaxation of the heart muscles (heartbeat), respiratory organ vibrations, movement of the upper and lower limbs in the knees, muscle movements during walking and moving etc. is very much useful to power these implantable devices [4].Various nanogenerators such as

B. P. Yalagala (✉) · J. Zhang · H. Heidari
University of Glasgow, Glasgow, UK
e-mail: Bhavani.Yalagala@glasgow.ac.uk

R. Das
Electronic Engineering, University of Exeter, Exeter, UK
e-mail: Rupam.das@glasgow.ac.uk

R. Das, H. Heidari (eds.), *Wireless Power Technologies for Biomedical Devices*,
Analog Circuits and Signal Processing, https://doi.org/10.1007/978-3-031-52834-7_2

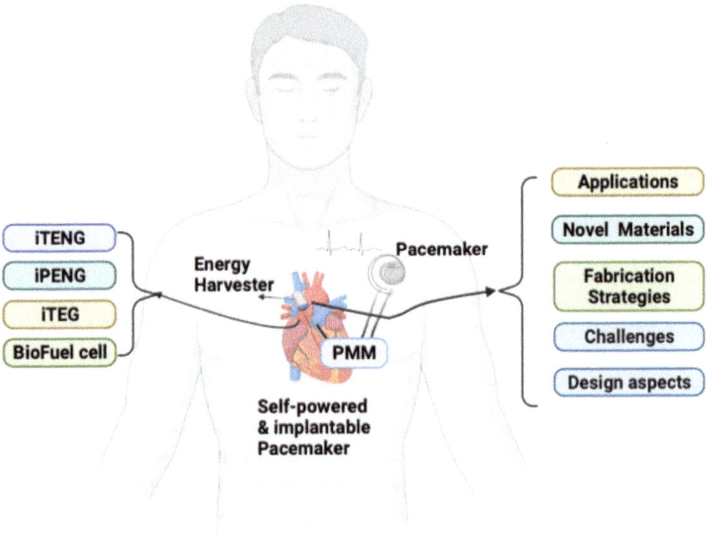

Fig. 2.1 Schematic representation of the graphical abstract of the self-powered implantable energy harvesters

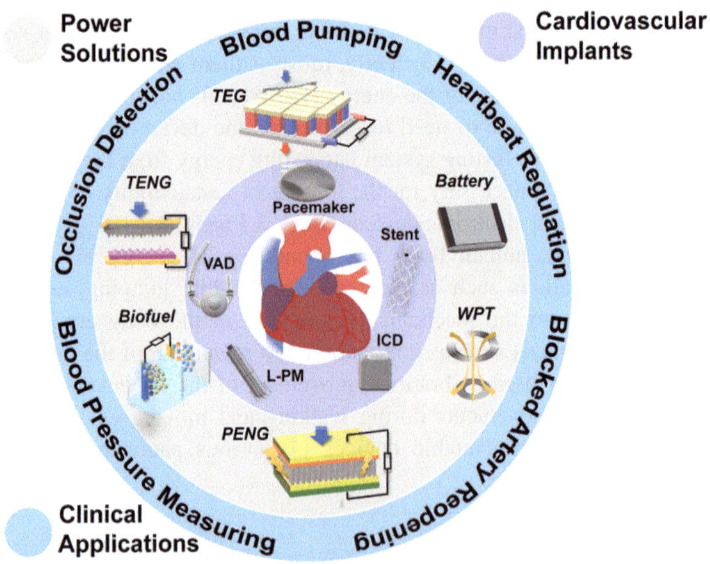

Fig. 2.2 Advancements in cardiovascular implantable devices: power strategies and medical applications using PENG, TENG, ICD (implantable cardioverter defibrillators), VAD (ventricular assist devices), L-PM (leadless pacemaker)

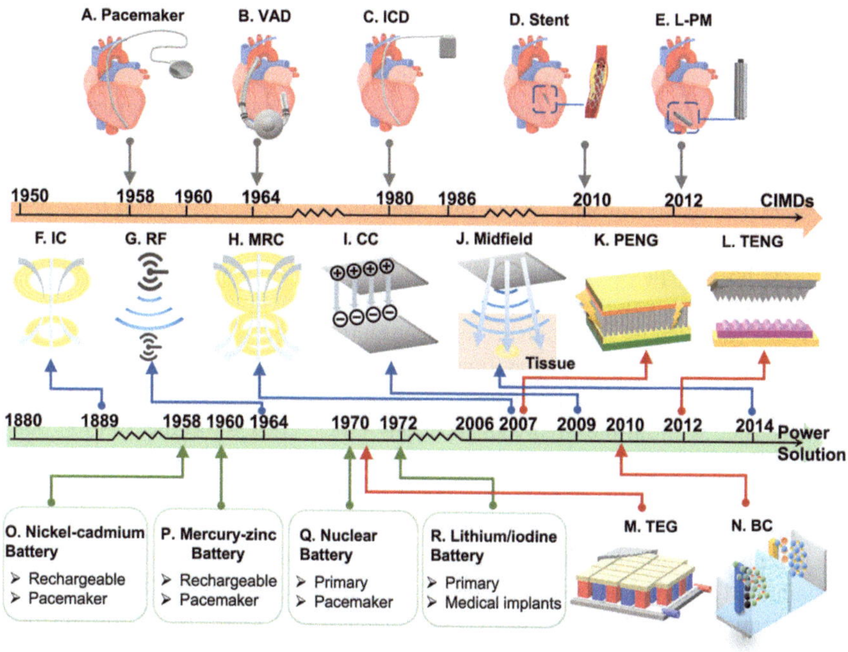

Fig. 2.3 Milestones in cardiovascular implantable devices and corresponding power strategies [6–22]

piezoelectric, triboelectric, pyroelectric, thermoelectric, biofuels etc. have already been explored self-powered, flexible, and wearable sensors and are expanding rapidly for implantable electronics as well. However, there is a still lot of research works need to be demonstrated in terms of the boosting the reliable and constant power, front end electronic circuitry to regulate the desired power and there is a trade-off between the performance, battery lifetime, and volume [5].

In this section, a brief overview on the different types of the energy harvesters and its working mechanism along with its applications in various implantable and flexible electronic devices (Fig. 2.3).

2.1 Implantable Bio-fuel Cell

Biofuel cells especially, the implantable bio-fuel cells are miniature devices designed to generate electrical energy within the human body using biochemical catalysed chemical reactions. The implantable biofuels mainly use human biofuel (glucose) and bioxidants (molecular oxygen) to drive the implantable devices [23]. These bio-fuel cells are typically constructed using biocompatible and environmentally friendly green materials and utilize enzymes or microorganisms to catalyse the electrochemical reactions involved in energy production. The concept behind implantable bio-fuel cells is to harness the energy derived from the metabolic processes of the body, such as glucose or other organic molecules, and convert it into electrical energy that can power various medical devices or implanted sensors [24]. By utilizing the body's own resources as a fuel source, these devices offer a potential alternative to traditional batteries or external power sources. The bio-fuel cell typically consists of an anode and a cathode, separated by a membrane. The anode is usually coated with an enzyme that can oxidize the biological fuel, while the cathode is coated with another enzyme that facilitates the reduction reaction [25]. As the fuel passes through the anode, it is oxidized, releasing electrons that travel through an external circuit to the cathode, generating an electric current. There are two primary mechanisms for electron transfer (ET) in bioelectrochemical systems such as direct electron transfer (DET) and mediated electron transfer (MET). DET involves the direct transfer of electrons between the biocatalyst and the electrode without the involvement of any other redox molecule. On the other hand, MET relies on a separate redox molecule, which can be a small molecule in solution, a redox active protein, or a redox polymer. This redox moiety acts as a mediator, shuttling electrons between the biocatalyst and the electrode, facilitating the ET process [26]. Implantable bio-fuel cells have the advantage of being self-sustaining if a continuous supply of the biological fuel is available in the body. They can potentially provide a long-lasting and reliable power source for various medical applications, such as pacemakers, neurostimulators, insulin pumps, or biosensors, eliminating the need for regular battery replacements or external connections.

Based on the biocatalyst nature, biofuels were broadly classified into three categories such as the enzymatic biofuel cells (EBFCs), microbial biofuel cells (MBFCs) and hybrid biofuel cells (HBFCs). The EBFCs employ enzymes as biocatalysts to facilitate the oxidation and reduction reactions involved in the generation of electrical energy [27]. These biofuel cells utilize the high catalytic activity and selectivity of enzymes to efficiently convert the chemical energy of a fuel source into electrical energy. EBFCs offers high catalytic efficiency, a wide range of fuel sources, and operation at mild conditions. However, they face challenges related to enzyme stability and cost. The second one MFCs are a type of biofuel cell that utilize microorganisms as biocatalysts to generate electrical energy from organic compounds. MBFCs harness the metabolic activity of microorganisms to convert the chemical energy stored in organic matter into usable electrical power. MBFCs utilize the complex fuels, self-sustaining operation, and potential for higher power output. Nevertheless, they have slower reaction rates and require management of microbial communities. HBFCs, also known as hybrid enzymatic-microbial fuel cells, are a type of biofuel cell that combines both enzymatic and microbial elements to enhance the overall performance and stability of the system. These cells leverage the advantages of both enzymatic and microbial biofuel cells to overcome limitations associated with individual types. The enzymes act as biocatalysts to facilitate the oxidation or reduction of the fuel molecules, while the microorganisms provide a continuous supply of fuel and contribute to the overall power output. However, they entail complex system design, electron transfer optimization, and stability concerns.

Recently, Mark et al. demonstrated a self-powered pacemaker using an implantable biofuel cell utilizing buckypaper electrodes modified with enzymes [28]. The biofuel cell generated an open circuit voltage (V_{oc}) of 470 mV and a short circuit current of 5 mA and is connected to the pacemaker via a charge pump and the variable voltage regulator. The power generated activated a pacemaker, demonstrating the potential for implantable medical devices powered by electricity harvested from the human body. Similarly, Lenka et al. has demonstrated an abiotic and enzyme based implantable biofuel cells operating in a small living organism like a snail [29]. The works demonstrate an excellent electrical output such as the V_{OC} and I_{SC} of 530 mV and 42.5 μA respectively with an optimal resistance of 20 kΩ. Moreover, it is observed that the overall electrical output produced from the different snail specimens was much higher than the output produced by the other biofuel cells implanted in the other animals such as rats, rabbits etc. Alon et al has successfully designed and tested an implanted biofuel cell in a snail, utilizing compressed carbon nanotubes as electrodes modified with specific enzymes for glucose oxidation and oxygen reduction [30]. This breakthrough allowed for the sustainable generation of electrical power from glucose produced metabolically. The team further extended their research to clams, assembling multiple implanted biofuel cells into batteries, leveraging the availability and affordability of clams for experimental purposes. The electrodes were composed of buckypaper, a compressed multiwalled carbon nanotube material, and the enzymes were immobilized on the electrodes using a cross-linker, enabling effective catalytic reactions. Guray et al. shown a successful application of direct electrochemical communication using nanostructured buckypaper composed of multi-wall CNTs is demonstrated. Enzymes immobilized onto buckypaper through a heterobifunctional cross-linker allow efficient communication between the CNTs and the enzyme, enabling the biofuel cell to operate in human serum [31]. Cyclic voltammetry and biofuel cell polarization were used to characterize the power generated from human serum, presenting a model system for biofuel cells in a physiological context. The maximum power density of 16.12 $\mu W\ cm^{-2}$ was succeeded in the human serum for lower than physiological concentrations of glucose.

2.2 Piezoelectric Nanogenerator

Piezoelectricity, coined from the Greek words "piezein" meaning "to press" and "ēlektron" meaning "amber," refers to the generation of electricity under mechanical stress. This phenomenon occurs in certain materials that can produce opposite electrical charges when subjected to external forces, causing a mechanical deformation [32]. The accumulation of these charges at different ends of the material creates an electric dipole, leading to the formation of a piezopotential. Non-centrosymmetric materials, such as polymers and biological systems, exhibit this effect due to the absence of a centre of symmetry in their crystalline structure. PENG devices are one of the promising nanogenerator which provide clean and efficient

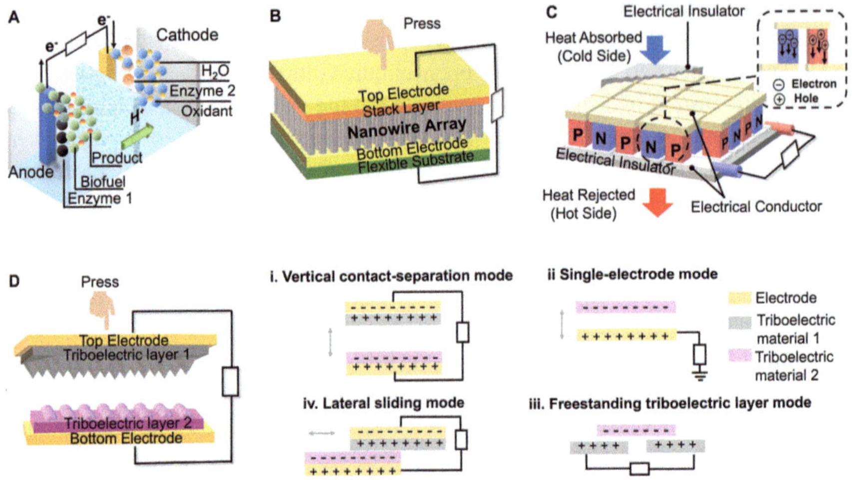

Fig. 2.4 Different energy harvesters and their device structures (**a**) Bio-fuel cell (**b**) PENG (**c**) TEG (**d**) TENG and its different modes of operations

energy by transducing the mechanical energy to electrical energy i.e., piezoelectric effect to power the microelectronic devices. These wearable PENG devices can harvest energy from different day-to-day activities like muscle movements, running, walking, and other activities from the human body. Moreover, these PENG devices have been extensively studied and developed for use as self-powered dynamic pressure sensors in applications such as e-skin, soft robotics, medical devices, and prosthetics [33]. They offer promising prospects for wearable and implantable technologies, enabling efficient and sustainable power generation while accurately detecting and measuring pressure changes (Fig. 2.4).

2.2.1 *Piezoelectric Materials*

Various materials such as polymers both natural and synthetic polymers (organic), metal oxides and nitrides (inorganic), and their composites (hybrid) were well explored as the piezoelectric layers for energy harvesting and sensing applications. Among different nanomaterials based on metal oxides and nitrides such as Titanium dioxide (TiO_2), Molybdenum disulfide (MoS_2), Tin Sulfide (SnS), Tungsten Selenide (WSe), Alluminium Nitride (AlN), Cadmium Sulfide (CdS), Barium titanate ($BaTiO_3$), Lead titanate ($PbTiO_3$), Potassiumniobate ($KNbO_3$), Sodium nitrate ($NaNbO_3$) etc. [34]. have been explored for the past decade towards the fabrication of PENG devices. Among all, Zinc oxide (ZnO) exhibits excellent piezoelectric properties due to its unique crystal structure and material properties. The crystalline structure of ZnO is non-centrosymmetric, meaning it lacks a center of symmetry and allows to generate electric charges when subjected to mechanical stress or pressure. Initially, the positive charges will be localised near the Zn atoms and the negative charge centres were localised on the oxygen and both balance each other [35]. As the external stress is applied then the dipoles get separated resulting in the generation of the potential. Additionally, ZnO has a large piezoelectric coefficient, which indicates its high sensitivity to mechanical deformation and its high electron mobility helps in the faster and high output performance along with excellent biocompatibility and non-toxicity. Further, different nanostructures nanoparticles, nanowires, nanotubes etc. also will be explored for various sensing and energy harvesting applications. As an example, the Wang et al. demonstrated a high-performance PENG device harvesting energy from tiny vibrations using the ZnO NW as a piezoelectric layer showing a good energy efficiency of 17–30% [36]. Similarly, Li et al. achieved a significant milestone by developing an implantable PENG device using a single ZnO nanowire [37]. This device was successfully attached to the diaphragm and heart of a living rat, enabling the harvesting of energy from the rat's breath and heartbeats. The PENG demonstrated outputs typically below 50 mV and 500 pA, showcasing its potential for energy generation in biomedical applications. Although, PENG devices using ZnO has demonstrated good output performance but still lot of enhancements needs to be done in terms of the device repeatability, cyclic stability along with the high currents and larger energy conversion efficiency [38]. Improving the piezoelectric properties of ZnO materials and optimizing device design and architecture are areas of active research to achieve higher power generation. Alternatively, the thin films of PZT and AlN also have exhibited a very good piezoelectric performance owing to its large piezoelectric coefficients. Among different piezopolymers such as Poly-1,1-difluoroethene (PVDF), Nylon, Polylactic acid (PLA) etc. PVDF and its co-derivatives like PVDF-trifluoroethylene (PVDF-TrFe) has proven to be excellent materials for exhibiting the high-performance piezoelectric and is well towards for neural and bone regeneration applications [39]. The PVDF offers numerous advantages in terms of the excellent piezoelectric response, flexibility, biocompatibility, mechanical durability, ease of processing, and compatibility with other materials to enhance the device performance. Recently, Wang et al. demonstrated a PVDF-TrFE nanofiber as an energy harvesting device for wound healing application [40]. The implanted NG generate a maximum voltage and current of 6 mV and 6 nA respectively during the cell proliferation studies and results show that the rate was increased 1.6-fold after the stimulation. Similarly, a recent study demonstrated a piezoelectric suture to harvest energy from valvular perturbations for self-powering a wireless sonomicrometry sensor using PVDF as the piezoelectric layer [41]. The study provides insights into the magnitude of valvular perturbations and estimates the levels of electrical energy that can be harvested using a simplified piezoelectric suture model. The results indicate that power generation can range from nano-watts to milli-watts, with the maximum power harvested from the leaflet plane [42] (Table 2.1).

On the other hand, piezoelectric biomaterials, such as proteins or biological polymers, are derived from organic sources, including animal tissues, plants, or

Table 2.1 Self-powered IMDs using PENG and piezoelectric transducers

Energy source & reference	IMDs	Piezoelectric materials	Transducer type	Max output (power/energy/V_{oc}/I_{sc})	Dimension/volume/thickness
Heartbeat 2012 [43]	Pacemaker	PZT-5A ceramics	Cantilever-type (zigzag)	1.3×10^{-2} μW/mm^2 (power density)	$27 \times 27 \times 6$ mm^3
Heartbeat 2014 [44]	Pacemaker (trigger energy of 1.1 μJ)	PMN-PT single crystal	Thin film	8.2 V, 145 μA 2.7 μJ	1.7 cm × 1.7 cm
Heartbeat 2015 [45]	Pacemaker	PZT ceramics	Nanogenerator	3 V	400 nm–600 nm (thickness)
Heartbeat 2017 [46]	Pacemaker	PZT ceramics	Cantilever-type (Fan-folded)	16.25 μW	1 cm^3
Heartbeat 2019 [47]	Pacemaker	Piezoelectric composites	Multiple layers structure	36 V, 22 μA, 33μW 20 V,8 μA 22 V,40 μA, 14μW 12 V, 15μA	100 μm (thickness)
Heartbeat 2020 [48]	Pacemaker	PDMS-infilled microporous P(VDF-TrFE)	Cantilever-type (multibeam)	3.22 ± 0.24 V	6 mm (thickness)
Heartbeat 2021 [49]	Pacemaker	Nanofibers of PVDF and hybrid nanofillers	Nanogenerator	24.52 ± 1.11 V, 19.39 ± 1.08 μA 0.487 μJ/heartbeat	1.5 cm × 1.5 cm
Ascending aorta pulse 2015 [47]	Pacemaker/ICD/Cardiac monitor	PVDF & Al layer	Thin film	10.3 V,400 nA, 681 nW 1.5 V, 300 nA, 20 nW	$56 \times 56 \times 0.2$ mm^3

microorganisms. These materials possess inherent piezoelectric properties due to the presence of polar biomolecules with dipole moments, such as amino acids or nucleotides [50]. Examples of natural bio-piezoelectric materials include collagen, silk, DNA, and certain types of proteins. Similarly, other natural biomaterials such as cellulose, chitosan, glycine, PGLA etc. also has demonstrated excellent piezoelectric response along with its superior mechanical, chemical and biological properties [51]. Piezoelectricity in organic biomaterials arises primarily from the reorientation of molecular dipoles, whereas in traditional inorganic materials, it stems from the disruption of structural symmetry. A piezoelectric biomaterial is subjected to stress, the molecular chains with permanent dipoles align in a specific direction, resulting in a change in net polarization and exhibiting piezoelectric behaviour, even in the absence of non-centrosymmetric. To exhibit the piezoelectric properties, they need to possess permanent molecular dipoles, the capability to align these dipoles, and the ability to maintain the alignment of dipoles [52]. However, the piezoelectric coefficient of the natural biomaterials is lower as compared with the inorganic metal oxides or nitrides but demonstrate excellent biocompatibility and biodegradability which are very useful for the medical and implantable devices. Further, the solution processability of these materials allows for their transformation into a liquid or gel-like state, which can be easily manipulated and shaped into desired forms. Typically, the natural piezoelectric material is dissolved in a suitable solvent, such as water or an organic solvent, to form a homogeneous solution or dispersion. This solution can then be applied using various techniques, such as spin-coating, dip-coating, or inkjet printing, onto substrates or templates to create thin films, fibers, or patterns. Further, it enables the fabrication of flexible and conformal devices, as well as the integration of these materials into complex structures or electronic systems. Additionally, it allows for the incorporation of other functional additives or biomolecules to enhance the performance or tailor the properties of the natural piezoelectric material. For instance, sumanta et al. demonstrated a work on the fabrication of the silk fibroin-based PENG device which demonstrated excellent mechanical flexibility and superior electrical performance. This literature review focuses on the design and characterization of a mechanically robust piezoelectric nanogenerator (PNG) utilizing spider silk [53]. The PENG exhibits impressive energy conversion efficiency (~66%), generating a high output voltage (~21.3 V) and current (~0.68 μA), accompanied by an instantaneous power density of approximately 4.56 $\mu W/cm^2$. Moreover, the PENG demonstrates biocompatibility and ultra-sensitivity in physiological signal monitoring, particularly arterial pulse response, opening some of the new possibilities for potential biomedical applications. Similarly, Sarkar et al. demonstrated as flexible piezoelectric nanogenerator based silk fibroin embedded with ZnO nanoparticles with a maximum output voltage of 25 V_{PP} under the higher concentration of ZnO [54]. The nanogenerator's output response exhibited a strong correlation with the bending angle, achieving a peak output power density of 6.67 mW cm^{-2}. The prototype demonstrated the ability to power commercial red LEDs and harvest energy from human body movement. Furthermore, the experimental determination of the piezoelectric coefficient (d_{33}) for the silk-ZnO composite film was carried out, providing valuable insights into its piezoelectric properties.

PENG devices have shown great potential for energy harvesting applications, but they also face several challenges that need to be addressed for their widespread adoption and improved performance. Some of the key challenges in PENG-based devices are (1) power generated by PENGs is relatively low, limiting their practical applications. Improving the energy conversion efficiency and power output of PENGs is crucial to make them more viable for powering electronic devices and systems (2) PENG devices need to withstand mechanical stresses and strains during operation. However, the mechanical durability and stability of PENGs are often compromised due to the fragility of the piezoelectric materials or the structural design. Enhancing the mechanical robustness and stability of PENGs is essential to ensure their long-term performance and reliability (3) scaling up the production of PENG devices while maintaining their performance and quality is a significant challenge. The fabrication techniques for PENGs need to be scalable, cost-effective, and compatible with large-scale manufacturing processes. Developing scalable manufacturing methods for PENGs is essential for their commercialization and widespread adoption (4) PENG devices often need to be integrated with other electronic components or systems. Achieving seamless integration and compatibility with existing electronics, sensors, or energy storage devices can be challenging due to differences in materials, electrical interfaces, or mechanical properties. Overcoming these integration challenges is crucial to enable the efficient utilization of PENGs in practical applications (5) environmental impact of PENG-based devices, particularly in terms of materials used and waste generated, is an important consideration. The development of sustainable and eco-friendly materials for PENGs, as well as strategies for recycling or disposing of these devices at the end of their lifespan, is essential to minimize their environmental footprint (6) designing PENG-based systems to maximize their energy harvesting performance requires careful optimization of various parameters such as material properties, device geometry, and operating conditions. Developing efficient and optimized PENG device designs is crucial to maximize energy conversion and overall system performance.

Various fabrication methods such as electrospinning of nanofibers, 3D printing, or additive manufacturing, thin film based physical vapor deposition such as sputtering, thermal evaporation etc., template assisted fabrication such as nanoimprint lithography, porous templates, hydrothermal/solvothermal methods etc. were well explored for the fabrication of the PENG devices [55]. For instance, Wang et al. demonstrated the vertically aligned ZnO nanowires were embedded between the two metal electrodes to fabricate a PENG device and the output voltage is generated as the nanowires get deformed by weak forces [56]. Further, they successfully demonstrated the conversion of nanoscale mechanical energy into electrical energy using piezoelectric ZnO nanowire (NW) arrays. By deflecting aligned NWs with a conductive atomic force microscope tip, the strain field and charge separation generated by the piezoelectric and semiconducting properties of zinc oxide resulted in electrical current generation through the rectifying characteristic of the Schottky barrier. The efficiency of this NW-based piezoelectric power generator was estimated to be between 17% and 30% [36]. Similarly, the 3D printing and inkjet printing methods have been used for the large area printing with low cost and ease of fabrication methodologies. Further, these printed PENG devices provide advantages in terms of design flexibility, rapid prototyping, integration capabilities, material variety, scalability, cost-effectiveness, and waste reduction. These novel fabrication strategies such as the microstructures, devices structures, chemical composites, core-cladding type electrospinning etc. are advancing the field of PENGs by offering new possibilities for device design, performance optimization, and integration into various applications [57]. Continued research and development in fabrication techniques will contribute to the realization of high-performance, scalable, and customizable PENG devices for energy harvesting and other emerging applications.

2.3 Triboelectric Nanogenerator

The current section focuses on the iTENGs and their potential applications in the field of biomedical implants. The growing demand for implantable medical devices has driven the need for alternative power sources that can provide sustained energy to these devices [58]. Conventional batteries have limitations such as finite lifetime and the need for replacement surgeries, which are invasive and inconvenient for patients. Therefore, the development of self-powered implantable devices has become an active area of research. TENGs have emerged as a promising technology for harvesting energy from mechanical movements and converting it into electrical energy. TENGs utilize the triboelectric effect and electrostatic induction to generate electricity. When two materials with different electronegativities come into contact and then separate, a charge transfer occurs due to the redistribution of electrons. This charge separation creates an electric potential difference between the two materials, which can be harvested as electrical energy. The contact and separation can be achieved through mechanical movements, such as tapping, sliding, or rubbing [59].

Researchers have developed iTENGs that can harvest energy from the mechanical movements of the heart. These TENGs are integrated into cardiac devices such as pacemakers or defibrillators and utilize the pulsation and contraction of the heart muscles to generate electrical energy, providing a sustainable power source for the implanted devices. For instance, Ma et al. demonstrated a TENG device for energy harvesting from respiratory movements in rats by implanting them under the left thoracic skin [60]. A TENG with a working surface of 0.8×0.8 cm^2 achieved an open circuit voltage (V_{oc}) of 3.75 V and a short circuit current (I_{sc}) of 0.14 μA. In a pig model, a TENG implanted between the pericardium and epicardium produced a V_{OC} of 10 V and an I_{sc} of 4 μA, enabling monitoring of ventricular premature contractions and atrial fibrillation, as well as estimation of blood flow parameters Similarly, in a study by Ouyang et al., developed a fully implanted symbiotic pacemaker based on a TENG was developed for correcting sinus arrhythmias in a large animal model [61] (Fig. 2.5).

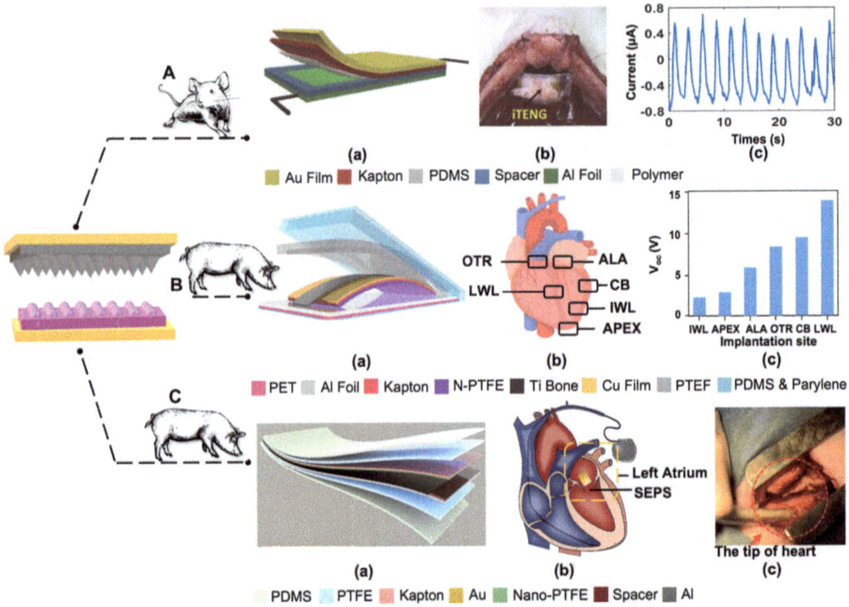

Fig. 2.5 Self-powered cardiovascular implant for in vivo recording demonstrated using TENG devices

The TENG generated a maximum V_{OC} of 65.2 V and an energy of 0.495 μJ during each cardiac motion cycle, surpassing the endocardial pacing threshold energy of 0.377 μJ. Another experiment using a five-staged TENG in a mongrel *in vivo* model demonstrated the ability to charge a lithium-ion battery by harnessing the biomechanical body and gravity motion at a subcutaneous implantation site. This enabled the pacemaker system to operate in the self-rechargeable ventricle pacing and sensing mode. TENGs have been utilized in implantable prosthetic devices to harvest energy from the natural movements of the human body. For example, TENGs integrated into prosthetic limbs can convert the mechanical energy generated during walking or running into electrical energy, which can be used to power sensors, actuators, or even rechargeable batteries within the prosthetic device, eliminating the need for external power sources. As an instance, Shlomy et al. have developed a practical tactile restoration device that address the limitations of existing neuro-prosthetic solutions using TENG based devices. The TENG technology converts mechanical energy into electricity through triboelectrification and electrostatic induction [62]. In this study, the researchers created an integrated tactile (IT) sensory restoration device (TENG-IT) and successfully demonstrated its functionality *in vivo*. Similarly, TENGs have been incorporated into implantable biosensors for continuous monitoring of physiological signals. These sensors can harvest energy from the body's movements or even internal processes such as blood flow or muscle contractions, enabling long-term, self-powered monitoring of vital signs, glucose levels, or other biomarkers. The harvested energy can power the sensors and wireless communication modules, eliminating the need for frequent battery replacements or external power sources. As an example, Liu et al., introduced a bionic cochlear auditory sensor utilizing a TENG to directly convert acoustic signals into electrical signals and achieve frequency selectivity [63]. The TENG based sensor design imitates the basement membrane of the cochlea and consists of a trapezoidal polytetrafluoroethylene (PTFE) film and nine small rectangular silver electrodes. When exposed to external sound signals, the PTFE film resonates, contacting and separating from the electrode layer, thereby generating a voltage signal. The different positions on the trapezoidal structure exhibit specific response frequency ranges spanning from 20 to 3000 Hz Previous studies have showcased the potential of TENG in various applications, including biomedical uses such as energy harvesting and nerve repair. TENGs have been explored in implantable drug delivery systems to provide a self-sustained power source for controlled drug release. The mechanical motions within the body, such as respiratory movements or blood flow, can be harnessed by TENGs to generate electrical energy, which can be used to drive drug release mechanisms, ensuring a continuous and controlled release of medication without the need for external power or manual intervention. For instance, in a pioneering study, researchers developed a self-powered controlled drug-release system using a triboelectric nanogenerator (TENG). The system consisted of a TENG for energy harvesting and a drug pumping system comprising an electrochemical pressure pump, a drug reservoir, and a long microtube. The output from the TENG was converted and rectified to generate direct current (15 V; 1.5 mA) for the electrochemical pressure pumping mechanism [64]. The process involved water-splitting, generating oxygen and hydrogen to pressurize the drug reservoir and propel the drug through the microtube for ocular drug delivery. The drug-releasing speed could be adjusted within a range of 5.3–40 $\mu L\ min^{-1}$, which correlated with the TENG output. TENGs have been investigated for use in implantable neurostimulation devices, such as deep brain stimulators or spinal cord stimulators. By harvesting energy from the body's movements or muscle contractions, TENGs can provide the necessary power for stimulating neural tissues, offering a self-powered and potentially more efficient alternative to conventional battery-powered devices.

iTENGs offer several advantages for powering implantable medical devices and provide a sustainable and renewable power source, eliminating the need for frequent battery replacements [65]. This improves the long-term reliability and reduces the risk of device failure. Next, iTENGs can harvest energy from various human body movements, such as heartbeat, respiration, and muscle contractions, enabling continuous power generation. Moreover, the compatibility of iTENGs with biocompatible materials allows for seamless integration into implantable devices, minimizing the risk of rejection or adverse reactions. The choice of materials and fabrication techniques play a crucial role in the development of iTENGs. Biocompatible materials, such as polymers, are commonly used as the triboelectric layers in iTENGs to ensure compatibility with the human body [66]. Various fabrication techniques, including 3D printing, thin-film deposition, and nanostructuring, are employed to create the desired device structures and optimize the performance of iTENGs. The selection of appropriate materials and fabrication methods is essential to enhance the energy conversion efficiency and durability of iTENGs.

iTENGs have the potential to power a wide range of implantable medical devices, including pacemakers, neurostimulators, drug delivery systems, and sensors. They can also be used for monitoring physiological signals and harvesting energy from biomechanical movements in the human body. However, there are several challenges that need to be addressed, such as device miniaturization, optimization of power output, long-term stability, and biocompatibility. Further research and development efforts are required to overcome these challenges and fully exploit the potential of implantable triboelectric nanogenerators in biomedical applications. TENGs offer great potential for various applications, but several challenges need to be addressed for their successful implementation. Some of the challenges associated with TENG devices, focusing on materials, fabrication methods, stability, cost, large-scale fabrication methods, and ease of processing are discussed below. (1) The choice of materials is crucial for TENG performance. Finding suitable materials with high triboelectric properties, mechanical flexibility, and durability is a challenge. Additionally, ensuring compatibility with the desired application, such as biocompatibility for medical applications, adds another layer of complexity. (2) The fabrication of TENG devices often requires precise patterning and integration of multiple layers or structures. Conventional fabrication techniques, such as photolithography or vacuum deposition, may not be suitable for large-area or flexible substrates. Developing scalable and cost-effective fabrication methods for TENGs, such as roll-to-roll printing or solution-based processes, is a challenge that needs to be overcome. (3) TENG devices should maintain stable performance over extended periods of operation. However, factors like mechanical wear, environmental conditions, and material degradation can affect the device stability. Ensuring long-term stability and reliability is a challenge that requires robust materials, protective coatings, and proper encapsulation techniques. (4) The cost of materials, fabrication processes, and device integration plays a crucial role in the commercial viability of TENG devices. Cost-effective manufacturing methods and the use of low-cost materials without compromising performance are essential challenges to address (5) Scaling up the fabrication of TENG devices to meet industrial or commercial demands presents challenges. Maintaining uniformity and high performance across large-area devices, while minimizing production costs, requires the development of scalable manufacturing techniques and equipment. (6) TENG devices should ideally be compatible with existing manufacturing processes and easily integrated into various applications. Compatibility with different substrates, compatibility with flexible or stretchable materials, and ease of integration with other electronic components pose challenges that need to be addressed. Addressing these challenges requires multidisciplinary efforts combining materials science, device engineering, fabrication techniques, and cost-effective manufacturing approaches. Overcoming these hurdles will enable the widespread adoption of TENG devices in various industries, ranging from energy harvesting and self-powered systems to healthcare and human-machine interfaces.

2.4 Thermoelectric Nanogenerator

Thermoelectric nanogenerators (TEGs) have emerged as a promising technology for harvesting waste heat and converting it into usable electrical energy. The underlying principle of TEGs is based on the seebeck effect, where a temperature gradient across a material generates an electric potential difference. This thermoelectric conversion process allows for the direct conversion of heat into electricity, offering a sustainable and efficient means of power generation [67]. The underlying principle of thermoelectric generation is the seebeck effect, where a temperature gradient across a material generates an electric potential difference. This effect is based on the behaviour of charge carriers in a material with different seebeck coefficients, leading to the creation of a voltage. The efficiency of a TEG is determined by the figure of merit, ZT, which is a measure of the material's ability to convert heat into electricity. ZT depends on several factors, including the seebeck coefficient, electrical conductivity, and thermal conductivity. Enhancing ZT is crucial for improving the performance of TEGs [68]. Various types of TEGs have been developed to cater to different applications and operating conditions. Bulk TEGs utilize high-performance thermoelectric materials and are suitable for applications with moderate temperature gradients. They are often used in industrial waste heat recovery systems to improve overall energy efficiency. Thin-film TEGs, on the other hand, employ flexible substrates and thin layers of thermoelectric materials, allowing for their integration into wearable devices and portable electronics. These TEGs can be fabricated using deposition techniques such as sputtering or chemical vapor deposition, enabling scalable and cost-effective production [69]. Nanostructured TEGs leverage the unique properties of nanomaterials to enhance performance. By reducing thermal conductivity and improving charge carrier transport, these TEGs exhibit improved efficiency and power output. Hybrid TEGs combine thermoelectric conversion with other energy conversion mechanisms, such as photovoltaics or piezoelectricity, to further enhance performance. These hybrid systems offer synergistic effects, allowing for improved energy harvesting capabilities. Research efforts are focused on developing novel materials, optimizing device engineering, and exploring innovative fabrication techniques to enhance the efficiency and commercial viability of TEGs. These advancements are crucial for enabling the widespread adoption of TEGs in various sectors, including automotive, aerospace, and wearable electronics, for sustainable energy harvesting and utilization of waste heat [70].

TEGs have shown great potential for a wide range of applications in the biomedical field, particularly in the development of implantable devices. The ability of TEGs to directly convert heat into electricity makes them suitable for harnessing thermal energy from the human body and powering various biomedical devices, thereby eliminating the need for traditional power sources such as batteries or external connections. One important application of TEGs in the biomedical field is in self-powered implantable devices. Implantable medical devices, such as pacemakers, neurostimulators, and drug delivery systems, require a continuous and reliable power source. TEGs can be integrated into these devices to harvest the body's heat and convert it into electrical energy to power the device. This eliminates the need for battery replacement surgeries or external connections, leading to longer device lifespan and improved patient comfort [71]. TEGs have been used to power implantable sensors for monitoring physiological parameters, such as temperature, pressure, and glucose levels, providing real-time data for medical diagnostics and personalized healthcare. Another important application of TEGs in the biomedical field is in tissue engineering and regenerative medicine. TEGs can be utilized to generate electrical stimulation for promoting tissue growth and regeneration. By incorporating TEGs into scaffolds or implants, electrical stimulation can be applied directly to the target tissue, enhancing cell proliferation, differentiation, and tissue integration. This has significant implications in areas such as bone regeneration, wound healing, and nerve regeneration. TEGs also find applications in wearable and portable medical devices. With the miniaturization and flexibility of TEGs, they can be integrated into wearable devices, such as smart garments or wearable patches, to harvest body heat and power various health monitoring systems. These devices can continuously monitor vital signs, such as heart rate, respiration rate, and body temperature, providing valuable data for healthcare professionals and individuals to track their health status. Moreover, TEGs have potential applications in targeted drug delivery systems. Implantable TEGs can generate electrical energy to drive drug release mechanisms, such as electrochemical pumps or electro-responsive hydrogels. This enables precise and controlled drug delivery, allowing for personalized and localized therapy. Further, the TEGs offer exciting opportunities in the field of biomedical and implantable devices. They provide a self-sustainable power source, enabling long-term operation of implantable devices, tissue regeneration, wearable health monitoring systems, and targeted drug delivery [72]. With further advancements in TEG materials, device design, and fabrication techniques, their integration into biomedical applications is expected to continue growing, leading to improved healthcare outcomes and enhanced patient care.

Various TEGs have been successfully demonstrated using various materials such as the composites of the 1D nanowires and nanotubes, 2D nanosheets, nanoparticles, micro belts, and their composites. In addition, different other transition metal alloys and its hybrids such as bismuth telluride, lead telluride, antimony telluride have demonstrated excellent performance owing to their high ZT coefficients (>2) but suffers from the toxicity and rigidity further exhibiting its non-suitability for the implantable and biomedical applications [73]. On the other hand, as compared to the conducting polymers such as PANI, PEDOT: PSS etc. have demonstrated a greater biocompatibility, biodegradability and mechanical flexibility but exhibits poor thermal stability and low conductivity. These organic polymeric materials exhibit 2–3 orders lower ZT values as compared to the inorganic metal alloys. Hence, an organic-inorganic hybrid composite offers multiple advantages and demonstrate excellent thermoelectric properties. In a recent study introduced a TEG prototype featuring a specialized boost circuit designed to power a pacemaker [74]. Additionally, they validated the efficiency of the energy harvesting circuit using a clock circuit with higher power consumption (145 μW) as a substitute for an actual pacemaker with the same pulse interval. When the temperature gradient reached a stable state of 0.5 °C in a rabbit, the measured output voltage was initially 20 mV, but it increased to 3.3 V upon passing through the boost circuit. Similarly, Bhatia et al. reported a TEG using doped bismuth telluride to power the pacemaker and 4000 TEGs were connected in series to drive the packemaker with the surface area of 6 cm^2. The TEG generate an V_{oc} of 4 V and has delivered an output power of 100 μW further demonstrating its potential for driving the future biomedical and implantable devices [75]. Similarly, there were several studies has been reported using Te nanowires/PEDOT: PSS, MoS_2 nanoclusters embedded in polyurethane composite film, Sb-doped ZnO microbelts or nanobelts encapsulated with PDMS etc.

Although, TEGs have shown great potential as energy conversion devices, they also have certain limitations that need to be addressed for their widespread implementation. The limitations of the TEGs were discussed as follows (1) The efficiency of TEGs is generally low compared to other energy conversion technologies. TEGs typically have conversion efficiencies in the range of 5–10%, which means a significant amount of the input heat is wasted. Improving the efficiency of TEGs is an ongoing challenge to maximize their energy conversion capabilities. (2) TEGs have a limited operating temperature range within which they can effectively convert heat into electricity. This restricts their applicability to specific temperature differentials. TEGs typically work best with high temperature differentials, but their performance decreases as the temperature difference decreases. (3) The efficiency of TEGs depends heavily on the choice of thermoelectric materials used. Currently available materials often have limitations in terms of their thermoelectric properties, such as low conversion efficiency, limited temperature range, and high cost. Finding new materials or optimizing existing ones with better thermoelectric properties is crucial for enhancing the performance of TEGs. (4) TEGs can be expensive to produce due to the high cost of thermoelectric materials and complex manufacturing processes. The cost factor limits their widespread adoption in various applications. Finding cost-effective manufacturing methods and exploring alternative low-cost materials are important for making TEGs more economically viable. (5) Scaling up TEGs for large-scale applications can be challenging. The production of TEG modules with sufficient power output requires the integration of many thermoelectric elements, which adds complexity and cost to the system. Developing scalable fabrication methods and improving module design are necessary for achieving practical and commercially viable TEG systems.

2.4.1 Applications of Energy Harvesters

Energy harvesters hold significant promise for a wide range of applications, particularly in the domains of wireless communications, implantable devices, and self-powered sensors. These innovative devices have the potential to revolutionize the way we power and operate various technologies. In the realm of wireless communications, energy harvesters offer an intriguing solution to address the challenge of limited battery life. By harvesting energy from the surrounding environment, such as solar power, vibrations, or radiofrequency signals, these devices can provide a continuous and sustainable power source for wireless communication systems. This opens exciting possibilities for long-term deployment of wireless sensors, IoT devices, and remote monitoring systems, without the need for frequent battery replacements or inconvenient wired connections. Implantable devices stand to benefit greatly from energy harvesters as well. Medical implants, such as pacemakers, neural stimulators, and drug delivery systems, often require a reliable and long-lasting power source. Energy harvesters can tap into the body's own energy, such as body heat or movement, to generate power for these implants. This not only eliminates the need for battery replacements or external power sources but also enhances patient comfort and safety by reducing invasive procedures. Self-powered sensor applications also stand to gain from energy harvesters. These sensors are increasingly utilized in various industries, including environmental monitoring, structural health monitoring, and wearable technology. By harnessing energy from the ambient environment, energy harvesters can provide a self-sustaining power source for these sensors, enabling continuous data collection and reducing maintenance efforts. This opens new possibilities for long-term and widespread deployment of sensors in remote or inaccessible locations. In conclusion, the potential applications of energy harvesters in wireless communications, implantable devices, and self-powered sensor applications are vast and promising. By tapping into ambient energy sources, these devices offer sustainable power solutions, reduce reliance on batteries, and enable continuous operation in diverse settings. As research and development in this field continue to advance, we can expect to witness ground-breaking innovations that will shape the future of these industries.

2.4.2 Challenges and Opportunities of Energy Harvesters

Performance Enhancement

Triboelectric and piezoelectric nanogenerators have emerged as promising technologies for harvesting mechanical energy from the environment and converting it into usable electrical power. These nanogenerators hold great potential for various applications, ranging from self-powered sensors to wearable electronics and even energy harvesting in large-scale systems. This chapter explores recent

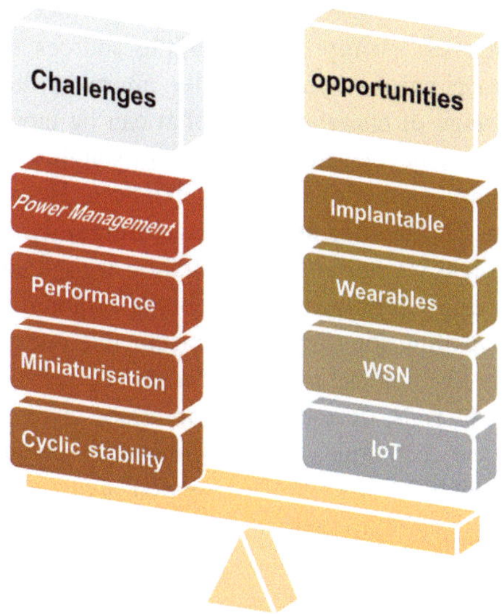

Fig. 2.6 Challenges and opportunities of the energy harvesting devices

advancements and strategies employed to enhance the performance of triboelectric and piezoelectric nanogenerators, focusing on improving their efficiency, output power, and reliability [76] (Fig. 2.6).

The choice of materials and device design plays a crucial role in the performance of nanogenerators. Researchers have been exploring novel materials with high triboelectric and piezoelectric properties to enhance energy conversion efficiency [77]. Furthermore, advancements in nanomaterials, such as graphene, CNTs, and nanostructured polymers, have shown improved performance in terms of charge generation and collection. Additionally, device design optimization, such as maximizing contact area, improving electrode configurations, and implementing three-dimensional structures, has led to increased power output and enhanced mechanical-to-electrical energy conversion. The interface between the triboelectric or piezoelectric materials and the electrodes significantly impacts the device performance. Surface modification techniques, such as functionalization, nanostructuring, and surface patterning, have been employed to enhance the charge transfer efficiency and reduce energy loss at the interfaces [78]. Introducing nanostructured materials and coatings on the electrode surfaces can improve charge collection and reduce surface energy dissipation, leading to enhanced device performance. Efficient energy harvesting requires the integration of nanogenerators with energy management systems. Several developments such as power conditioning circuits, energy storage devices, and power management systems to optimize the power output and utilization of nanogenerators. These systems enable efficient power transfer, voltage regulation, and energy storage, allowing for the seamless integration of nanogenerators into various applications [79]. To further enhance performance, various studies from the literature have explored the concept of hybrid and multifunctional nanogenerators. These systems combine different energy harvesting mechanisms, such as triboelectric, piezoelectric, and photovoltaic, to harvest energy from multiple sources simultaneously. The integration of these mechanisms synergistically improves the overall power output and expands the range of energy sources that can be tapped into, providing greater versatility and efficiency. Ensuring long-term stability and reliability is crucial for the practical implementation of nanogenerators. However, there are still some challenges that needs attention such as device durability, environmental robustness, and mechanical stress. Protective encapsulation techniques, material engineering, and mechanical design improvements are being explored to enhance device stability and reliability, enabling long-term and sustainable energy harvesting. TENG and PENG devices hold immense potential for energy harvesting applications [80]. Advancements in materials, device design, interface engineering, system integration, and hybridization techniques have significantly enhanced their performance, efficiency, and reliability. Continued research and development in these areas will unlock new opportunities and applications for these nanogenerators, paving the way for a sustainable and self-powered future.

Device Miniaturisation

Miniaturization plays a crucial role in the successful integration of energy harvesting technologies, such as TENGs and PENGs, into implantable devices. Achieving small form factors is essential to ensure seamless integration, minimize the device's physical footprint, and maintain the functionality and comfort of the implant. Here, we delve into the device miniaturization aspects for TENGs and PENGs in implantable device applications. Component size and integration reducing the size of the individual components within TENG and PENG systems is vital for achieving miniaturization [81]. This includes downsizing the energy harvesting units, such as the triboelectric or piezoelectric materials, electrodes, and structural components. Advances in nanofabrication techniques enable the precise manufacturing of miniature components with high precision and reliability. Additionally, integrating multiple components into a compact unit or module helps optimize the overall size of the energy harvesting system. Flexible and thin-film configurations are advantageous for miniaturization in implantable devices. By employing flexible substrates and thin-film deposition techniques, the overall thickness and size of the energy harvesting device can be significantly reduced. Flexible designs also facilitate conformal integration with the target implant, allowing the device to adapt to the irregular surfaces and contours of the human body. Microfabrication techniques is crucial for achieving miniaturization in TENG and PENG systems [82]. Microelectromechanical systems (MEMS) fabrication methods, such as photolithography, etching, and deposition, enable the precise fabrication of microscale features and structures. These techniques aid in creating miniature energy harvesting devices with high efficiency, reliability, and reproducibility. System integration efficient integration of TENGs and PENGs with other components of the implantable device is vital for miniaturization. This includes integrating the energy harvesting system with the sensor, control circuitry, data processing unit, and wireless communication modules. Compact and intelligent integration reduces the overall size of the implantable device, allowing for easy and unobtrusive implantation while maintaining seamless functionality [83]. Power management and storage systems are crucial for miniaturized TENG and PENG implantable devices. Efficient power management circuitry ensures optimal energy transfer from the energy harvester to the device components while minimizing energy losses. Miniaturized energy storage technologies, such as thin-film batteries or supercapacitors, can store harvested energy and provide a stable power supply for the implantable device during periods of low energy generation. Miniaturization should not compromise the biocompatibility of the energy harvesting system. It is crucial to select materials and fabrication techniques that are biocompatible and non-toxic to ensure long-term implant viability [84]. Furthermore, the miniaturized device should undergo rigorous biocompatibility testing to ensure it meets the necessary safety and regulatory standards. Finally, miniaturization is a critical aspect in the successful integration of TENG and PENG systems into implantable devices. By reducing component size, employing flexible and thin-film designs, utilizing microfabrication techniques, optimizing system integration, implementing efficient power management and storage, and considering biocompatibility aspects, energy harvesting systems can be miniaturized for seamless integration into implantable devices. These advancements pave the way for the development of compact and efficient self-powered implantable devices that can revolutionize healthcare practices.

Long Term Cyclic Stability

The long-term cyclic stability of TENGs and PENGs is a critical consideration for their successful implementation in implantable device applications. Implantable devices often require consistent and reliable power sources over extended periods to ensure continuous operation and minimize the need for maintenance or replacement. Therefore, understanding and addressing the cyclic stability aspects of TENGs and PENGs is crucial for their long-term performance and viability in such applications. One key aspect affecting the long-term cyclic stability of TENGs and PENGs is the durability of the materials used. The nanogenerator components, including the triboelectric or piezoelectric materials and the electrodes, should exhibit excellent mechanical and electrical stability [85]. The materials must withstand repeated cycles of strain, stress, and electrical potential without significant degradation or fatigue. Robust materials with high resistance to wear, corrosion, and mechanical fatigue are desirable to ensure long-term reliability and cyclic stability. The structural integrity and adhesion of TENG and PENG components are crucial for maintaining stable performance over an extended period. Ensuring strong bonding between the triboelectric or piezoelectric layer and the electrodes is essential to prevent delamination or detachment, which can lead to performance degradation or failure. Proper encapsulation or protective coatings can be employed to enhance the long-term stability of the device, safeguarding against environmental factors such as moisture, temperature variations, and chemical exposure. TENGs and PENGs may experience performance degradation due to various factors [86]. For TENGs, factors such as wear, surface contamination, and changes in the triboelectric properties of materials can lead to reduced charge transfer efficiency and output power. Similarly, for PENGs, factors such as mechanical fatigue, polarization fatigue, and material aging can impact the piezoelectric properties and energy conversion efficiency. Understanding the degradation mechanisms specific to each nanogenerator type is crucial for devising strategies to mitigate performance degradation and ensure long-term cyclic stability. To assess the long-term cyclic stability of TENGs and PENGs, reliability testing and accelerated aging studies are commonly employed. These tests subject the nanogenerators to accelerated stress conditions, simulating prolonged use, and cycling [87]. By monitoring the performance parameters, such as output voltage, current, power density, and energy conversion efficiency, over an extended period, the cyclic stability and potential degradation trends can be evaluated. Reliability testing helps identify weak points or failure modes and allows for optimization of the device design and materials to improve long-term performance. To ensure the suitability of TENGs and PENGs for implantable device applications, in vivo and long-term monitoring studies are crucial. Implanting the nanogenerators in animal models or human subjects allows for the assessment of their performance, stability, and biocompatibility in a real physiological environment over an extended period. Monitoring factors such as output power, voltage, and charge transfer efficiency can provide insights into the long-term cyclic stability and performance of the devices in practical applications.

Long Term Self-Powering Capacity

The long-term self-powering capacity of TENGs and PENGs holds significant promise for implantable device applications. This aspect refers to the ability of these energy harvesters to sustainably generate power over extended periods, thereby eliminating or reducing the need for frequent battery replacements or external power sources. TENGs and PENGs utilize renewable energy sources present in the surrounding environment, such as mechanical vibrations, body movements, or ambient light [88]. These energy sources are virtually infinite and readily available, allowing for continuous power generation without relying on finite resources like traditional batteries. TENGs and PENGs are designed to efficiently convert mechanical energy into electrical energy through the triboelectric or piezoelectric effect, respectively. Their ability to harness and convert ambient energy ensures a sustainable power supply for implantable devices over the long term, without depleting or relying on external power sources. Implantable devices traditionally rely on batteries as their power source. However, batteries have limited lifespans and require periodic replacement or recharging, which can be inconvenient and pose risks during invasive procedures. By integrating TENGs or PENGs into implantable devices, the reliance on batteries can be reduced or eliminated, ensuring continuous operation, and minimizing the need for maintenance or battery-related procedures [89]. The long-term self-powering capacity of TENGs and PENGs contributes to the longevity and reliability of implantable devices. By providing a continuous power source, these energy harvesters support uninterrupted device functionality, enabling long-term monitoring, diagnosis, or treatment without interruptions caused by battery depletion. This enhances the overall lifespan and effectiveness of implantable devices, improving patient outcomes and reducing the need for device replacements or interventions. TENGs and PENGs can be miniaturized and seamlessly integrated into implantable devices due to their compact size and flexible design. This allows for direct integration with the target device, minimizing additional space requirements and enabling efficient energy harvesting without compromising the device's functionality or aesthetics. TENGs and PENGs can be engineered using biocompatible materials, ensuring compatibility with the human body, and minimizing the risk of adverse reactions [90]. Moreover, as these energy harvesters generate power through non-invasive means, they offer a safe and reliable power source for long-term implantable device applications. In summary, the long-term self-powering capacity of TENGs and PENGs for implantable device applications offers numerous advantages. By utilizing renewable energy sources and providing sustainable power generation, these energy harvesters reduce dependence on batteries, enhance device longevity, and support continuous operation. Their miniaturization and biocompatibility further enable seamless integration into implantable devices, ensuring safe and reliable long-term power supply. As research and development progress, these aspects will continue to be refined, paving the way for the widespread adoption of TENGs and PENGs in the field of implantable devices.

2.4.3 Power Management Circuitry

Power management for TENGs and PENGs in implantable device applications poses several challenges that need to be addressed for their successful integration. TENGs and PENGs generate relatively low power outputs, especially in miniaturized and implantable configurations. The power generated may not be sufficient to meet the energy demands of complex implantable devices, such as pacemakers or neurostimulators. Ensuring a stable and consistent power output from these nanogenerators is crucial for the reliable operation of the implantable devices. Power Conversion Efficiency: The conversion efficiency of TENGs and PENGs, i.e., the ability to convert mechanical energy into electrical energy, needs improvement. Higher efficiency is required to maximize the energy harvesting capabilities of these nanogenerators and minimize power losses. Strategies such as optimizing materials, device designs, and interface engineering need to be explored to enhance the conversion efficiency [91]. Implantable devices have strict size and weight constraints. Thereforc, miniaturizing TENGs and PENGs while maintaining their power generation capabilities is a challenge. Additionally, integrating these nanogenerators seamlessly with the implantable devices without compromising their functionality and performance poses further difficulties. Implantable devices require materials that are biocompatible and safe for long-term use within the body. The materials used in TENGs and PENGs should meet these requirements to ensure the device's compatibility with living tissues. Additionally, the nanogenerators should demonstrate long-term stability and durability to withstand the harsh physiological conditions and mechanical stresses experienced within the body. Efficient power management circuitry is essential to regulate, store, and distribute the harvested energy from TENGs and PENGs to power the implantable devices. The power management circuitry should be designed to maximize energy extraction, minimize losses, and ensure optimal utilization of the available energy. It should also incorporate appropriate energy storage elements, such as supercapacitors or rechargeable batteries, to provide a stable power supply when the nanogenerator output is insufficient or intermittent [92]. Implantable devices powered by TENGs and PENGs must adhere to stringent safety and regulatory standards. The design and implementation of the power management system should ensure safe operation within the human body and comply with relevant medical device regulations. Robust safety measures, such as overvoltage and overcurrent protection, should be incorporated to prevent any potential harm to the patient. Addressing these challenges requires interdisciplinary collaboration between materials scientists, engineers, biologists, and medical professionals. Continued research and development efforts are necessary to overcome these hurdles and unlock the full potential of TENGs and PENGs as reliable and sustainable power sources for implantable devices.

2.4.4 Opportunities of Energy Harvester

Energy harvesters, also known as energy scavengers or energy harvesters, are devices that capture and convert ambient energy from the environment into usable electrical energy. These devices have gained significant attention in recent years due to their potential to provide sustainable and renewable power sources for various applications. Here are some of the opportunities and potential applications of energy harvesters. Energy harvesters can power small electronic devices and sensors used in IoT applications [93]. They can scavenge energy from ambient sources such as light, vibration, thermal gradients, or radio waves to enable self-powered and wireless operation of IoT devices, eliminating the need for batteries or frequent maintenance. Energy harvesters can be integrated into wearable devices, such as smartwatches, fitness trackers, and medical sensors. They can utilize body movements, body heat, or ambient light to generate power, extending the battery life or even eliminating the need for batteries altogether. Energy harvesters can be employed in remote or inaccessible locations to power environmental monitoring systems [94]. They can extract energy from the surrounding environment, allowing continuous operation of sensors for measuring parameters like temperature, humidity, air quality, or water quality. Energy harvesters can be integrated into infrastructure and structural components, such as bridges, buildings, or aircraft wings, to monitor their health and detect structural damage. The harvesters can convert vibrations or mechanical strain into electrical energy to power sensors that monitor stress levels, cracks, or other structural abnormalities. Energy harvesters can be used to power wireless sensor networks (WSNs) that are deployed in remote areas or harsh environments where frequent battery replacements are impractical. The harvested energy can be used to power the sensor nodes and transmit data wirelessly, enabling long-term and autonomous operation of the network [95]. Energy harvesters can contribute to the development of smart cities by powering various infrastructure and devices. For example, they can harness energy from foot traffic, vehicles, or solar radiation to generate electricity for streetlights, traffic monitoring systems, or public charging stations for electric vehicles. Energy harvesters can be used in portable electronic devices, such as smartphones or tablets, to provide supplementary power. By capturing and converting energy from ambient sources like solar radiation or kinetic energy from user interactions, these harvesters can extend the battery life and enhance the device's usability. Energy harvesters can be particularly beneficial in remote or off-grid areas where access to a reliable power grid is limited. They can provide a sustainable power source for low-power devices, such as wireless communication systems, monitoring equipment, or small appliances, reducing the reliance on traditional energy sources like batteries or fossil fuels.

2.5 Conclusion

In conclusion, energy harvesters have emerged as a promising power source for implantable cardiac applications. These innovative devices have the potential to revolutionize the field of cardiology by addressing the critical challenge of long-term power supply for cardiac implants. By converting ambient energy sources, such as mechanical vibrations or thermal gradients, into electrical energy, energy harvesters offer a sustainable and self-sufficient solution for powering implantable devices. The use of energy harvesters in implantable cardiac applications brings several significant advantages. Firstly, it eliminates the need for frequent battery replacements or recharging procedures, reducing the invasiveness and inconvenience associated with traditional power sources. This not only enhances patient comfort but also minimizes the risk of infections and complications related to repeated surgeries. Moreover, energy harvesters offer a reliable and continuous power supply, ensuring uninterrupted operation of vital cardiac devices. This is particularly crucial in critical situations where sudden power failures could have life-threatening consequences. By harnessing and storing energy from the patient's own body or the surrounding environment, these devices can provide a dependable source of power, improving the safety and reliability of implantable cardiac systems. Additionally, energy harvesters contribute to sustainability and cost-effectiveness in healthcare. By reducing or eliminating the reliance on batteries, they minimize the environmental impact associated with their disposal and reduce the overall healthcare waste. Furthermore, the long-lasting nature of energy harvesters reduces the need for frequent medical interventions, resulting in cost savings for both patients and healthcare systems. Although energy harvesters for implantable cardiac applications are still in the early stages of development, the progress made thus far is promising. Ongoing research and technological advancements will continue to refine these devices, improving their efficiency, miniaturization, and integration with existing cardiac implants. With further advancements, energy harvesters have the potential to enhance the lives of countless individuals with cardiac conditions, offering a sustainable and reliable power source for their implantable devices. In conclusion, energy harvesters represent an exciting avenue for powering implantable cardiac devices, revolutionizing the field and providing a sustainable, self-sufficient, and reliable solution for long-term power supply. Their potential to improve patient comfort, ensure uninterrupted operation, contribute to sustainability, and reduce healthcare costs makes them a compelling choice for the future of implantable cardiac applications.

References

1. Y. Zou, L. Bo, Z. Li, Recent progress in human body energy harvesting for smart bioelectronic system. Fundam. Res. **1**(3), 364–382 (2021)
2. S.-Y. Yang, V. Sencadas, S.S. You, N.Z.-X. Jia, S.S. Srinivasan, H.-W. Huang, A.E. Ahmed, J.Y. Liang, G. Traverso, Powering implantable and ingestible electronics. Adv. Funct. Mater. **31**(44), 200–289 (2021)
3. V. Leonov, R.J.M. Vullers, Wearable electronics self-powered by using human body heat: the state of the art and the perspective. J. Renew. Sustain. Energy **1**(6), 62–701 (2009)
4. F. Hartmann, M. Baumgartner, M. Kaltenbrunner, Becoming sustainable, the new frontier in soft robotics. Adv. Mater. **33**(19), 200–413 (2009)
5. Y. Yang, S. Bremner, C. Menictas, M. Kay, Modelling and optimal energy management for battery energy storage systems in renewable energy systems: a review. Renew. Sustain. Energy Rev. **167**, 112–671 (2022)
6. E.E. Van der Wall, Milestones in cardiovascular medicine: 10 or more? Neth. Hear. J. **21**(12), 527–529 (2013)
7. B.P. Yalagala, S.A. Sankaranarayanan, A.K. Rengan, S.R.K. Vanjari, Biocompatible, flexible, and high-performance nanowelded silver nanowires on silk fibroin for transparent conducting electrodes toward biomemristor application. ACS Sustain. Chem. Eng. **10**(14), 4473–4485 (2022)
8. C.J. Borleffs, L. Willem, R.J. van Erven, E.T. van Bommel, E.E. van der Velde, J.J. van der Wall, F.R. Bax, Rosendaal, and Martin J. Schalij., Risk of failure of transvenous implantable cardioverter-defibrillator leads. Circ. Arrhythm. Electrophysiol. **2**(4), 411–416 (2009)
9. M. Spiecker, R. Erbel, H.-J. Rupprecht, J. Meyer, Emergency angioplasty of totally occluded left main coronary artery in acute myocardial infarction and unstable angina pectoris—institutional experience and literature review. Eur. Heart J. **15**(5), 602–607 (1994)
10. N.E.G. Beurskens, K.T.N. Breeman, K.J. Dasselaar, A. Chris Meijer, A.-F.B.E. Quast, F.V.Y. Tjong, R.E. Knops, Leadless cardiac pacing systems: current status and future prospects. Expert Rev. Med. Devices **16**(11), 923–930 (2019)
11. H.G. Mond, G. Freitag, The cardiac implantable electronic device power source: evolution and revolution. Pacing Clin. Electrophysiol. **37**(12), 1728–1745 (2014)

12. K. Jeffrey, V. Parsonnet, Cardiac pacing, 1960–1985: a quarter century of medical and industrial innovation. Circulation **97**(19), 1978–1991 (1998)
13. V. Parsonnet, A.D. Berstein, G.Y. Perry, The nuclear pacemaker: is renewed interest warranted? Am. J. Cardiol. **66**(10), 837–842 (1990)
14. B.D. McCauley, A.F. Chu, Leadless cardiac pacemakers: the next evolution in pacemaker technology. R. I. Med. J. **100**(11), 31–34 (2017)
15. S. El Ichi, A. Zebda, J.-P. Alcaraz, A. Laaroussi, J. François Boucher, N.R.-B. Boutonnat, et al., Bioelectrodes modified with chitosan for long-term energy supply from the body. Energ. Environ. Sci. **8**(3), 1017–1026 (2015)
16. J. Dong, F.R. Fan, Z.-Q. Tian, Droplet-based nanogenerators for energy harvesting and self-powered sensing. Nanoscale **13**(41), 17290–17309 (2021)
17. X.D. Wang, C.S. Jun Zhou, J.H. Lao, N.S.X. Song, Z.L. Wang, In situ field emission of density-controlled ZnO nanowire arrays. Adv. Mater. **19**(12), 1627–1631 (2007)
18. A. Kurs, A. Karalis, R. Moffatt, J.D. Joannopoulos, P. Fisher, M. Soljacic, Wireless power transfer via strongly coupled magnetic resonances. Science **317**(5834), 83–86 (2007)
19. M.A. Hannan, S. Mutashar, S.A. Samad, A. Hussain, Energy harvesting for the implantable biomedical devices: issues and challenges. Biomed. Eng. Online **13**(1), 1–23 (2014)
20. W.C. Brown, The history of power transmission by radio waves. IEEE Trans. Microwave Theory Tech. **32**(9), 1230–1242 (1984)
21. X. Shi, C. Qi, Q. Meiling, S. Ye, G. Wang, L. Sun, Y. Zhiping, Effects of coil shapes on wireless power transfer via magnetic resonance coupling. J. Electromagn. Waves Appl. **28**(11), 1316–1324 (2014)
22. B.P. Yalagala, P. Sahatiya, C.S.R. Kolli, S. Khandelwal, V. Mattela, S. Badhulika, V2O5 nanosheets for flexible memristors and broadband photodetectors. ACS Appl. Nano Mater. **2**(2), 937–947 (2019)
23. R. Ibrahim, N. Shaari, A.H.M. Aman, Bio-fuel cell for medical device energy system: a review. Int. J. Energy Res. **45**(10), 14245–14273 (2021)
24. V. Nishaa, B.V. Spoorthi, B.T. Soumya, U.S. Meda, V.S. Desai, Powering implantable medical devices with biological fuel cells. ECS Trans. **107**(1), 19197 (2022)
25. J.A. Cracknell, K.A. Vincent, F.A. Armstrong, Enzymes as working or inspirational electrocatalysts for fuel cells and electrolysis. Chem. Rev. **108**(7), 2439–2461 (2008)
26. F.A.V. Pires, Disposable immunosensor for diagnosis of human cytomegalovirus congenital infection. PhD diss., Universidade da Beira Interior (Portugal) (2013)
27. R.A. Bullen, T.C. Arnot, J.B. Lakeman, F.C. Walsh, Biofuel cells and their development. Biosens. Bioelectron. **21**(11), 2015–2045 (2006)
28. K. MacVittie, J. Halámek, L. Halámková, M. Southcott, W.D. Jemison, R. Lobel, E. Katz, From "cyborg" lobsters to a pacemaker powered by implantable biofuel cells. Energ. Environ. Sci. **6**(1), 81–86 (2013)
29. L. Halámková, J. Halámek, V. Bocharova, A. Szczupak, L. Alfonta, E. Katz, Implanted biofuel cell operating in a living snail. J. Am. Chem. Soc. **134**(11), 5040–5043 (2012)
30. A. Szczupak, J. Halámek, L. Halámková, V. Bocharova, L. Alfonta, E. Katz, Living battery–biofuel cells operating in vivo in clams. Energ. Environ. Sci. **5**(10), 8891–8895 (2012)
31. C. Agnès, M. Holzinger, A. Le Goff, B. Reuillard, K. Elouarzaki, S. Tingry, S. Cosnier, Supercapacitor/biofuel cell hybrids based on wired enzymes on carbon nanotube matrices: autonomous reloading after high power pulses in neutral buffered glucose solutions. Energ. Environ. Sci. **7**(6), 1884–1888 (2014)
32. B.P. Yalagala, A.S. Dahiya, R. Dahiya, ZnO nanowires based degradable high-performance photodetectors for eco-friendly green electronics. Opto-Electron Adv **6**(2), 220020–220021 (2023)
33. J. Li, R. Bao, J. Tao, Y. Peng, C. Pan, Recent progress in flexible pressure sensor arrays: from design to applications. J. Mater. Chem. **C6**(44), 11878–11892 (2018)
34. N. Sezer, M. Koç, A comprehensive review on the state-of-the-art of piezoelectric energy harvesting. Nano Energy **80**, 105567 (2021)

35. Z. Liang, C.-F. Yan, S. Rtimi, J. Bandara, Piezoelectric materials for catalytic/photocatalytic removal of pollutants: recent advances and outlook. Appl. Catal. Environ. **241**, 256–269 (2019)
36. C.N.R. Rao, F.L. Deepak, G. Gundiah, A. Govindaraj, Effect of Mn doping on solvothermal synthesis of CdS nanowires. Prog. Solid State Chem. **31**(5) (2003)
37. Z. Li, G. Zhu, R. Yang, A.C. Wang, Z.L. Wang, Muscle-driven in vivo nanogenerator. Adv. Mater. **22**(23), 2534–2537 (2010)
38. X. Li, R. Liang, J. Tao, Z. Peng, X. Qiming, X. Han, X. Wang, et al., Flexible light emission diode arrays made of transferred Si microwires-ZnO nanofilm with piezo-phototronic effect enhanced lighting. ACS Nano **11**(4), 3883–3889 (2017)
39. M.T. Chorsi, E.J. Curry, H.T. Chorsi, R. Das, J. Baroody, P.K. Purohit, H. Ilies, T.D. Nguyen, Piezoelectric biomaterials for sensors and actuators. Adv. Mater. **31**(1), 1802084 (2019)
40. A. Wang, Z. Liu, M. Hu, C. Wang, X. Zhang, B. Shi, Y. Fan, Y. Cui, Z. Li, K. Ren, Piezoelectric Nanofibrous Scaffolds as in vivo energy harvesters for modifying fibroblast alignment and proliferation in wound healing. Nano Energy **43**, 63–71 (2018)
41. M.E.E. Alahi, A. Nag, S.C. Mukhopadhyay, Self-powered implantable energy harvesters for medical electronics, in *Flexible Sensors for Energy-Harvesting Applications*, (Springer, Cham, 2022), pp. 169–184
42. S.H. Kondapalli, Y. Alazzawi, M. Malinowski, T. Timek, S. Chakrabartty, Feasibility of self-powering and energy harvesting using cardiac valvular perturbations. IEEE Trans. Biomed. Circuits Syst. **12**(6), 1392–1400 (2018)
43. M. Amin Karami, D.J. Inman, Powering pacemakers from heartbeat vibrations using linear and nonlinear energy harvesters. Appl. Phys. Lett. **100**(4), 042901 (2012)
44. G.T. Hwang, H. Park, J.H. Lee, S. Oh, K.I. Park, M. Byun, H. Park, G. Ahn, C.K. Jeong, K. No, Adv. Mater. **26**, 4880 (2014)
45. B. Lu, Y. Chen, O. Dapeng, H. Chen, L. Diao, W. Zhang, J. Zheng, W. Ma, L. Sun, X. Feng, Ultra-flexible piezoelectric devices integrated with heart to harvest the biomechanical energy. Sci. Rep. **5**(1), 16065 (2015)
46. C. Hu, K. Behdinan, R. Moradi-Dastjerdi, PVDF energy harvester for prolonging the battery life of cardiac pacemakers. Actuators **11**(7), 187 (2022) MDPI
47. N. Li, Z. Yi, Y. Ma, F. Xie, Y. Huang, Y. Tian, X. Dong, et al., Direct powering a real cardiac pacemaker by natural energy of a heartbeat. ACS Nano **13**(3), 2822–2830 (2019)
48. B.P. Yalagala, S. Deswal, S.R.K. Vanjari, R. Dahiya, Flexible and ultra-fast bioresorbable nanofibers of silk fibroin-PVA composite, in *2021 IEEE international conference on flexible and printable sensors and systems (FLEPS)*, (IEEE, 2021), pp. 1–4
49. E. Samimi-Sohrforozani, S. Azimi, A. Abolhasani, S. Malekian, S. Arbab, M. Zendehdel, M.M. Abolhasani, N.Y. Nia, Development of porous polyacrylonitrile composite fibers: new precursor fibers with high thermal stability. Electron. Mater. **2**(4), 454–465 (2021)
50. T. Li, L. Xian-Mao, M.-R. Zhang, H. Kuan, Z. Li, Peptide-based nanomaterials: self-assembly, properties and applications. Bioact. Mater. **11**, 268–282 (2022)
51. J. Li, Y. Long, F. Yang, X. Wang, Degradable piezoelectric biomaterials for wearable and implantable bioelectronics. Curr. Opin. Solid State Mater. Sci. **24**(1), 100806 (2020)
52. Y.S. Choi, S.K. Kim, M. Smith, F. Williams, M.E. Vickers, J.A. Elliott, S. Kar-Narayan, Unprecedented dipole alignment in α-phase nylon-11 nanowires for high-performance energy-harvesting applications. Sci. Adv. **6**(24), eaay5065 (2020)
53. S.K. Karan, S. Maiti, O. Kwon, S. Paria, A. Maitra, S.K. Si, Y. Kim, J.K. Kim, B.B. Khatua, Nature driven spider silk as high energy conversion efficient bio-piezoelectric nanogenerator. Nano Energy **49**, 655–666 (2018)
54. L. Sarkar, M.V. Sushma, B.P. Yalagala, A.K. Rengan, S.G. Singh, S.R.K. Vanjari, ZnO nanoparticles embedded silk fibroin—a piezoelectric composite for nanogenerator applications. Nanotechnology **33**(26), 265403 (2022)
55. D. Wang, Y. Zhang, L. Xi, Z. Ma, C. Xie, Z. Zheng, Chemical formation of soft metal electrodes for flexible and wearable electronics. Chem. Soc. Rev. **47**(12), 4611–4641 (2018)
56. G. Liu, Y. Tang, A.M. Soomro, P. Shen, L. Shiqiang, Y. Cai, H. Wang, et al., Vertically aligned ZnO nanoarray directly orientated on Cu paper by h-BN monolayer for flexible and transparent piezoelectric nanogenerator. Nano Energy **109**, 108265 (2023)

57. M. Qi, W. Tingting, Z. Wang, Z. Wang, B. He, H. Zhang, Y. Liu, et al., Progress in metafibers for sustainable radiative cooling and prospects of achieving thermally drawn metafibers. Adv. Energy Sustainability Res. **3**(5), 2100168 (2022)
58. S. Parandeh, N. Etemadi, M. Kharaziha, G. Chen, A. Nashalian, X. Xiao, J. Chen, Advances in triboelectric nanogenerators for self-powered regenerative medicine. Adv. Funct. Mater. **31**(47), 2105169 (2021)
59. M.E. Karagozler, I. Poupyrev, G.K. Fedder, Y. Suzuki, Paper generators: harvesting energy from touching, rubbing and sliding, in *Proceedings of the 26th Annual ACM Symposium on User Interface Software and Technology*, (2013), pp. 23–30
60. Y. Ma, Q. Zheng, Y. Liu, B. Shi, X. Xue, W. Ji, Z. Liu, et al., Self-powered, one-stop, and multifunctional implantable triboelectric active sensor for real-time biomedical monitoring. Nano Lett. **16**(10), 6042–6051 (2016)
61. H. Ouyang, Z. Liu, N. Li, B. Shi, Y. Zou, F. Xie, Y. Ma, et al., Symbiotic cardiac pacemaker. Nat. Commun. **10**(1), 1821 (2019)
62. I. Shlomy, S. Divald, K. Tadmor, Y. Leichtmann-Bardoogo, A. Arami, B.M. Maoz, Restoring tactile sensation using a triboelectric nanogenerator. ACS Nano **15**(7), 11087–11098 (2021)
63. X. Qu, Y. Liu, Z. Liu, Z. Li, Assistive devices for the people with disabilities enabled by triboelectric nanogenerators. J. Phys.: Mater. **4**(3), 034015 (2021)
64. P. Wang, H. Mengmeng, H. Wang, Z. Chen, Y. Feng, J. Wang, W. Ling, Y. Huang, The evolution of flexible electronics: from nature, beyond nature, and to nature. Adv. Sci. **7**(20), 2001116 (2020)
65. S. Roy, A.N.M. Wasekul Azad, S. Baidya, M.K. Alam, F. Khan, Powering solutions for biomedical sensors and implants inside the human body: a comprehensive review on energy harvesting units, energy storage, and wireless power transfer techniques. IEEE Trans. Power Electron. **37**(10), 12237–12263 (2022)
66. E. McGlynn, B.P. Yalagala, H. Heidari, Microfabrication of implantable, flexible neural probes towards bidirectional interfacing in the deep brain, in *2022 29th IEEE International Conference on Electronics, Circuits and Systems (ICECS)*, (IEEE, 2022), pp. 1–4
67. X. Zhang, L.-D. Zhao, Thermoelectric materials: energy conversion between heat and electricity. J. Materiomics **1**(2), 92–105 (2015)
68. L. Yang, Z.-G. Chen, M.S. Dargusch, J. Zou, High performance thermoelectric materials: progress and their applications. Adv. Energy Mater. **8**(6), 1701797 (2018)
69. P.-X. Hou, F. Zhang, L. Zhang, C. Liu, H.-M. Cheng, Synthesis of carbon nanotubes by floating catalyst chemical vapor deposition and their applications. Adv. Funct. Mater. **32**(11), 2108541 (2022)
70. Q. Zhang, K. Deng, L. Wilkens, H. Reith, K. Nielsch, Micro-thermoelectric devices. Nat. Electron. **5**(6), 333–347 (2022)
71. S.H. Kwon, L. Dong, Flexible sensors and machine learning for heart monitoring. Nano Energy **102**, 107632 (2022)
72. A.A. Mathew, A. Chandrasekhar, S. Vivekanandan, A review on real-time implantable and wearable health monitoring sensors based on triboelectric nanogenerator approach. Nano Energy **80**, 105566 (2021)
73. C. Sanchez, E.M. Maria, S. Bartoletti, B.P. Yalagala, B.C. Garofani, A. Capodiferro, E. Russell, et al., Bioresorbable insertion aids for brain implantable flexible probes: A comparative study on silk fibroin, alginate and disaccharides. Adv. NanoBiomed Res. **2**(8), 601–612 (2023)
74. M. Zhu, Z. Yi, B. Yang, C. Lee, Making use of nanoenergy from human–nanogenerator and self-powered sensor enabled sustainable wireless IoT sensory systems. Nano Today **36**, 101016 (2021)
75. N. Bhatia, M. El-Chami, Leadless pacemakers: a contemporary review. J. Geriatr. Cardiol. **15**(4), 249 (2018)
76. B. Yalagala, J. Shivam Khandelwal, Deepika, and Sushmee Badhulika., Wirelessly destructible MgO-PVP-graphene composite based flexible transient memristor for security applications. Mater. Sci. Semicond. Process. **104**, 104673 (2019)

77. R. Bagherzadeh, S. Abrishami, A. Shirali, A.R. Rajabzadeh, Wearable and flexible electrodes in nanogenerators for energy harvesting, tactile sensors, and electronic textiles: novel materials, recent advances, and future perspectives. Mater. Today Sustainability **20**, 100233 (2022)
78. S.K. Nemani, R.K. Annavarapu, B. Mohammadian, A. Raiyan, M. Jamie Heil, A. Haque, A. Abdelaal, H. Sojoudi, Surface modification of polymers: methods and applications. Adv. Mater. Interfaces **5**(24), 1801247 (2018)
79. S. Zheng, H. Wang, P. Das, Y. Zhang, Y. Cao, J. Ma, S. Liu, W. Zhong-Shuai, Multitasking MXene inks enable high-performance printable microelectrochemical energy storage devices for all-flexible self-powered integrated systems. Adv. Mater. **33**(10), 2005449 (2021)
80. J. Li, X. Wang, Materials perspectives for self-powered cardiac implantable electronic devices toward clinical translation. Acc. Mater. Res. **2**(9), 739–750 (2021)
81. A. Haroun, X. Le, S. Gao, B. Dong, T. He, Z. Zhang, F. Wen, X. Siyu, C. Lee, Progress in micro/nano sensors and nanoenergy for future AIoT-based smart home applications. Nano Express **2**(2), 022005 (2021)
82. Q. Zheng, Q. Tang, Z.L. Wang, Z. Li, Self-powered cardiovascular electronic devices and systems. Nat. Rev. Cardiol. **18**(1), 7–21 (2021)
83. R. Herbert, H.-R. Lim, S. Park, J.-H. Kim, W.-H. Yeo, Recent advances in printing technologies of nanomaterials for implantable wireless systems in health monitoring and diagnosis. Adv. Healthc. Mater. **10**(17), 2100158 (2021)
84. D.C. Aduba Jr., H. Yang, Polysaccharide fabrication platforms and biocompatibility assessment as candidate wound dressing materials. Bioengineering **4**(1), 1 (2017)
85. Y. Sun, L. Yun, X. Li, Y. Zheyin, S. Zhang, H. Sun, Z. Cheng, Flexible hybrid piezo/triboelectric energy harvester with high power density workable at elevated temperatures. J. Mater. Chem. A **8**(24), 12003–12012 (2020)
86. L. Wang, F. Xuemei, J. He, X. Shi, T. Chen, P. Chen, B. Wang, H. Peng, Application challenges in fiber and textile electronics. Adv. Mater. **32**(5), 1901971 (2020)
87. B. Yalagala, P. Sahatiya, V. Mattela, S. Badhulika, Ultra-low cost, large area graphene/MoS2-based piezotronic memristor on paper: a systematic study for both direct current and alternating current inputs. ACS Appl. Electron. Mater. **1**(6), 883–891 (2019)
88. X. Pu, H. Weiguo, Z.L. Wang, Toward wearable self-charging power systems: the integration of energy-harvesting and storage devices. Small **14**(1), 1702817 (2018)
89. Y. Long, J. Li, F. Yang, J. Wang, X. Wang, Wearable and implantable electroceuticals for therapeutic electrostimulations. Adv. Sci. **8**(8), 2004023 (2021)
90. H. Xiang, Y. Zeng, X. Huang, N. Wang, X. Cao, Z.L. Wang, From triboelectric nanogenerator to multifunctional triboelectric sensors: a chemical perspective toward the interface optimization and device integration. Small **18**(43), 2107222 (2022)
91. V. Raghunathan, A. Kansal, J. Hsu, J. Friedman, M. Srivastava, Design considerations for solar energy harvesting wireless embedded systems, in *IPSN 2005. Fourth international Symposium on Information Processing in Sensor Networks*, (2005), pp. 457–462. IEEE
92. K. Dong, Z.L. Wang, Self-charging power textiles integrating energy harvesting triboelectric nanogenerators with energy storage batteries/supercapacitors. J. Semicond. **42**(10), 101601 (2021)
93. A. Lazaro, R. Villarino, D. Girbau, A survey of NFC sensors based on energy harvesting for IoT applications. Sensors **18**(11), 3746 (2018)
94. M. Cerezo-Sanchez, B. Yalagala, E. McGlynn, E. Russell, F. Walton, H. Heidari, Flexible neural probe modelling for optimal microelectrode-tissue interaction, in *2022 29th IEEE International Conference on Electronics, Circuits and Systems (ICECS)*, (IEEE, 2022), pp. 1–4
95. T. Wu, W. Fan, J.-M. Redoute, M.R. Yuce, An autonomous wireless body area network implementation towards IoT connected healthcare applications. IEEE Access **5**, 11413–11422 (2017)

Chapter 3
Magnetoelectric Composites for Implantable Wireless Power Transfer in Biomedical Applications

Eve McGlynn

3.1 Introduction

The magnetoelectric effect is a reversible coupling between a piezoelectric phase and a magnetostrictive phase, meaning, a changing magnetic field may be used to generate a potential difference or vice versa. Formally, this is more accurately described as a change in polarisation with a change in the magnetic field. While this effect has been observed in selected materials, it is more commonly seen in composites which combine a piezoelectric material with a magnetostrictive one, with coupling occurring at the interface. Over the past decades, research in this area has come in and out of fashion with the advent of new materials, geometries and fabrication processes, each of which bring magnetoelectric devices closer to the point of usefulness (Fig. 3.1).

Magnetic stimulation in the human body has been used extensively to treat neurological disorders with techniques such as transcranial magnetic stimulation (TMS) [1]. Safety protocols, in line with other high frequency devices such as personal telephones, ensure that the magnetic field does not cause harm to the body. Turning the benefits of magnetoelectric composites to the field of biomedical engineering, these devices may be used for wireless power transfer [2], energy harvesting [3], sensing and communication [4].

The most common configurations include a Helmholtz coil to generate an AC magnetic field, and a bias magnet, which can be adapted for wearable or implantable use cases [5]. So-called giant or colossal magnetoelectric effects have been reported for miniaturised composites, with high performance metrics or magnetoelectric coefficients as per the equation below. While the output power or impedance of

E. McGlynn (✉)
University of Glasgow, Glasgow, UK
e-mail: Eve.McGlynn@glasgow.ac.uk

R. Das, H. Heidari (eds.), *Wireless Power Technologies for Biomedical Devices*,
Analog Circuits and Signal Processing, https://doi.org/10.1007/978-3-031-52834-7_3

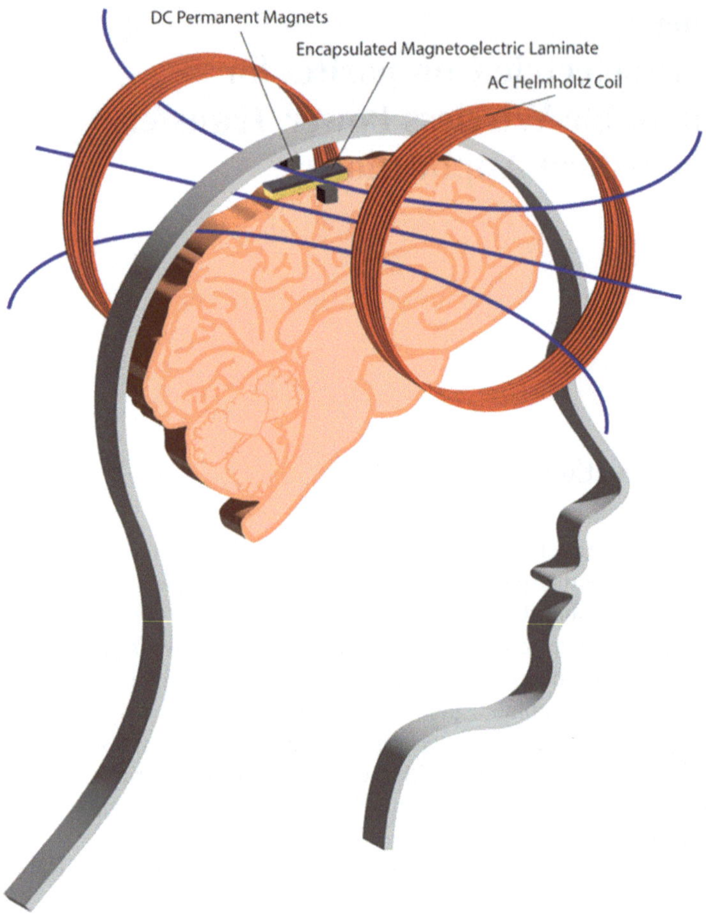

Fig. 3.1 Representative figure of an implanted magnetoelectric composite, powered by an external AC Helmholtz coil with a small implanted permanent magnet

these devices are also of interest, the magnetoelectric coefficient encompasses the three most important measurements: the voltage output (V); the thickness of the piezoelectric phase in cm; and the magnitude of the stimulating AC field in Oersted (Oe).

$$\alpha_{ME} = \frac{dE}{dH} = \frac{dV}{t \times dH}$$

Modern developments in this field have produced a number of exciting research avenues. Self-biased systems omit the bias magnet while maintaining magnetoelectric coefficients which can compete with, or even outstrip, biased systems.

Core-shell magnetoelectric nanoparticles improve the interface between the two constituent phases, encapsulating a magnetostrictive particle in a piezoelectric sheath [6]. Polymer-based composites are ideal for implantation into the body, as they conform to tissue and prevent toxic effects from materials such as PZT, which contains lead. Nano-scale magnetoelectric antennas, as developed by the NaN laboratory of Tsinghua University, marry the field of nanofabrication with mechano-magnetoelectric effects to create antennas operating at medical frequencies [4].

This chapter endeavours to provide both a theoretical and practical overview of the current state of the art in magnetoelectric composites. From the materials and designs to the fabrication techniques and experimental validation, each step in the process of developing a new magnetoelectric device will be described. In particular, this chapter will focus on how magnetoelectric devices may be implemented in vivo, discussing biocompatibility concerns and frequency optimisation.

3.2 Material Choice

While PZT has the highest piezoelectric charge coefficient, and as such is the most attractive in terms of voltage output, it is commonly found as a rigid ceramic plate, which is less suitable for implantable and wearable devices than flexible alternatives. The fact that PZT contains lead means that it poses a health risk if not properly encapsulated. PVDF is an inherently piezoelectric polymer, which means it can either be loaded with magnetostrictive particles to create a conformable polymer sheet, or spin-coated onto a magnetostrictive material such as Metglas to achieve a thin coating. $BaTiO_3$ is often employed as part of core-shell magnetoelectric nanoparticles, thanks to its high piezoelectric charge coefficient (Tables 3.1 and 3.2).

The most important compromise in magnetostrictive materials is the saturation magnetostriction versus the level of DC bias required, which accounts for the popularity of Metglas as a constituent of simple laminates. The magnetic permeability of Metglas ($\mu_r = 50{,}000$) is 5000 times higher than alternatives such as Terfenol-D [15]. Clearly, the higher the saturation magnetostriction, the higher the maximum strain coupled to the piezoelectric. DC bias may be applied by a permanent magnet or a Helmholtz coil, and which may have consequences for the size or material of the

Table 3.1 Key parameters for piezoelectric materials commonly used in magnetoelectric composites

Piezoelectric material	Piezoelectric charge coefficient (pC/N)	Piezoelectric voltage coefficient Vm/N	Density kg/m^3	Relative dielectric constant
PZT-5A	225–590 [7]	-11.6×10^{-3} [8]	7800 [8]	1800 [8]
PVDF	24–34 [7]	216×10^{-3} [8]	1780 [8]	12 [8]
BaTiO3	191 [7]	11×10^{-3}	6020	4000–6000 [9]
AlN	5.5 [10]	0.06 [11]	3260	8.5 [12]

Table 3.2 Key parameters for magnetostrictive materials commonly used in magnetoelectric composites

Magnetostrictive material	Piezomagnetic coefficient	Density kg/m^3	Saturation magnetostriction (ppm)
Galfenol	19.53 T/GPa [7]	7800 [8]	120–240 [7]
Terfenol-D	18.42 T/GPa [7]	9250 [7]	800–1200 [7]
Metglas	25 nm/A [13]	7180 [7]	27 [14]
CoFeO		8400 [7]	80–140 [7]
FeGaB	12 ppm/Oe [14]		75 [14]

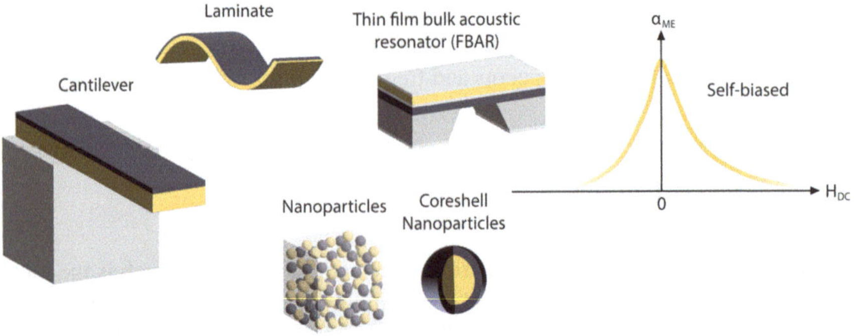

Fig. 3.2 Geometries of magnetoelectric composites

magnet, or the power requirements of the Helmholtz coil. In many cases, a DC bias field of a few Oe is sufficient. Consideration has already been given to encapsulation of neodymium magnets in vivo, concluding that parylene C was the most promising coating in terms of both cytotoxicity and degradation.

3.3 Composite Geometries (Fig. 3.2)

3.3.1 Laminates

Typically exhibiting resonance peaks in the order of 10–100 kHz, magnetoelectric laminates are sheets of magnetoelectric material, for example an amorphous metal such as Metglas, bonded to a piezoelectric plate. The ratio of length to thickness is high, even for mm-scale laminates, since the thickness tends to fall in the range of microns. Polymeric piezoelectric phases can also be used in laminates, for example after spin coating on a magnetostrictive film.

Individually, the MS and PE elements exhibit resonance. That is, a specific frequency of magnetic field will elicit a significantly higher deformation in the magnetostrictive phase. Vibrations at a specific frequency will create much higher voltage output in the piezoelectric. While these frequency values are disparate in the individual phases, the average acoustic velocity in the bilayer is such that the piezoelectric and magnetostrictive resonances occur at the same frequency [16]. With increasing frequency, the mode number for magnetoelectric resonance increases. The first few modes, termed bending modes [17], do not engender significant deformation of the film, but can be easily modelled.

Every bilayer falls into one of four categories. Both the piezoelectric element and magnetostrictive element can be in either a longitudinal or transverse orientation, which determines the direction of the deformation under excitation. Piezoelectric plates are most commonly poled in their thickness direction, a term which will be expanded in Sect. 4.1. This means that the potential difference will also occur across the thickness direction. Materials such as Metglas, being amorphous, exhibit the magnetoelectric effect regardless of the orientation [8].

Significant research has been carried out to determine the optimum length to thickness ratio for the piezoelectric phase. Further to this, the number of layers in the laminate have been increased to investigate the optimum combination of phases. A bilayer refers to a single piezoelectric layer coupled with a single magnetostrictive layer. An MPM trilayer is a piezoelectric plate sandwiched between two magnetostrictive phases, with PMP trilayers being less common. Extensive modelling performed by Martins et al. [18] aims to determine best practice for the design of multi-layered laminates, focusing on the thickness limits for the piezoelectric phase.

As expected, the polymer PVDF shows less consistent trends than the ceramic PZT due to its semicrystalline nature, and its performance is therefore less predictable. Since it is a polymer, however, its thickness can be carefully controlled, which is especially important considering the optimum thickness ratio between the magnetostrictive and piezoelectric phases. While the magnetostrictive material may be purchase off-the-shelf with a set thickness, the PVDF may be spin coated or otherwise cast to the recommended thickness ratio of 0.2.

3.3.2 Nanoparticles

Piezoelectric and magnetostrictive nanoparticles may be combined in a polymer to produce flexible, conformable, and often biocompatible films. Or the polymer itself may exhibit piezoelectricity, and be filled with magnetostrictive particles only. The most common piezoelectric nanoparticle, BaTiO3, is attractive in that it is lead-free [19], and has been an integral component in the creation of devices which exhibit 'giant' and 'colossal' magnetoelectric effects. Core-shell nanoparticles, briefly mentioned in Sect. 1, have a magnetostrictive core and piezoelectric shell, with a significantly improved interfacial coefficient when compared to standard mixtures of nanoparticles. The magnetic field, which penetrates the piezoelectric outer layer, causes deformation of the inner magnetostrictive particle, generating a voltage on the outside of the PE.

Core-shell particles range from tens to thousands of nanometres, typically prepared by a sol-gel method which will be discussed in Sect. 4.3. Core-shell films exhibit magnetoelectric coefficients orders of magnitude higher than standard nanoparticle films. For example, using $BaTiO_3$ and $CoFe_2O_4$, Wang et al. achieved a magnetoelectric coefficient of 5 V/cmOe [6], whereas Omelyanchik et al. illustrated a maximum of 18.5 mV/cmOe [20].

3.3.3 Nanomechanical ME Devices

Magnetoelectric cantilevers have been used extensively as sensors [21]. The mechanical aspect of these devices firmly place them in the MEMS category, which also includes nanomechanical magnetoelectric antennas. The thin-film bulk acoustic resonator antenna developed by Nan et al. is a very promising alternative to low-frequency laminates and polymer films. By significantly reducing the footprint of the magnetoelectric composite element, the resonant frequency is tailored to the GHz range, specifically the 2.4 GHz medical frequency. The same nanofabrication techniques and similar materials (FeGaB and AlN) are also used to fabricate a Nanoplate Resonator (NPR) [4]. An array of NPR antennas has been incorporated into the NanoNeuroRFID communication and sensing system [3].

3.3.4 Self-Biased ME Composites

Self-biased MEs (SMEs) operate without the need for bulky bias magnets, which reduces the total volume of the system. The non-zero magnetization at H_{DC} of 0 Oe occurs through a variety of mechanisms, some of which may be directly controlled during the fabrication process. For example, the magnetization grading approach involves layering several different magnetostrictive materials, which creates a small internal magnetic field as a result of their disparate saturation magnetization values. In [22], Metglas and nickel are combined with a PZT ceramic in a five-layer structure. The MS materials must be pre-magnetized before testing the SME output voltage. For an AC magnetic field of 100 μT, following IEEE Safety Standards, the ME outputs 160 μW, suitable for a biomedical device. Nickel foil is also used in [23], sprayed with PZT to create a thin film with an ME coefficient as high as 3.15 V/cmOe. The shape of the laminate significantly impacts the demagnetization in the Ni, and therefore the effective magnetic field. The thinner the Ni, the lower the demagnetization. This work takes advantage of the remanent magnetic field in the material. Internal stress may be introduced to the composite through the manufacturing process as described in [24], especially considering that thin-film stress, difference in thermal expansion coefficient, and poor mechanical coupling are difficult to prevent during fabrication. This 'mechanical pre-load' may provide a starting point with a non-zero magnetization at zero bias field.

3.4 Fabrication

3.4.1 *Poling*

Poling is a high-voltage process, which aligns the dipole moments within piezoelectric crystals to point in the same direction until the constant electric field is removed, at which point some imperfections are observed in the alignment. While this change is permanent, piezoelectric materials suffer degradation with repeated mechanical loading or extremes of temperature. For example, the Curie temperature of a piezoelectric is the maximum temperature of operation: beyond this, the material loses its piezoelectric properties.

This process is directly related to the modality of the composite, as the location of the contacts and the direction of the applied electric field determine the direction of the dipoles in the material. Poling may be performed with a handful of techniques such as contact or corona poling. Contact poling is the simplest, in which physical connections are made between the voltage source and the electrodes of the piezoelectric. Corona poling requires only physical contact with the ground, ionising the air above the sample [25]. The piezoelectric constant, phase angle, and electromechanical coupling factor may be tuned with careful selection of the DC voltage (range of kV/mm), time (tens of minutes), and temperature (20–50 °C) [26].

3.4.2 *Epoxy*

Simply fabricated piezoelectric laminates are bonded together with epoxy, which impacts the interfacial constant in the ME coefficient equation. Most examples use hard, non-conductive epoxy to couple the two phases together. However, the use of conductive epoxy may reduce fabrication complexity [27, 28]. The location of the wires which interface with a measurement setup is particularly important: increased resonant voltage has been observed when wires are placed in the centre of the laminate versus at the edge [5].

3.4.3 *Solgel*

As the name suggests, the sol-gel method involves the gelation of a solution before drying into powder or film through the application of heat. For the creation of sol-gel core-shell magnetoelectric materials, a reaction will occur between a solution and the magnetostrictive nanoparticle, adhering the piezoelectric outer later to the core.

3.4.4 *Milling*

Small particles interface with one another more effectively, especially when combining two phases with different particle sizes. It is common to mill the larger of these phases to match the smaller particle diameter. Ball milling uses balls made from steel, ceramic or rubber in a rotating drum to break particles up into smaller pieces on impact. In a notable example, milling increased the magnetic coercive field of the samples [29].

3.4.5 *Spin Coating and Doctor Blading*

The thickness of a spin-coated layer can be easily controlled by changing the speed of rotation. Above 1000 rpm, the process is reliable and repeatable, and if the density of the liquid is known, the theoretical thickness may be calculated before verification against an experimental spin-speed curve. Polymer-based piezoelectric materials, for example PVDF, may be spin-coated directly onto a magnetoelectric thin sheet.

Doctor blading is a simple alternative to spin coating. The distance between the blade and the sample surface is the height of the liquid coating, after the excess has been removed. This can be controlled to tens of microns [30].

3.5 Experimental Setup

In most cases, DC bias magnets are used to improve the voltage output of the ME composite. Permanent magnets or coils of an electromagnet achieve the same end. A Helmholtz coil is used to produce the AC perturbation necessary to yield an ME voltage output. The Helmholtz coil creates a magnetic field in the centre of the two coils which mimics one continuous cylindrical coil, therefore the ME composite is placed in the centre of both the AC and DC fields. In the simplest case, a laminate composite, the resonant frequency is correlated to the length of the plate. From calculation, simulation, or simply a frequency sweep, the resonant voltages are identified as the voltage increases sharply. The optimum DC bias corresponds to the highest sensitivity of the ME: that is, the largest voltage output per Tesla applied DC bias (Fig. 3.3).

In simple terms, a typical experiment proceeds as follows:

1. Fabrication of the magnetoelectric composite.
    (a) Bond together the two phases.
    (b) Metallisation to create electrodes on the piezoelectric phase.
    (c) Attach one wire to each face of the piezoelectric.

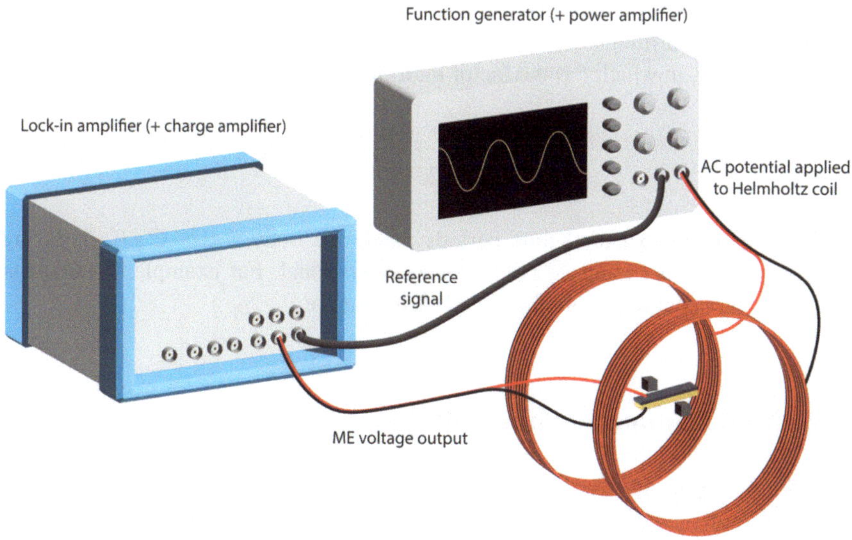

Fig. 3.3 Measurement setup including lock-in amplifier, function generator, and Helmholtz coil to produce the necessary AC field

2. Setup of the magnetic coils.
   (a) Position DC bias magnets close to the composite.
   (b) Apply AC voltage to Helmholtz coil.
3. Setup of lock-in amplifier (LIA).
   (a) Attach the same AC signal (to Helmholtz coil) to the reference input of the LIA.
   (b) Attach the signal and ground wires from the ME composite to the LIA.
4. Sweep AC frequency and note voltage output.
   (a) Identify the resonant voltage by noting the voltage peak and change of phase.

Safety regulations limiting the magnetic field applied to specific areas of the body have been published by: the Institute of Electrical and Electronics Engineers; the American Conference of Governmental Industrial Hygienists; the UK Health and Safety Executive; the European Commission; the International Commission on Non-Ionizing Radiation Protection. The World Health Organisation collects data on these standards, which typically differentiate between 'workers' and 'public.'

If the implant is intended for biomedical applications, biocompatibility and ageing tests are crucial to ensure the safety of future patients. Beginning with in vitro cytotoxicity assays, in which the viability of target cells is assessed in the short term, these experiments will expand to in vivo implantation, for example in rat models, to investigate the chronic effects. Accelerated ageing may be performed on the bench top, in phosphate buffered saline or another liquid medium which mimics the inside of the body (e.g. artificial cerebrospinal fluid). The liquid is heated according to the equation below, in which f is the factor multiplied by the time.

$$f = 2^{\frac{T - T_{ref}}{10}}$$

This simulates extended time periods inside the body at 37 °C (T_{ref}), during which the implant should be continuously operational. For example, 1 month at 87 °C is equivalent to 32 months at body temperature.

3.6 Modifications for Implantation

3.6.1 Voltage Multiplication

Despite the fact that a well-designed ME device can output useful voltages, they are commonly combined with voltage multipliers to improve the range of outputs. Circuits are selected for their footprint, efficiency and maximum power output. Both the Dickson and Cockcroft-Walton voltage multipliers can be adapted to have several stages, and increase the voltage at the expense of the number of necessary diodes. While full-wave bridge rectifiers, or even simple single-diode circuits perform the necessary rectification (to power stimulators, or even apply stimulation directly to tissue), they do not have the added benefit of voltage multiplication.

In [31] the authors compare a full-wave bridge rectifier to voltage multipliers with varying stages, concluding that the highest output power density was achieved with the Dickson 2-stage circuit. At the optimal load the output power was 12 μW, which is of the order required for implantable biomedical devices. Depending on the application, the circuit design may be chosen to favour (comparatively) high voltage, high current, or high power density.

Active diodes are generally designed for low frequencies around 100 Hz, but the design described in [32] operates in the range of 100 kHz in which ME laminates typically have their resonant frequencies. At high frequencies, the switching time can significantly increase the loss coefficient in the diode and reduce the overall efficiency of the voltage doubler circuit. The authors note that the footprint of a DC-DC converter is too large for an implantable device, and reducing the number of components is paramount in biomedical applications. The output power achieved with this circuit was 680 μW.

Integrating a Synchronous Electric Charge Extraction circuit with a 6-stage Dickson charge pump, Mukherjee and Mallick increased the output voltage of their ME laminate to 2.4 V, which would be sufficient to power a DBS stimulator, applying only 2.5 mT to the ME.

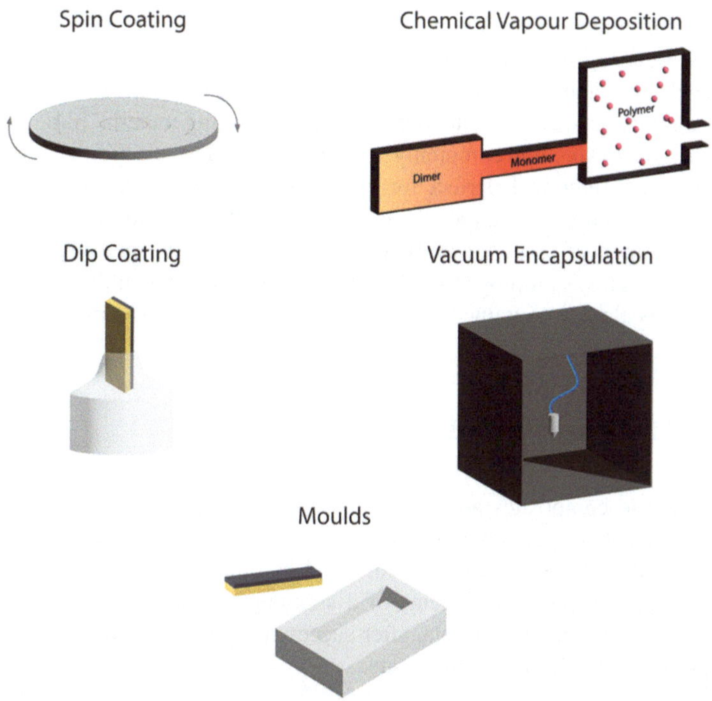

Fig. 3.4 Methods for polymer or epoxy encapsulation including spin coating, chemical vapour deposition, dip coating, vacuum encapsulation and moulds

3.6.2 Encapsulation

Parylene C is a common substrate and encapsulation material for neural probes and other similar implantable devices. Its single layer thickness is on the order of microns, carefully controlled during chemical vapor deposition. In two notable instances, parylene C is used to encapsulate an ME device for implantation in the brain. Singer et al. [5] noted that the encapsulation increased the resonant voltage of the Metglas/PZT laminate, an effect which was attributed to improved coupling between the two phases. During cytotoxicity tests, healthy cell proliferation was noted. The NanoNeuroRFID sytem [3], which utilises micron-scale ME antennas for data transfer is also encapsulated in parylene C and intended for insertion under the skull (Fig. 3.4).

Two other candidates for polymer encapsulation include polydimethylsiloxane (PDMS) [33] and polyimides (PI) [5]. The stiffness of PDMS can be tuned by altering the ratio of curing agent, and curing may be carried out at room temperature, or in an oven at around 60 °C. This is important when considering the Curie temperature of the constituent ME materials, above which they lose their ferroelectric properties. By contrast, most polyimides require curing in a nitrogen environment above 300 °C, and have a higher Young's modulus compared to PDMS. Benzocyclobutane (BCB) has been employed as an interlayer between the piezoelectric and magnetostrictive phases of a laminate, and thanks to its biocompatibility, is another contender for an encapsulating material [34].

Epoxies and resins are commonly used to encapsulate electronics, creating a robust and watertight seal perfect for in vivo experiments such as those carried out in [2]. Vacuum encapsulation is used to fill all the gaps in and around small electronic components, with no voids or bubbles which could jeopardize the efficacy of the coating. This is employed in [21, 35], in which an ME cantilever for sensing applications is already contained within a metallic chamber, and can move freely.

3.7 In Vivo Case Studies

Magnetoelectric composites are incorporated into implantable systems in one of two ways: the voltage generated by the magnetoelectric directly stimulates tissue in the body; or the ME is used purely for wireless power transfer, with its voltage conditioned to suit the requirements of a stimulator. In each of these cases, the technology is tested in vivo in rats, after successful stimulation in models such as PBS, agarose, and ex vivo porcine tissue.

3.7.1 Magnetoelectrically Powered Stereotrode in Moving Rats [5]

Combining Metglas with either PVDF polymer or PZT ceramic, Singer et al. fabricated simple bilayer ME laminates with resonant frequencies between 250 and 400 kHz. By altering the length of the laminate, they could selectively power multiple laminates with different resonant frequencies. One of the most promising results of the study confirmed that the action potentials observed in the rat brain were not a result of the applied AC magnetic field, as at non-resonant frequencies, there was no such neuron activity. Moving from a benchtop system, the 8–9 mT bias field was applied with a small permanent magnet worn by the rat, inside a cage fitted with a Helmholtz coil. With the intention of performing DBS on hemi-Parkinsonian rats, the voltage from the ME laminate was rectified and modulated to fall within the 'therapeutic band' of 100–200 Hz. During stimulation with the ME-powered stereotrode, rats exhibited improved motor function, no longer favouring one paw and moving more freely. The entire implantable system was enclosed within 175 mm^3 with the AC coils requiring only 7 W of power. This system proves the usefulness of simple ME laminates for wireless power transfer of medical devices.

3.7.2 *ME-BIT Mm-Size Neural Stimulator in Pigs [2]*

Incorporated with an application-specific integrated circuit (ASIC), a basic ME laminate can perform double duty: both wireless power transfer, and data transfer through amplitude and frequency modulation. The MagnetoElectric-powered Bio Implant (ME-BIT) can stimulate the sciatic nerve of a rat at a distance of 4 cm, and is resistant to changes in orientation, which is ideal for an implantable system intended for in vivo tests. Once more, the laminate was fabricated with Metglas and PZT, bonded with epoxy. After it is encased in non-conductive epoxy, the full-size implant has a volume of 95.46 mm^3. When implanted along the sciatic nerve of a rat, the miniaturised ME-BIT (6.2 mm^3) evoked compound muscle action potentials (CMAPs), accompanied by the rat kicking its leg. To confirm that this result was directly produced from the ME stimulation and not the AC magnetic field, the DC bias was removed (rendering the ME inactive), and no leg kicks were observed. The femoral nerve of a pig was similarly stimulated, with the transmitter placed 1.5 cm away, outside the body. The ME-BIT implant itself was placed in the femoral artery, and both CMAPs and nerve action potentials observed. Moving from rats to pigs, the ME-BIT brings ME technology one step closer to the clinic.

3.7.3 *Coreshell Magnetoelectric Hydrogel for Stimulation in Rat Brain [36]*

Hydrogels are commonly employed in tissue engineering, and the inclusion of magnetoelectric nanoparticles (MENPs) in a biomimetic matrix illustrates their efficacy in stimulating cell regrowth. In this instance, Fe_3O_4 nanoparticles are synthesised with the sol-gel method, followed by a hydrothermal reaction to create $BaTiO_3$ from TiO_2 on the surface of the nanoparticles. Hydrogels with MENP concentrations between 0 and 1% w/v were compared, with the 0.5% w/v gel emerging with the highest cell proliferation in the hydrogel. In vivo tests required that an incision was made in the spinal cord of rats, and the hydrogel injected around the injury. The rats which underwent magnetic stimulation showed improved cell proliferation and healing when compared to the sham trials (PBS injection and MENP hydrogel injection with no stimulation). The stimulated group showed reduced scarring, "more nerve fibers and enhance[d] axonal regeneration." If necessary, such hydrogels can be made biodegradable, which indicates that MENPs may be an exciting short-term treatment during the recovery from spinal cord injury.

3.7.4 *Magnetoelectric Nanoparticles Invoke Cortical Activity in Mice [37]*

The most important benefit of MENPs over larger laminates is their reduced invasiveness. Nguyen et al. illustrate how 30 nm coreshell MENPs may achieve

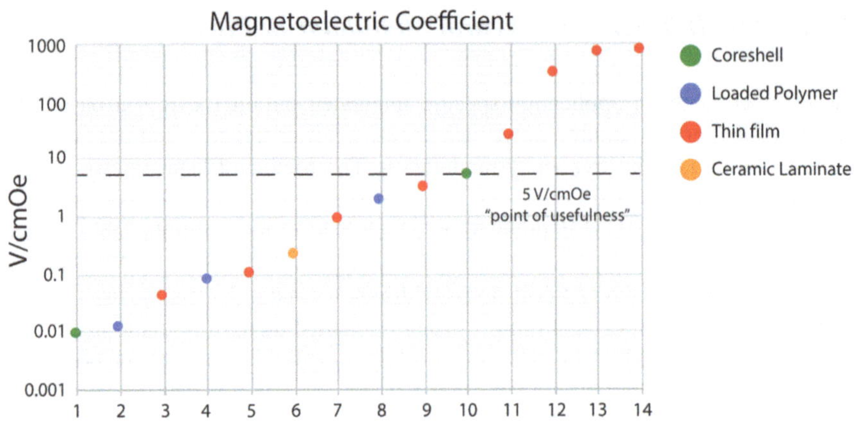

Fig. 3.5 Magnetoelectric constant values for various composites in the literature, with corresponding colour code to denote different geometries. Exact values and references are provided in Table 3.3

Table 3.3 Magnetoelectric coefficient values for composites from literature

	ME coefficient (V/cmOe)	Type	PE	MS	Refs.
1	9.18E-03	Coreshell	BTO	CFO	[39]
2	1.14E-02	Nanoparticle loaded polymer	PVD-TrFe	NiFe2O4	[33]
3	4.05E-02	Thin film	PZT	Ni	[40]
4	8.00E-02	Nanowire loaded polymer	PDVF-TrFe	Fe, Ni, Galfenol	[41]
5	1.00E-01	Thin film			[42]
6	2.20E-01	Ceramic laminate	PZT	Ni	[43]
7	8.80E-01	Thin film	PZT	Metglas	[44]
8	1.84E+00	Coreshell loaded polymer	PVDF-HFP, BaTiO3	CoFe2O4	[45]
9	3.15E+00	Thin film	PZT	Ni	[23]
10	5.00E+00	Coreshell			[37]
11	2.50E+01	Thin film	PZT matrix	CFO	[46]
12	3.10E+02	Thin film	PVDF	Metglas	[47]
13	6.86E+02	Thin film	PVDF ink	Metglas	[48]
14	7.37E+02	Thin film	PVDF-TrFe	Metglas	[35]

a magnetoelectric coefficient as high as 5 V/cmOe, which has been highlighted as a 'point of usefulness' by Rupp et al. [8]. First injected intravenously, the MENPs were guided through the bloodstream and across the blood brain barrier (BBB), something that becomes possible for small molecules and particles, with permeability significantly increased for particles between 50 and 100 nm [38]. After the arrival of MENPs into the rat brain, stimulation was possible for at least 24 hours before the volume of MENPs fell below the threshold to stimulate the cortical tissue. Thanks to the minimally invasive delivery method and small size of the nanoparticles, there was no observable microgliosis, and no observable change in microglial activation. The activity of the neurons was determined with calcium imaging to avoid noise from the outer electromagnet. Latency in neuronal responses to ME stimulation was explained in two ways: randomised placement of the MENPs means that some neurons may have a much greater minimum distance to the stimulation site; or that within one area of the cortex, the excitability of the neurons, and therefore their response time, will vary. When compared to DBS surgery, for example, the intravenous delivery of MENPs has minimal recovery time, which may significantly improve patient outcomes if the technology proves reliable in human trials (Fig. 3.5).

3.8 Conclusion

Magnetoelectric wireless power transfer is an exciting alternative to traditional electromagnetic coils, with proven efficacy in animal trials. The 'point of usefulness' has been identified by Rupp [8] as 4.68 V/cmOe, while Bichurin [49] cites 4 V/cmOe as suitable for 'practical devices.' Composites are the clear frontrunner in terms of high magnetoelectric coefficients, with near-unlimited combinations of materials, geometries, and polarities. Coreshell MENPs are especially interesting from a healthcare standpoint, where minimally invasive treatments are prioritised. Through careful design, considering the exposure limits set by international organisations, it is possible to create ME systems for long-term implantation which can withstand the environment inside the body and do not pose a risk to the patient. Miniaturised antennas such as FBAR and NPR antennas are capable of WPT and data transfer, with dimensions on the order of hundreds of microns. Despite the long history of ME experimentation, it is just emerging as a modality for implantable wireless power transfer in the body. Miniaturisation, improved biocompatibility, and exciting advancements in efficiency will allow this technology to advance from the laboratory bench to the clinic.

References

1. A. Valero-Cabré, J.L. Amengual, C. Stengel, A. Pascual-Leone, O.A. Coubard, Transcranial magnetic stimulation in basic and clinical neuroscience: a comprehensive review of fundamental principles and novel insights. Neurosci. Biobehav. Rev. **83**, 381–404 (2017)
2. J.C. Chen et al., A wireless millimetric magnetoelectric implant for the endovascular stimulation of peripheral nerves. Nat. Biomed. Eng. **6**, 706 (2022). https://doi.org/10.1038/s41551-022-00873-7
3. M. Zaeimbashi et al., NanoNeuroRFID: a wireless implantable device based on magnetoelectric antennas. IEEE J. Electromagn. RF Microw Med. Biol. **3**(3), 206–215 (2019)

4. T. Nan et al., Acoustically actuated ultra-compact NEMS magnetoelectric antennas. Nat. Commun. **8**(296), 1–8 (2017)
5. A. Singer et al., Magnetoelectric materials for miniature, wireless neural stimulation at therapeutic frequencies. Neuron **107**, 631 (2020)
6. P. Wang et al., Colossal magnetoelectric effect in coreshell magnetoelectric nanoparticles. Nano Lett. **20**, 5765 (2020)
7. F. Narita, M. Fox, A review on piezoelectric, magnetostrictive, and magnetoelectric materials and device technologies for energy harvesting applications. Adv. Eng. Mater. **20**(5), 1700743 (2018). https://doi.org/10.1002/adem.201700743
8. T. Rupp, B.D. Truong, S. Williams, S. Roundy, Magnetoelectric transducer designs for use as wireless power receivers in wearable and implantable applications. Materials **12**(3), 512 (2019)
9. N.H. Yusoff, R.A.M. Osman, M.S. Idris, K.N.D.K. Muhsen, N.I.M. Nor, Dielectric and structural analysis of hexagonal and tetragonal phase BaTiO3, Putrajaya, Malaysia, 2020, p. 020038. https://doi.org/10.1063/1.5142130
10. Y. Lu, *Development and Characterization of Piezoelectric AlScN-Based Alloys for Electroacoustic Applications* (2019). https://doi.org/10.6094/UNIFR/149771
11. S.A. Anggraini, M. Uehara, K. Hirata, H. Yamada, M. Akiyama, Enhancement in piezoelectric responses of AlN thin films by co-addition of Mg and Ta. Mater. Chem. Phys. **276**, 125394 (2022). https://doi.org/10.1016/j.matchemphys.2021.125394
12. A. Witek et al., Synthesis of oxygen-free aluminium nitride ceramics. J. Mater. Sci. **33**(13), 3321–3324 (1998). https://doi.org/10.1023/A:1013281212757
13. D. Mukherjee, D. Mallick, Magnetoelectric wireless power transfer system for biomedical implants, in *2021 IEEE International Midwest Symposium on Circuits and Systems (MWSCAS), Lansing, MI, USA*, Aug. 2021, pp. 356–359. https://doi.org/10.1109/MWSCAS47672.2021.9531861
14. C. Dong et al., Characterization of magnetomechanical properties in FeGaB thin films. Appl. Phys. Lett. **113**(26), 262401 (2018). https://doi.org/10.1063/1.5065486
15. L. Chen, Y. Wang, T. Luo, Y. Zou, Z. Wan, The study of magnetoimpedance effect for magnetoelectric laminate composites with different magnetostrictive layers. Materials **14**(21), 6397 (2021). https://doi.org/10.3390/ma14216397
16. J.G. Wan, Z.Y. Li, Y. Wang, M. Zeng, G.H. Wang, J.-M. Liu, Strong flexural resonant magnetoelectric effect in Terfenol-D/epoxy-Pb(Zr,Ti)O3 bilayer. Appl. Phys. Lett. **86**(20), 202504 (2005). https://doi.org/10.1063/1.1935040
17. J. Zhai, Z. Xing, S. Dong, J. Li, D. Viehland, Magnetoelectric laminate composites: an overview. J. Am. Ceram. Soc. **91**(2), 351–358 (2008). https://doi.org/10.1111/j.1551-2916.2008.02259.x
18. P. Martins, C.S.L. Fernandez, D. Silva, S. Lanceros-Méndez, Theoretical optimization of magnetoelectric multilayer laminates. Compos. Sci. Technol. **204**, 108642 (2021). https://doi.org/10.1016/j.compscitech.2020.108642
19. M. Acosta et al., $BaTiO_3$ -based piezoelectrics: fundamentals, current status, and perspectives. Appl. Phys. Rev. **4**(4), 041305 (2017). https://doi.org/10.1063/1.4990046
20. A. Omelyanchik et al., Boosting magnetoelectric effect in polymer-based nanocomposites. Nano **11**(5), 1154 (2021). https://doi.org/10.3390/nano11051154
21. S. Zuo et al., Ultrasensitive magnetoelectric sensing system for pico-tesla magnetomyography. IEEE Trans. Biomed. Circuits Syst. **14**, 971 (2020)
22. O. Saha, E. Andersen, S. Roundy, Wireless power transfer by self-biased magnetoelectric laminate for biomedical implants, in *2021 IEEE 20th International Conference on Micro and Nanotechnology for Power Generation and Energy Conversion Applications (PowerMEMS), Exeter, United Kingdom*, Dec. 2021, pp. 36–39. https://doi.org/10.1109/PowerMEMS54003.2021.9658356
23. H. Palneedi et al., Enhanced self-biased magnetoelectric coupling in laser-annealed Pb (Zr, Ti) O3 thick film deposited on Ni foil. ACS Appl. Mater. Interfaces **10**(13), 11018–11025 (2018)

24. Y. Zhou, D. Maurya, Y. Yan, G. Srinivasan, E. Quandt, S. Priya, Self-biased magnetoelectric composites: an overview and future perspectives. Energy Harvesting Syst. **3**(1), 1–42 (2016). https://doi.org/10.1515/ehs-2015-0003
25. M. Rotan, M. Zhuk, J. Glaum, Activation of ferroelectric implant ceramics by corona discharge poling. J. Eur. Ceram. Soc. **40**(15), 5402–5409 (2020). https://doi.org/10.1016/j.jeurceramsoc.2020.06.058
26. V. Bijalwan, I. Sokolov, P. Tofel, Poling procedures and piezoelectric response of (Ba $_{0.85}$ Ca$_{0.15}$ Zr$_{0.1}$ T$_{0.9}$) O$_3$ ceramics. J. Asian Ceramic Soc. **9**(1), 229–236 (2021). https://doi.org/10.1080/21870764.2020.1860438
27. T.C. Darvade, P.S. Kadhane, B.G. Baraskar, A.R. James, V.R. Reddy, R.C. Kambale, Structural and multiferroic properties of trilayer green magnetoelectric Co0.8Ni0.2Fe2O4/K0.25Na0.75NbO3/Co0.8Ni0.2Fe2O4 composite structure. J. Magn. Magn. Mater. **505**, 166733 (2020). https://doi.org/10.1016/j.jmmm.2020.166733
28. R. Stampfli, N.U. Huynh, G. Youssef, Long-term converse magnetoelectric response of actuated 1-3 multiferroic composite structures. Magnetochemistry **7**(4), 55 (2021). https://doi.org/10.3390/magnetochemistry7040055
29. S. Lather et al., Effect of mechanical milling on magnetic, dielectric and magneto-electric properties of Z-type (Ba, Sr) hexaferrites. J. Alloys Compd. **902**, 163807 (2022). https://doi.org/10.1016/j.jallcom.2022.163807
30. B. Sapkota, M.T. Hasan, A. Martin, R. Mahbub, J.E. Shield, V. Rangari, Fabrication and magnetoelectric investigation of flexible PVDF-TrFE/cobalt ferrite nanocomposite films. Mater. Res. Express **9**(4), 046302 (2022). https://doi.org/10.1088/2053-1591/ac6151
31. S. Reis et al., Electronic optimization for an energy harvesting system based on magnetoelectric Metglas/poly(vinylidene fluoride)/Metglas composites. Smart Mater. Struct. **25**(8), 085028 (2016). https://doi.org/10.1088/0964-1726/25/8/085028
32. J.M. Sanchez-Chiva, D. Galayko, A. Rhouni, A CMOS active voltage doubler for a low voltage \$100–1000\ \mu \mathrm{W}\$ range magnetoelectric energy transducer, in *2021 28th IEEE International Conference on Electronics, Circuits, and Systems (ICECS), Dubai, United Arab Emirates*, Nov. 2021, pp. 1–6. https://doi.org/10.1109/ICECS53924.2021.9665515
33. S.K. Ghosh et al., Rollable magnetoelectric energy harvester as a wireless IoT sensor. ACS Sustain. Chem. Eng. **8**(2), 864–873 (2019)
34. Z. Xiao et al., Enhanced magnetoelectric coupling in a composite multiferroic system via interposing a thin film polymer. AIP Adv. **8**(5), 055907 (2018)
35. A. Kulkarni et al., Giant magnetoelectric effect at low frequencies in polymer-based thin film composites. Appl. Phys. Lett. **104**(2), 022904 (2014)
36. Y. Zhang et al., Magnetoelectric nanoparticles incorporated biomimetic matrix for wireless electrical stimulation and nerve regeneration. Adv. Healthc. Mater. **10**(16), 2100695 (2021). https://doi.org/10.1002/adhm.202100695
37. T. Nguyen et al., In vivo wireless brain stimulation via non-invasive and targeted delivery of magnetoelectric nanoparticles. Neurotherapeutics **18**(3), 2091–2106 (2021). https://doi.org/10.1007/s13311-021-01071-0
38. T.D. Brown, N. Habibi, D. Wu, J. Lahann, S. Mitragotri, Effect of nanoparticle composition, size, shape, and stiffness on penetration across the blood–brain barrier. ACS Biomater Sci. Eng. **6**(9), 4916–4928 (2020). https://doi.org/10.1021/acsbiomaterials.0c00743
39. B.N. Rao, P. Kaviraj, S.R. Vaibavi, A. Kumar, S.K. Bajpai, A. Arockiarajan, Investigation of magnetoelectric properties and biocompatibility of CoFe2O4-BaTiO3 core-shell nanoparticles for biomedical applications. J. Appl. Phys. **122**(16), 164102 (2017)
40. S. Babu, A. Siddeshwar, S. Kuchipudi, S. Suryanarayana, T. Bhimasankaram, Magnetoelectric properties of Ni/PZT/Ni layered composite for low field applications. J. Mater. Sci. **44**, 3948–3951 (2009). https://doi.org/10.1007/s10853-009-3534-4
41. M.Y. Alnassar, Y.P. Ivanov, J. Kosel, Flexible magnetoelectric nanocomposites with tunable properties. Adv. Electron. Mater. **2**(6), 1600081 (2016)
42. A. Amirov, I. Baraban, L. Panina, V. Rodionova, Direct magnetoelectric effect in a sandwich structure of PZT and magnetostrictive amorphous microwires. Materials **13**(4), 916 (2020)

43. R. Gupta et al., Giant magnetoelectric effect in PZT thin film deposited on nickel. Energy Harvest. Syst. **3**(2), 181 (2016)
44. H. Palneedi et al., Tailoring the magnetoelectric properties of Pb (Zr, Ti) O3 film deposited on amorphous metglas foil by laser annealing. J. Am. Ceram. Soc. **99**(8), 2680–2687 (2016)
45. L. Zhou et al., Self-assembled core-shell CoFe2O4@ BaTiO3 particles loaded P (VDF-HFP) flexible films with excellent magneto-electric effects. Appl. Phys. Lett. **111**(3), 032903 (2017)
46. A. McDannald et al., Switchable 3-0 magnetoelectric nanocomposite thin film with high coupling. Nanoscale **9**(9), 3246–3251 (2017)
47. J. Zhai, S. Dong, Z. Xing, J. Li, D. Viehland, Giant magnetoelectric effect in Metglas/polyvinylidene-fluoride laminates. Appl. Phys. Lett. **89**(8), 083507 (2006)
48. A.A. Chlaihawi et al., Novel screen printed flexible magnetoelectric thin film sensor. Procedia Eng. **168**, 684–687 (2016)
49. M. Bichurin, V. Petrov, *Modeling of Magnetoelectric Effects in Composites*, vol 201 (Springer, 2014)

Chapter 4
Triboelectric Nanogenerators Competency to Wireless Device Applications

P. Ravi Sankar, P. Supraja, K. Prakash, R. Rakesh Kumar, and K. Uday Kumar

4.1 Introduction

The rapid progress in the field of electronics observes an obvious trend of functionality, miniaturization, and portability. This is evident from the development of handheld mobile phones, one of the most popular and widely used examples of portable electronics that come with large functionalities. The present and future goals would be to deploy a range of sensors, including but not limited to gas, chemical, biological, motion, and navigation, to improve the functionality of electronics that are already in use. Many attempts are being undertaken to create devices far smaller than a mobile phone's size so that each person may typically own many electronic gadgets that serve various human needs. These small-sized electronic gadgets are generally operated on ultralow-power levels, allowing us to power them by harvesting energy from our living environment [1]. With the application of nanomaterials and nanotechnology for harvesting energy for powering micro/nano-power systems, new technologies capable of harvesting energy from the environment and thereby functioning as sustainable and self-sufficient small-scale power sources are constantly coming out and being researched in the field of nano energy. With the growing popularity of portable electronics, the challenge remains to develop energy storage-related technologies, as most are still running on batteries. As sensors are employed in different electronic devices, it will become more and more challenging to power such a tremendous network consisting of trillions of sensors around the world. One cannot opt for batteries due to their

P. R. Sankar · K. Prakash (✉)
Flexible Electronics Lab, National Institute of Technology, Warangal, India
e-mail: kprakash@nitw.ac.in

P. Supraja · R. R. Kumar · K. U. Kumar
Energy Materials and Devices Lab, National Institute of Technology, Warangal, India

R. Das, H. Heidari (eds.), *Wireless Power Technologies for Biomedical Devices*, Analog Circuits and Signal Processing, https://doi.org/10.1007/978-3-031-52834-7_4

drawbacks, such as replacing them, checking their orientation, and inspecting them from time to time.

A sustainable power source capable of harvesting energy from the environment can be seen as one of the possible solutions to the problem mentioned. Thus, there is a need for nano energy and dependent power sources which are available, efficient, and stable [2]. Triboelectric Nanogenerator (TENG) is a new type of technology, first demonstrated in 2012 by Prof. Zhong Lin Wang's group at the Georgia Institute of Technology [3, 4]. TENG Working in conjunction with the triboelectric effect and electrostatic induction, it is an energy harvesting device that converts tiny amounts of mechanical energy into electrical energy. TENG has a wide range of applications due to its ability to produce a range of output. This chapter briefly introduces TENGs, a detailed literature review, their fundamental modes of operation, their working principle, and potential applications in self-powered medical devices and health monitoring.

4.2 Detailed Literature Review

Figure 4.1 represents the detailed literature statistics of the TENG. We have used the search term Triboelectric nanogenerator **Or** Triboelectric generator keywords in the Web of Science database. A total of 4677 articles are found, and there is an exponential growth of articles published from 2012 (6 articles) to 2021 (1116 articles), as shown in Fig. 4.1a. Figure 4.1a indicates the potential of the TENG technology. Further analyses were carried out with the top ten results in each category with relative percentages. The significant contribution to TENG comes from the Chinese academy of sciences institute (1355 articles), and china country (3201 articles). The majority of the articles were published in Nano Energy (Elsevier publications), and Prof Z. L Wang has published the highest number of articles. The paper "Flexible triboelectric generator" is cited the highest number of times [5].

4.3 Triboelectric Effect and Working of TENG

The triboelectric effect is a contact-induced electrification phenomenon in which the material happens to be electrically charged once it contacts another material through friction. The sign of charges carried by the material depends on its relative polarity compared to the material it's brought into contact with. The effect can be generally understood as the consequence of the flow of charges from one material to another material via a physical bond formed between some parts of the two surfaces, often referred to as adhesion, when brought into contact by the application of mechanical force. The transferred charges could be either electrons or ions/molecules [6–8]. When separated, some bonded atoms retain extra electrons

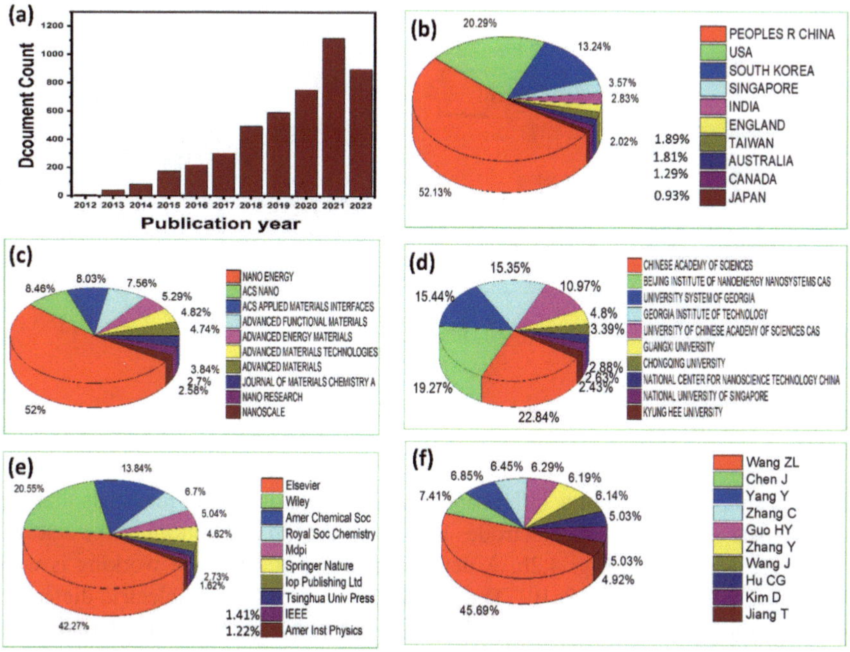

Fig. 4.1 Detailed literature review statistics (Source: Web of Science, as on Oct, 1 2022) (**a**) Total numbers documents published in each year, Top ten entries contributing to the TENG research by (**b**) Countries, (**c**) Journal, (**d**) Institute, (**e**) Publishers, (**f**) Author

while others surrender them, leading to the likely development of triboelectric charges on surfaces. Surprisingly, materials with a strong triboelectric effect are probably the ones with less conductivity or insulators. As a result, they can capture the charges transferred and hold them for a long period, thereby building up the triboelectric charges.

The working mechanism of TENG is presented in Fig. 4.2a. Two main components constituting TENG are dielectric materials that demonstrate a strong triboelectric effect and electrodes to balance the induced surface charges. Triboelectric charges of opposite polarity are induced on the surfaces of two different materials when they are made to contact each other. A potential difference is created between the top and bottom electrodes when they are separated. The generated potential difference causes a flow of electrons from one electrode to another. When the triboelectric layers move close to each other under force, electrons flow in the reverse direction from one electrode to another. Further, TENG has four basic modes, as described in Fig. 4.2b: (1) vertical contact-separation (VCS), (2) Lateral sliding (LS), (3) single-electrode (SE), (4) free-standing triboelectric (FST).

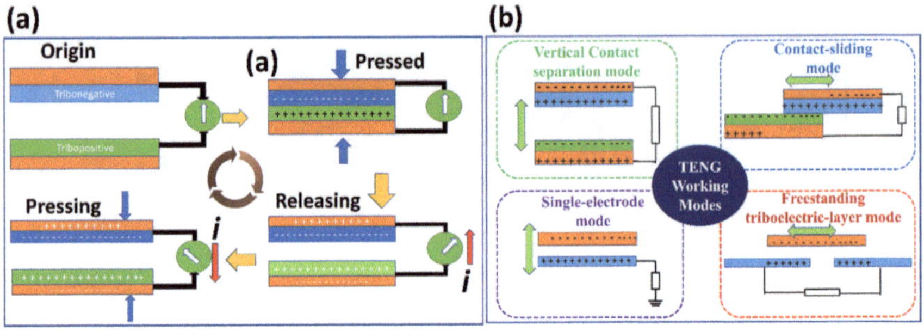

Fig. 4.2 (**a**) Basic working of TENG, (**b**) Different working modes of TENG

4.3.1 CS Mode

It is the simplest design of a TENG and composes two dissimilar materials that are face to face with each, with electrodes located on the top and bottom of the stacked structure, as shown in Fig. 4.2b. A detailed analysis of CS mode working is well described in the literature [9, 10]. The triboelectric layers can be dielectric-dielectric or metal-dielectric, or metal-metal. Physical contact by applying an external force causes charges of opposite signs to be induced on the surfaces of dissimilar metals. Once the two surfaces are separated under an external lifting force, a potential drop is created. Free electrons are propelled to another electrode to build an opposite potential to balance the induced electrostatic field if the setup is joined to an external load. Further, when the gap is closed, the potential due to triboelectric charges disappears, and electrons flow back.

4.3.2 LS Mode

The structure is quite akin to that of vertical contact mode, generates the difference being that a relative gliding parallel to the surfaces generates triboelectric charges on the surfaces in lateral sliding mode [11–13] in contrast to the touch and leave type in vertical contact mode depicted in Fig. 4.2b. Here, a lateral polarization induces the electric potential that causes the electrons to flow from one electrode to another to balance this potential. An AC output is achieved by periodic sliding and closing.

4.3.3 SE Mode

The previous two modes mentioned above see two electrodes interconnected to a load. A TENG is of that type suitable for free mobile applications and fails to be effective when some object that is part of TENG cannot be coupled with a load since it is a mobile object, such as the human motion of walking on the floor. One of the electrodes is grounded to harvest such energy, as seen in Fig. 4.2b [14, 15]. Given the finite size of TENG, separating the top object from the bottom would cause a local distribution of the electrical field so that there is an exchange in electrons between the bottom electrode top ground to balance the potential change of the electrode. This strategy can be used to harvest energy in Contact Sliding Mode [16] and contact Separation Mode [17].

4.3.4 FST Mode

A moving object by nature is naturally charged due to its contact with air or other objects. These charges are retained on the surface for a long time, and the charge density reaches a maximum value, thereby contact or friction unnecessary. This mode works on balancing the local potential distribution initiated by electron flow between electrodes due to asymmetric charge distribution [17, 18]. This is obtained by placing a pair of symmetric electrodes underneath a dielectric layer with the size of electrodes and creating a gap separation between two electrodes in the same order as the size of the moving object; therefore, the object's approach to or departing from the electrodes generates asymmetric charge distribution [19]. One added advantage of this model is its extension of the durability of the TENGs.

4.3.5 TENG Power Output

Power density is an essential parameter in quantifying the TENG. The power density first reported TENG is ~10.4 mW/cm^3 and increased rapidly over the years to a few hundred watts/m^2 [20, 21]. In recent times, TENGs achieved a remarkable milestone of achieving power density and conversion efficiency of 500 Wm^{-2} and >50%, respectively [22]. W. Tang et al. [23] have demonstrated an enhanced area power density as high as 500 Wm^{-2}, a volume power density of 15 MWm^{-3}, and instantaneous conversion efficiency of 70%. Y. Xie et al. [24] have also conducted experiments to show that the fast energy conversion efficiency may reach up to 80%.

4.3.6 Material Choices for TENG

Almost all materials we are aware of exhibit a triboelectrification effect, including metals, dielectrics, semiconductors, insulators, and polymers. A detailed review of material choices for TENG can be found in the literature [25, 26]. There are a lot of material options for TENG fabrication because all of the materials are potential candidates. But a material's capacity for receiving or shedding electrons relies on its polarity. In addition to selecting a material from the triboelectric series, physical techniques can change surface morphologies by generating micro- or nanopatterns based on pyramids, squares, or hemispheres. These patterns are helpful in increasing the contact area and, perhaps triboelectrification.

4.4 Applications

TENG, unlike traditional electromagnetic generators (EMGs), is particularly well-suited for collecting low-frequency ambient mechanical stimuli, and its output efficiency is [27] much higher [28]. The promising advantages of TENGs have now been applied to diverse fields of science, technology, engineering, and medicine. The application of TENGs can be summarized into four areas: (i) as a micro/nano power source in powering electronic devices [29–34], (ii) as a self-powered sensor system [35–42], (iii) as a blue energy harvesting system [43], and (iv) as a high-voltage (HV) power source [44]. The development of TENGs in these areas focused on the fabrication of TENGs based on material choice.

4.4.1 TENG as a Bio-mechanical Energy Harvester

A TENG device's primary function is to gather mechanical energy from the surrounding environment. Any type of mechanical energy is possible, including vibration, vehicle motion, and bodily motion in humans. A TENG device looks at the construction of wearable TENG and its applications to successfully scavenge biomechanical energy and use it gainfully. Figure 4.3a, b represents ZnO and ZnS nanosheet films based TENGs schematically [45, 46]. Figure 4.3c, d shows the TENG electrical response against hand slapping force. Further, these TENG devices are used to power the LEDs, portable electronic devices, as shown in Fig. 4.3e, f.

4.4.2 Powering Implantable Medical Devices by TENG

According to the World Health Organization (WHO), cardiovascular anomalies are the leading cause of death worldwide [47]. Various respiratory issues cause chronic respiratory diseases, such as asthma, allergies, lung diseases, and pulmonary hypertension due to increased air pollution has caused. Therefore, it is crucial to keep track of cardiac and respiratory parameters in real-time, especially heart rate and rhythmic motion, to diagnose underlying diseases. The most critical implantable medical device is the pacemaker, and the function of the pacemaker is to regulate the heartbeat with electric stimulation of heart muscles. The energy source of

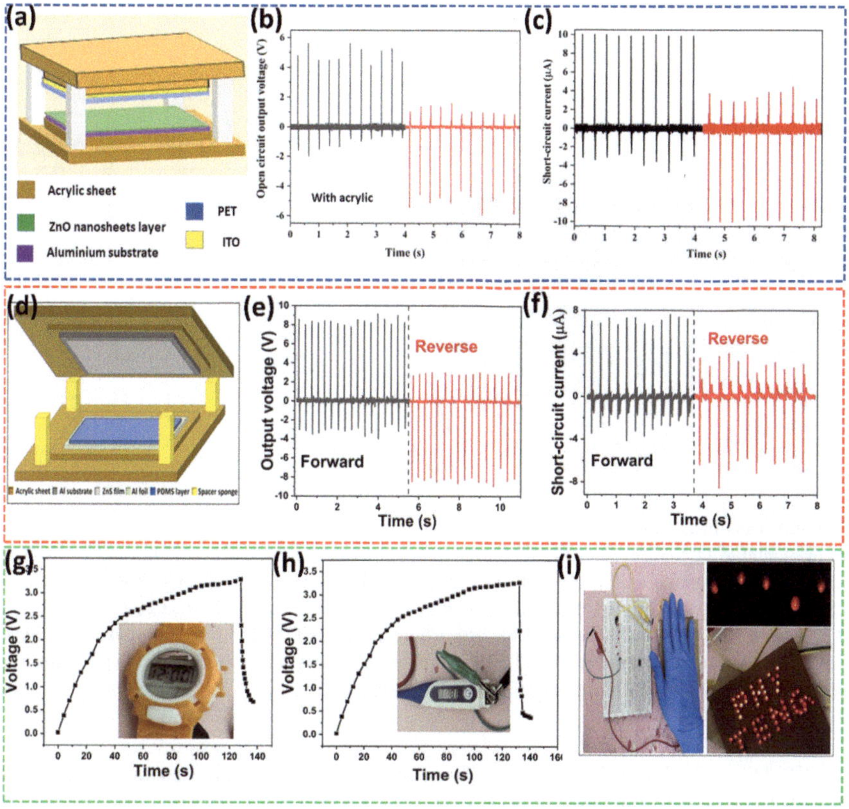

Fig. 4.3 Biomechanical energy harvesting with TENG made up of (**a**) with ZnO nanosheets, (**b**) ZnS nanosheets, (**c**) a Digital watch, thermometer, and LEDs powered with TENG (Reproduced with permission from [45, 46])

the pacemaker is the battery which needs periodic replacement due to its limited capacity. Replacement of battery is necessary as the abnormalities in the rate of heartbeats lead to death in some cases. The replacement needs additional surgery, which is expensive and causes inconvenience to patients due to the long recovery time. Therefore, self-powered pacemakers are attracted attention in recent days. The first report on powering pacemakers by TENG is reported in 2014 [48]. The proposed TENG is implanted under the thoracic skin and collects the periodic breathing energy in a living rat, and converted energy was utilized to power the pacemaker. In this TENG design, patterned PDMS and nanosurface-modified Al act as triboelectric layers with PET spacers. The breath-driven TENG generated an output power density of 8.44 mW/m^2, short-circuit current of 0.14 μA, and an open-circuit voltage of 3.73 V. A capacitor stored the electrical energy generated by the TENG to drive the pacemaker to regulate the heartbeat. The concept of driving the

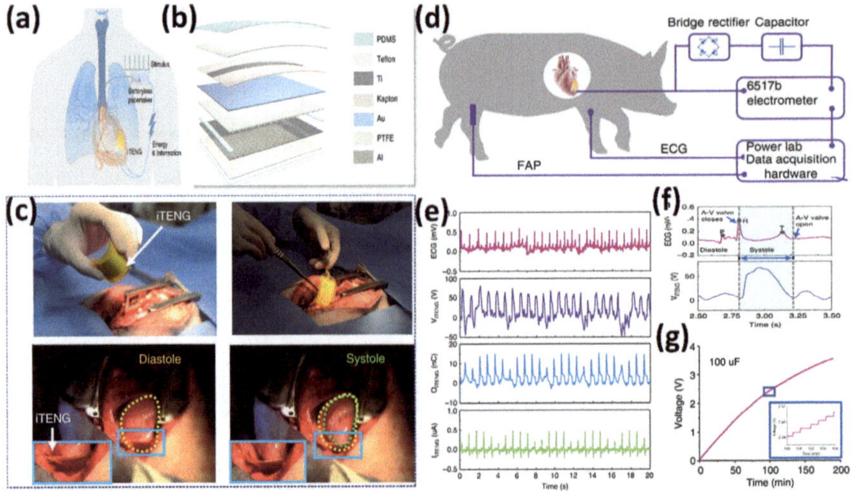

Fig. 4.4 (**a**) Illustration of symbiotic cardiac pacemaker system, (**b**) Schematic structure diagram of implantable TENG, (**c**) The iTENG implantation process in animal experiments and TENG was driven by the diastole and systole of the heart, (**d**) Schematic of in vivo experimental electrical characterizations, (**e**) In vivo open-circuit voltage, transferred charge, short-circuit current of the TENG and simultaneously recorded electrocardiography (ECG), (**f**) In vivo output open-circuit voltage and simultaneously recorded ECG signals, (**g**) Charging curve of a 100 μF capacitor charged by TENG. (Reproduced with permission from [49])

TENG from regular and irregular body movements has brought advancement in the field of self-powered lifetime IMDs.

Further, in 2019, H. Ouyang reported an implantable symbiotic pacemaker (SPM) powered by implanted TENG on an adult Yorkshire male porcine heart [49]. TENG achieves energy harvesting and storage and cardiac pacing on a large-animal scale. The source of energy and target of SPM is the heart. The energy harvested from each cardiac motion cycle is 0.495 μJ, higher than the required endocardial pacing threshold energy (0.377 μJ). Figure 4.4a shows the schematic of the placement of the TENG and SPM inside porcine body. The TENG is fabricated using nanostructured PTFE film and Al foil as triboelectric layers with additional spacer, encapsulation layers, as shown in Fig. 4.4b. The TENG was placed between the heart and pericardium, and the PTFE side faced the left ventricular wall. Cardiac motion results in regular contact and separation of the two triboelectric layers. In vivo open-circuit voltage, transferred charge, short-circuit current of the TENG, and simultaneously recorded electrocardiography (ECG), presented in Fig. 4.4e. In vivo mode, TENG produced a V_{OC} ~ 65.2 V, Q_{SC} ~ 13.6 nC, and the I_{SC} ~ 0.5 μA. The electrical output of TENG is completely synchronized with the corresponding ECG signals (Fig. 4.4f). Therefore, TENG output can be used to monitor the heartbeat rate. Under blood pressure of ~100/70 mmHg and heart rate of ~77 bpm, TENG output was stored in a capacitor through a rectifier. The voltage of the capacitor could be charged from 0 to 3.55 V within 190 min (Fig. 4.4g). The sinus arrhythmia is corrected, and deterioration is prevented using the symbiotic pacemaker by inducing myocardial contraction and regulating the heart rate. The applications can be extended to in vivo symbiotic bioelectronics. Recently, a self-rechargeable cardiac pacemaker system was demonstrated using an inertia-driven TENG based on body motion and gravity [50]. In this TENG, five-stacked TENG devices converted mechanical energy into electricity and stored it to power the pacemaker. The self-rechargeable cardiac pacemaker system's ventricle pacing and sensing operation mode is demonstrated.

4.4.3 Self-Powered Wireless Cardiac Monitoring via Implantable TENG

Initially, TENG was used to power implantable medical devices, and later it was used as a self-powered, wireless healthcare monitoring system. The first report on wireless healthcare monitoring using TENG is reported in 2016 [51]. In this report, TENG is implanted in the heart of an adult Yorkshire porcine, and TENG is made up of nanostructured PTFE film and Aluminium foil. The implanted TENG produced an output voltage of 14 V, and a current of 5 μA. The TENG responses directly indicate the physiological heart-beating behaviour. A self-powered wireless data transmission system was fabricated for real-time remote cardiac monitoring by connecting it with an implantable wireless transmitter. The real-time heartbeat of an adult Yorkshire porcine was monitored successfully through the wireless transmitted signal.

4.4.4 TENG Itself a Self-Powered Implantable Cardiac and Pulse Sensor

The first report on real-time monitoring of vital signs in vivo mode using TENG is reported in 2016 [52]. This report uses nanostructured PTFE and Kapton film as triboelectric layers. TENG is attached to the pericardium of the living porcine model to monitor the heart rate and rhythm (Fig. 4.5a). Figure 4.5b shows the R waves of the electrocardiogram (ECG) and TENG output voltage signals against the contraction and relaxation of the heart. The frictional layers repeatedly undergo contact and separation due to the periodic heartbeat (motion), which results in alternating current and voltage pulses due to contact electrification and electrostatic induction (Fig. 4.5c). Further heartbeat was calculated using the time gap between the two consecutive peaks, as shown in Fig. 4.5d.

The heartbeat value of 89.4 bpm and 88.6 bpm was observed with ECG and TENG, which confirm the high accuracy of heartbeat monitoring using TENG. Further, the heartbeat is altered using epinephrine and amiodarone, and the response

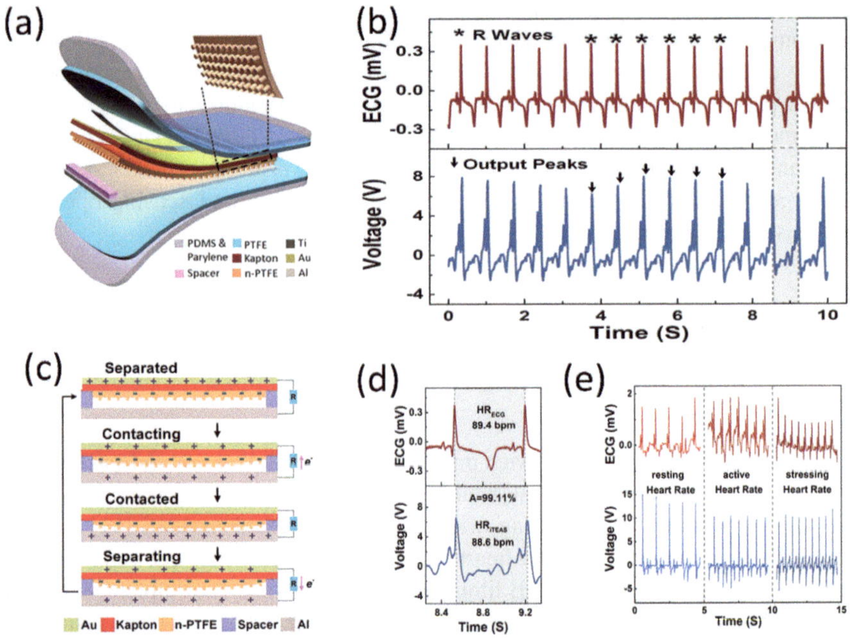

Fig. 4.5 (**a**) Schematic illustration of the TENG structure, (**b**) R waves of ECG (asterisks) absolutely corresponded to peaks of voltage output of the TENG (arrows), (**c**) Proposed working mechanism of TENG, (**d**) time intervals between two enlarged contiguous pairs of the R wave and output voltage of TENG, (**e**) The consistency between R waves and voltage peaks remained under resting (~60 bpm, rHR), active (~90 bpm, aHR), and stressed (~120 bpm, sHR) conditions. (Reproduced with permission from [52])

of ECG and TENG signal is recorded, as shown in Fig. 4.5e. Figure 4.5e further confirms the high accuracy of TENG in monitoring the heartbeat. The implanted TENG was used to estimate blood flow pressure and velocity.

Recently, non-invasive multi-indicators cardiovascular monitoring is reported using TENG implanted at different arterial pulse sites of the human body, as shown in Fig. 4.6a [53]. TENG's design is made up of microstructure FEP and Polyamide (PA) film using 3D elastic sponge spacers (Fig. 4.6b. Figure 4.6d shows the TENG response and ECG signals' response in detecting the arterial pulse wave generated by pumping the blood by the heart. The accurate matching of both signals indicates that implanted TENG is a potential candidate for cardiovascular monitoring. Further pulse wave signals detected from different arteries, including (i) carotid artery, (ii) brachial artery, (iii) radial artery, and (iv) fingertip artery, are presented in Fig. 4.6e. Simultaneous acquisition of multiple arterial pulse signals can fully extract human physiological information. The interval between two adjacent peaks is regarded as the single heartbeat cycle used for real-time heart rate (HR) monitoring. Further, LabVIEW program is developed for HR monitoring strategy, which alerts when the HR is abnormal (HR > 100) (Table 4.1).

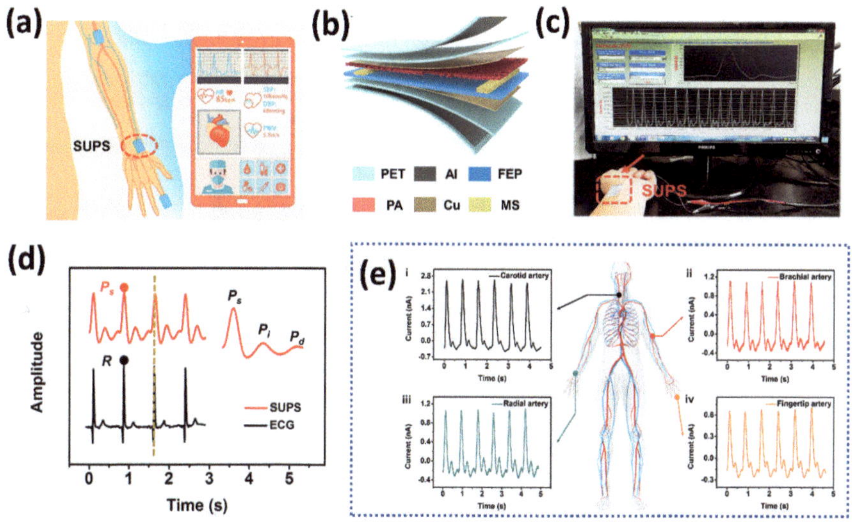

Fig. 4.6 (**a**) Illustration of the integration of self-powered ultrasensitive pulse sensor (SUPSs) for non-invasive multi-indicators cardiovascular monitoring, (**b**) Schematic structure diagram of the SUPs, (**c**) The real-time pulse wave signal detected by the SUPS attached on the radial artery, (**d**) R wave peak in ECG corresponds to the Ps in pulse wave signal accurately, (**e**) Pulse wave signals detected from different arteries including (i) Carotid artery, (ii) Brachial artery, (iii) Radial artery, and (iv) Fingertip artery (Reproduced with permission from [53])

Table 4.1 Different types of TENG Implantation and its Applications

Sl. No	TENG implanted place	Mode	Type of sensing	References
1	Swine's heart	In-vivo	Endocardial pressure sensor	[54]
2	Various artery position on a human body	In-vitro	Ultrasensitive pulse sensor	[55]
3	In the Mask	In-Vitro	Self-Powered Breath Monitoring Masks	[56]
4	Artery position on the hand	In-Vivo	Human Heart-Rate Monitoring	[57]
5	Attached on wrist	In-Vitro	Parkinson's disease tremor sensor	[58]

4.4.5 TENG-Based Wireless Power Transmission

Wireless electric energy transmission plays an important role where interconnecting wires is inconvenient, hazardous, or not possible. In 2017, a significant breakthrough in the wireless transmission of harvested energy is reported using TENG [59]. The proposed consists of 3D-printed eco-friendly graphene polylactic acid (gPLA) nanocomposite and Teflon as triboelectric layers. The proposed TENG generates very high output voltages >2 kV with a strong electric field that enables the wirelcss transmission of harvested energy over a distance of 3 m. The need for hard-wiring or

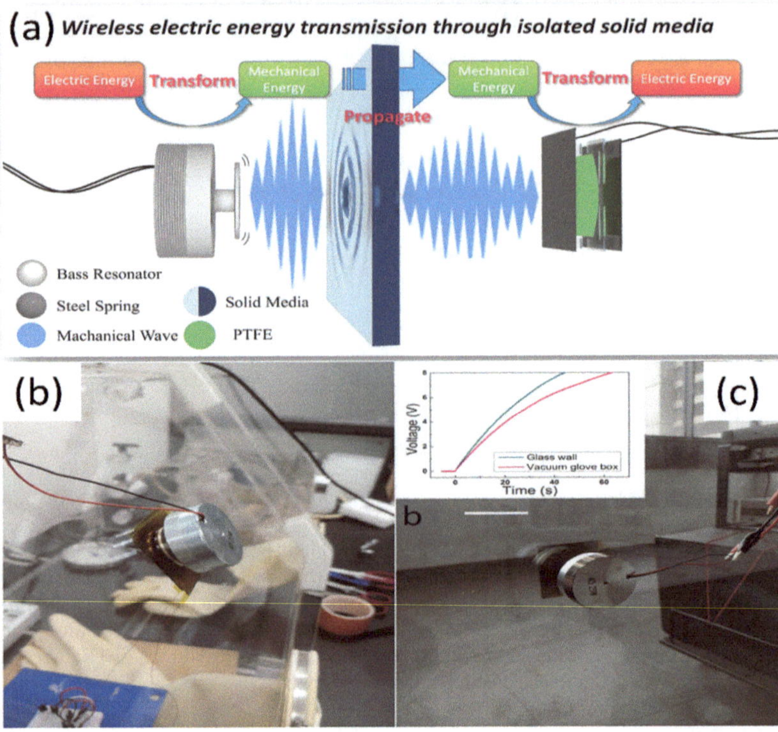

Fig. 4.7 (**a**) Schematic of the wireless electric energy transmission technology based on TENG, Photographs of the wireless charging technology working on (**b**) Vacuum glove box and (**c**) Glass wall (inset Corresponding voltage curve of the capacitor charged by the V-TENG in both situations). (Reproduced with permission from [60])

additional circuitry is avoided in charging the capacitor in this report. In addition to the above, wireless and secure actuation of smart-home applications such as smart tint windows, temperature sensors, liquid crystal displays, and security alarms are demonstrated.

In the same year, wireless electric energy transmission through various isolated solid media based on TENG is presented [60]. In this report, electric energy is first transformed into mechanical vibration energy in a mechanical wave that can pass well through the solid medium, and then the vibration energy is harvested by a TENG, as shown in Fig. 4.7a. The wireless energy supply through a metal chamber by the electromagnetic induction technique is not possible but possible with TENG technology. TENG is attached to the solid support to receive the mechanical wave and vibrate along with the solid support. A free-standing PTFE layer vibrates between the two conducting electrodes to produce electrical energy. The wireless charging technology with vacuum glovebox and glass wall, as shown in Fig. 4.7b, c.

Further, a rotating-disk-based hybridized electromagnetic–triboelectric nanogenerator was demonstrated for wirelessly transmitting energy using coils [61]. The sliding mode of TENG is utilized in harvesting rotational energy. When a supercapacitor bank is applied for storing the generated electricity from the nanogenerator, the wireless transmitted power can charge a mobile phone located 100 cm away from the transmitter. Recently, wireless transmission of solar and mechanical energy was demonstrated using a new type of hybrid nanogenerator consisting of a series connected TENG and photovoltaic cells [62]. The energy collected by both solar and mechanical energy was transmitted wirelessly by integrating a mechanical switch design based on LC oscillation.

4.5 Conclusions

In summary, focusing on the trend of downsizing, energy harvesting, and self-powered electronics, TENG is being developed as a sustainable, wireless, wearable power source. The ability to power electronic devices using waste mechanical energy from the environment has been feasible by the rapid improvement of energy harvesting nanogenerators. TENG development has evolved to large-scale manufacture, and the concept has begun to be marketed. TENG's further evolution would make it simpler to harvest mechanical energy from human motions according to the flexible structure. Building TENGs as a versatile power source would open up possibilities for a more dependable and sustainable power source and the potential to quickly substitute conventional batteries.

References

1. Z.L. Wang, W. Wu, Nanotechnology-enabled energy harvesting for self-powered micro–/nanosystems. Angew. Chem. Int. Ed. **51**, 11700 (2012). https://doi.org/10.1002/anie.201201656
2. H.H. Singh, N. Khare, Flexible ZnO-PVDF/PTFE based piezo-tribo hybrid nanogenerator. Nano Energy **51**, 216–222 (2018). https://doi.org/10.1016/j.nanoen.2018.06.055
3. Z.L. Wang, Triboelectric nanogenerators as new energy technology and self-powered sensors – principles, problems and perspectives. Faraday Discuss. **176**, 447 (2014). https://doi.org/10.1039/c4fd00159a
4. Z. Wen, Q. Shen, X. Sun, Nanogenerators for self-powered gas sensing. Nanomicro Lett. **9**, 1–19 (2017). https://doi.org/10.1007/s40820-017-0146-4
5. F.R. Fan, Z.Q. Tian, Z. Lin Wang, Flexible triboelectric generator. Nano Energy **1**, 328–334 (2012). https://doi.org/10.1016/j.nanoen.2012.01.004
6. Z.L. Wang, A.C. Wang, On the origin of contact-electrification. Mater. Today **30**, 34–51 (2019). https://doi.org/10.1016/j.mattod.2019.05.016
7. F.A. Vick, F.P. Inst, *Theory of Contact Electrification* (n.d.)
8. W. Zhang, Y. Shi, Y. Li, X. Chen, H. Shen, A review: contact electrification on special interfaces. Front. Mater. **9** (2022). https://doi.org/10.3389/fmats.2022.909746

9. H. Chen, Y. Xu, J. Zhang, W. Wu, G. Song, Theoretical system of contact-mode triboelectric nanogenerators for high energy conversion efficiency. Nanoscale Res. Lett. **13**, 346 (2018). https://doi.org/10.1186/s11671-018-2764-2
10. S. Niu, S. Wang, L. Lin, Y. Liu, Y.S. Zhou, Y. Hu, Z.L. Wang, Theoretical study of contact-mode triboelectric nanogenerators as an effective power source. Energy Environ. Sci. **6**, 3576–3583 (2013). https://doi.org/10.1039/c3ee42571a
11. S. Niu, Y. Liu, S. Wang, L. Lin, Y.S. Zhou, Y. Hu, Z.L. Wang, Theory of sliding-mode triboelectric nanogenerators. Adv. Mater. **25**, 6184–6193 (2013). https://doi.org/10.1002/adma.201302808
12. G. Zhu, J. Chen, Y. Liu, P. Bai, Y.S. Zhou, Q. Jing, C. Pan, Z.L. Wang, Linear-grating triboelectric generator based on sliding electrification. Nano Lett. **13**, 2282 (2013). https://doi.org/10.1021/nl4008985
13. S. Wang, L. Lin, Y. Xie, Q. Jing, S. Niu, Z.L. Wang, Sliding-triboelectric nanogenerators based on in-plane charge-separation mechanism. Nano Lett. **13**, 2226 (2013). https://doi.org/10.1021/nl400738p
14. S. Niu, Y. Liu, S. Wang, L. Lin, Y.S. Zhou, Y. Hu, Z.L. Wang, Theoretical investigation and structural optimization of single-electrode triboelectric nanogenerators. Adv. Funct. Mater. **24**, 3332–3340 (2014). https://doi.org/10.1002/adfm.201303799
15. M. Zhang, L. Xia, S. Dang, L. Shi, A. Cao, Q. Deng, C. Du, A flexible single-electrode-based triboelectric nanogenerator based on double-sided nanostructures. AIP Adv. **9** (2019). https://doi.org/10.1063/1.5114884
16. Y. Yang, Y.S. Zhou, H. Zhang, Y. Liu, S. Lee, Z.L. Wang, A single-electrode based triboelectric nanogenerator as self-powered tracking system. Adv. Mater. **25**, 6594–6601 (2013). https://doi.org/10.1002/adma.201302453
17. S. Wang, Y. Xie, S. Niu, L. Lin, Z.L. Wang, Free-standing triboelectric-layer-based nanogenerators for harvesting energy from a moving object or human motion in contact and non-contact modes. Adv. Mater. **26**, 2818–2824 (2014). https://doi.org/10.1002/adma.201305303
18. S. Niu, Y. Liu, X. Chen, S. Wang, Y.S. Zhou, L. Lin, Y. Xie, Z.L. Wang, Theory of free-standing triboelectric-layer-based nanogenerators. Nano Energy **12**, 760–774 (2015). https://doi.org/10.1016/j.nanoen.2015.01.013
19. M.H. Yeh, L. Lin, P.K. Yang, Z.L. Wang, Motion-driven electrochromic reactions for self-powered smart window system. ACS Nano **9**, 4757 (2015). https://doi.org/10.1021/acsnano.5b00706
20. J.W. Lee, B.U. Ye, J.M. Baik, Research update: recent progress in the development of effective dielectrics for high-output triboelectric nanogenerator. APL Mater. **5** (2017). https://doi.org/10.1063/1.4979306
21. J.M. Baik, J.P. Lee, Strategies for ultrahigh outputs generation in triboelectric energy harvesting technologies: from fundamentals to devices. Sci. Technol. Adv. Mater. **20**, 927–936 (2019). https://doi.org/10.1080/14686996.2019.1655663
22. W. Tang, T. Jiang, F.R. Fan, A.F. Yu, C. Zhang, X. Cao, Z.L. Wang, Liquid-metal electrode for high-performance triboelectric nanogenerator at an instantaneous energy conversion efficiency of 70.6%. Adv. Funct. Mater. **25**, 3718–3725 (2015). https://doi.org/10.1002/adfm.201501331
23. Y. Xie, S. Wang, S. Niu, L. Lin, Q. Jing, J. Yang, Z. Wu, Z.L. Wang, Grating-structured freestanding triboelectric-layer nanogenerator for harvesting mechanical energy at 85% total conversion efficiency. Adv. Mater. **26**, 6599 (2014). https://doi.org/10.1002/adma.201402428
24. H. Yang, Y. Pang, T. Bu, W. Liu, J. Luo, D. Jiang, C. Zhang, Z.L. Wang, Triboelectric micromotors actuated by ultralow frequency mechanical stimuli. Nat. Commun. **10**, 2309 (2019). https://doi.org/10.1038/s41467-019-10298-7
25. D.W. Kim, J.H. Lee, J.K. Kim, U. Jeong, Material aspects of triboelectric energy generation and sensors. NPG Asia Mater. **12** (2020). https://doi.org/10.1038/s41427-019-0176-0
26. R. Zhang, H. Olin, Material choices for triboelectric nanogenerators: a critical review. EcoMat **2**, 1–13 (2020). https://doi.org/10.1002/eom2.12062
27. T. Zhao, S. Cao, S. Yang, R. Guo, S. Sang, H. Zhang, A self-powered counter/timer based on a clock pointer-like frequency-tunable triboelectric nanogenerator for wind speed detecting. Nano Energy **65**, 104025 (2019). https://doi.org/10.1016/j.nanoen.2019.104025

28. J. Zhong, Q. Zhong, F. Fan, Y. Zhang, S. Wang, B. Hu, Z.L. Wang, J. Zhou, Finger typing driven triboelectric nanogenerator and its use for instantaneously lighting up LEDs. Nano Energy **2**, 491 (2013). https://doi.org/10.1016/j.nanoen.2012.11.015
29. L. Chen, C. Wen, S.L. Zhang, Z.L. Wang, Z. Bin Zhang, Artificial tactile peripheral nervous system supported by self-powered transducers. Nano Energy **82**, 105680 (2021). https://doi.org/10.1016/j.nanoen.2020.105680
30. J.W. Lee, H.J. Cho, J. Chun, K.N. Kim, S. Kim, C.W. Ahn, I.W. Kim, J.Y. Kim, S.W. Kim, C. Yang, J.M. Baik, Robust nanogenerators based on graft copolymers via control of dielectrics for remarkable output power enhancement. Sci. Adv. **3**, 1–10 (2017). https://doi.org/10.1126/sciadv.1602902
31. L. Zhao, K. Chen, F. Yang, M. Zheng, J. Guo, G. Gu, B. Zhang, H. Qin, G. Cheng, Z. Du, The novel transistor and photodetector of monolayer MoS2 based on surface-ionic-gate modulation powered by a triboelectric nanogenerator. Nano Energy **62**, 38–45 (2019). https://doi.org/10.1016/j.nanoen.2019.05.012
32. X. Chen, Y. Wu, A. Yu, L. Xu, L. Zheng, Y. Liu, H. Li, Z. Lin Wang, Self-powered modulation of elastomeric optical grating by using triboelectric nanogenerator. Nano Energy **38**, 91–100 (2017). https://doi.org/10.1016/j.nanoen.2017.05.039
33. X. Wang, B. Yang, J. Liu, Y. Zhu, C. Yang, Q. He, A flexible triboelectric-piezoelectric hybrid nanogenerator based on P(VDF-TrFE) nanofibers and PDMS/MWCNT for wearable devices. Sci. Rep. **6**, 1–10 (2016). https://doi.org/10.1038/srep36409
34. F.R. Fan, L. Lin, G. Zhu, W. Wu, R. Zhang, Z.L. Wang, Transparent triboelectric nanogenerators and self-powered pressure sensors based on micropatterned plastic films. Nano Lett. **12**, 3109–3114 (2012). https://doi.org/10.1021/nl300988z
35. Y. Yang, H. Zhang, Z.H. Lin, Y.S. Zhou, Q. Jing, Y. Su, J. Yang, J. Chen, C. Hu, Z.L. Wang, Human skin based triboelectric nanogenerators for harvesting biomechanical energy and as self-powered active tactile sensor system. ACS Nano **7**, 9213–9222 (2013). https://doi.org/10.1021/nn403838y
36. M. Xu, S. Wang, S.L. Zhang, W. Ding, P.T. Kien, C. Wang, Z. Li, X. Pan, Z.L. Wang, A highly-sensitive wave sensor based on liquid-solid interfacing triboelectric nanogenerator for smart marine equipment. Nano Energy **57**, 574 (2019). https://doi.org/10.1016/j.nanoen.2018.12.041
37. Z.H. Lin, G. Cheng, Y. Yang, Y.S. Zhou, S. Lee, Z.L. Wang, Triboelectric nanogenerator as an active UV photodetector. Adv. Funct. Mater. **24**, 2810–2816 (2014). https://doi.org/10.1002/adfm.201302838
38. P.K. Yang, Z.H. Lin, K.C. Pradel, L. Lin, X. Li, X. Wen, J.H. He, Z.L. Wang, Paper-based origami triboelectric nanogenerators and self-powered pressure sensors. ACS Nano **9**, 901 (2015). https://doi.org/10.1021/nn506631t
39. X. Pu, H. Guo, J. Chen, X. Wang, Y. Xi, C. Hu, Z.L. Wang, Eye motion triggered self-powered mechnosensational communication system using triboelectric nanogenerator. Sci. Adv. **3**, e1700694 (2017). https://doi.org/10.1126/sciadv.1700694
40. A. Yu, M. Song, Y. Zhang, Y. Zhang, L. Chen, J. Zhai, Z.L. Wang, Self-powered acoustic source locator in underwater environment based on organic film triboelectric nanogenerator. Nano Res. **8**, 765–773 (2015). https://doi.org/10.1007/s12274-014-0559-z
41. X. Wen, W. Yang, Q. Jing, Z.L. Wang, Harvesting broadband kinetic impact energy from mechanical triggering/vibration and water waves. ACS Nano **8**, 7405 (2014). https://doi.org/10.1021/nn502618f
42. R. Lei, Y. Shi, Y. Ding, J. Nie, S. Li, F. Wang, H. Zhai, X. Chen, Z.L. Wang, Sustainable high-voltage source based on triboelectric nanogenerator with a charge accumulation strategy. Energy Environ. Sci. **13**, 2178 (2020). https://doi.org/10.1039/d0ee01236j
43. G. Zhu, W.Q. Yang, T. Zhang, Q. Jing, J. Chen, Y.S. Zhou, P. Bai, Z.L. Wang, Self-powered, ultrasensitive, flexible tactile sensors based on contact electrification. Nano Lett. **14**, 3208 (2014). https://doi.org/10.1021/nl5005652
44. Z. Zhang, T. He, M. Zhu, Z. Sun, Q. Shi, J. Zhu, B. Dong, M.R. Yuce, C. Lee, Deep learning-enabled triboelectric smart socks for IoT-based gait analysis and VR applications. NPJ Flexible Electron. **4** (2020). https://doi.org/10.1038/s41528-020-00092-7

45. P. Supraja, R.R. Kumar, S. Mishra, D. Haranath, P.R. Sankar, K. Prakash, N. Jayarambabu, T.V. Rao, K.U. Kumar, A simple and low-cost triboelectric nanogenerator based on two dimensional ZnO nanosheets and its application in portable electronics. Sens. Actuators A Phys. **335**, 113368 (2022). https://doi.org/10.1016/j.sna.2022.113368
46. S. Mishra, S. Potu, R.S. Puppala, R.K. Rajaboina, P. Kodali, H. Divi, A novel ZnS nanosheets-based triboelectric nanogenerator and its applications in sensing, self-powered electronics, and digital systems. Mater. Today Commun. **31**, 103292 (2022). https://doi.org/10.1016/j.mtcomm.2022.103292
47. A.A. Mathew, A. Chandrasekhar, S. Vivekanandan, A review on real-time implantable and wearable health monitoring sensors based on triboelectric nanogenerator approach. Nano Energy **80**, 105566 (2021). https://doi.org/10.1016/j.nanoen.2020.105566
48. Q. Zheng, B. Shi, F. Fan, X. Wang, L. Yan, W. Yuan, S. Wang, H. Liu, Z. Li, Z.L. Wang, In vivo powering of pacemaker by breathing-driven implanted triboelectric nanogenerator. Adv. Mater. **26**, 5851–5856 (2014). https://doi.org/10.1002/adma.201402064
49. H. Ouyang, Z. Liu, N. Li, B. Shi, Y. Zou, F. Xie, Y. Ma, Z. Li, H. Li, X. Qu, Y. Fan, Z.L. Wang, H. Zhang, Z. Li, Symbiotic cardiac pacemaker. Nat. Commun **10**, 1–10 (2019). https://doi.org/10.1038/s41467-019-09851-1
50. H. Ryu, H.M. Park, M.K. Kim, B. Kim, H.S. Myoung, T.Y. Kim, H.J. Yoon, S.S. Kwak, J. Kim, T.H. Hwang, E.K. Choi, S.W. Kim, Self-rechargeable cardiac pacemaker system with triboelectric nanogenerators. Nat. Commun. **12**, 4374 (2021). https://doi.org/10.1038/s41467-021-24417-w
51. Q. Zheng, H. Zhang, B. Shi, X. Xue, Z. Liu, Y. Jin, Y. Ma, Y. Zou, X. Wang, Z. An, W. Tang, W. Zhang, F. Yang, Y. Liu, X. Lang, Z. Xu, Z. Li, Z.L. Wang, In Vivo Self-Powered Wireless Cardiac Monitoring via Implantable Triboelectric Nanogenerator. ACS Nano **10**, 6510–6518 (2016). https://doi.org/10.1021/acsnano.6b02693
52. Y. Ma, Q. Zheng, Y. Liu, B. Shi, X. Xue, W. Ji, Z. Liu, Y. Jin, Y. Zou, Z. An, W. Zhang, X. Wang, W. Jiang, Z. Xu, Z.L. Wang, Z. Li, H. Zhang, Self-powered, one-stop, and multifunctional implantable triboelectric active sensor for real-time biomedical monitoring. Nano Lett. **16**, 6042–6051 (2016). https://doi.org/10.1021/acs.nanolett.6b01968
53. L. Xu, Z. Zhang, F. Gao, X. Zhao, X. Xun, Z. Kang, Q. Liao, Y. Zhang, Self-powered ultrasensitive pulse sensors for noninvasive multi-indicators cardiovascular monitoring. Nano Energy **81**, 105614 (2021). https://doi.org/10.1016/j.nanoen.2020.105614
54. Z. Liu, Y. Ma, H. Ouyang, B. Shi, N. Li, D. Jiang, F. Xie, D. Qu, Y. Zou, Y. Huang, H. Li, C. Zhao, P. Tan, M. Yu, Y. Fan, H. Zhang, Z.L. Wang, Z. Li, Transcatheter self-powered ultrasensitive endocardial pressure sensor. Adv. Funct. Mater. **29** (2019). https://doi.org/10.1002/adfm.201807560
55. H. Ouyang, J. Tian, G. Sun, Y. Zou, Z. Liu, H. Li, L. Zhao, B. Shi, Y. Fan, Y. Fan, Z.L. Wang, Z. Li, Self-powered pulse sensor for antidiastole of cardiovascular disease. Adv. Mater. **29** (2017). https://doi.org/10.1002/adma.201703456
56. A. Rajabi-Abhari, J.N. Kim, J. Lee, R. Tabassian, M. Mahato, H.J. Youn, H. Lee, I.K. Oh, Diatom bio-silica and cellulose nanofibril for bio-triboelectric nanogenerators and self-powered breath monitoring masks. ACS Appl. Mater. Interfaces **13**, 219–232 (2021). https://doi.org/10.1021/acsami.0c18227
57. Z. Lin, J. Chen, X. Li, Z. Zhou, K. Meng, W. Wei, J. Yang, Z.L. Wang, Triboelectric nanogenerator enabled body sensor network for self-powered human heart-rate monitoring. ACS Nano **11**, 8830–8837 (2017). https://doi.org/10.1021/acsnano.7b02975
58. J.N. Kim, J. Lee, H. Lee, I.K. Oh, Stretchable and self-healable catechol-chitosan-diatom hydrogel for triboelectric generator and self-powered tremor sensor targeting at Parkinson disease. Nano Energy **82**, 105705 (2021). https://doi.org/10.1016/j.nanoen.2020.105705
59. S.S.K. Mallineni, Y. Dong, H. Behlow, A.M. Rao, R. Podila, A wireless triboelectric nanogenerator. Adv. Energy Mater. **8** (2018). https://doi.org/10.1002/aenm.201702736
60. G. Liu, J. Chen, Q. Tang, L. Feng, H. Yang, J. Li, Y. Xi, X. Wang, C. Hu, Wireless electric energy transmission through various isolated solid media based on triboelectric nanogenerator. Adv. Energy Mater. **8** (2018). https://doi.org/10.1002/aenm.201703086

61. Y. Chen, Y. Cheng, Y. Jie, X. Cao, N. Wang, Z.L. Wang, Energy harvesting and wireless power transmission by a hybridized electromagnetic-triboelectric nanogenerator, Energy. Environ. Sci. **12**, 2678–2684 (2019). https://doi.org/10.1039/c9ee01245a
62. H. Shi, H. Lu, X. Liu, X. Wang, Y. Wu, H. Zheng, Structure design and wireless transmission application of hybrid nanogenerators for swinging mechanical energy and solar energy harvesting. Nanoscale **14**, 10972–10979 (2022). https://doi.org/10.1039/d2nr02833f

Chapter 5
Photovoltaic Energy Scavenging for the Sustainable Implantable Medical Device

Jinwei Zhao

5.1 Overview of Photovoltaic Energy Harvesters

Harvesting energy from light has been used for powering portable consumer products. Here, photovoltaic (PV) cells are used to convert light or the Sun's energy into useful electricity. Ultimately, semiconductor materials are commonly used for the purpose of producing currents and voltages as a result of the absorption of light, which is a phenomenon known as the photovoltaic effect. The Photovoltaic Effect is discovered in 1839 by Alexandre-Edmund Becquerel, when he observed that electric current is generated from a light induced chemical Reaction. Following that Adam and Day discovered the photovoltaic effect in selenium in 1876 [14]. The theory of PV effect is developed by Plank and Wilson, who donated the quantum nature of light and quantum nature of the solid structure [14]. However, this phenomenon was not noticed by the industry until the first PV cell with an efficiency of 6% was invented in 1954 [14].

The first generation of PV cell is made of crystalline semiconductors which is composed of Mono-crystalline and poly-crystalline silicon. In its most basic form, a PV cell consists of a pn-junction diode. Typical PV cell efficiencies range from 18% for polycrystalline to 24% for highly efficient monocrystalline technologies [17]. This technology constantly developed with improved capability and efficiency over a few decades. The mono-crystalline structure occupies 80% of PV cell market, which essentially apply p-n junctions (Shown in Fig. 5.1a). Nowadays, the efficiency of the Crystalline semiconductor is limited by the energy induced by photons because it decreases in higher wavelength [14]. In addition, radiation with higher wavelength will heat up the temperature of the crystalline PV cells

J. Zhao (✉)
Imperial Brands, Wellington House, Liverpool, UK
e-mail: Jinwei.Zhao@impbrands.com

R. Das, H. Heidari (eds.), *Wireless Power Technologies for Biomedical Devices*,
Analog Circuits and Signal Processing, https://doi.org/10.1007/978-3-031-52834-7_5

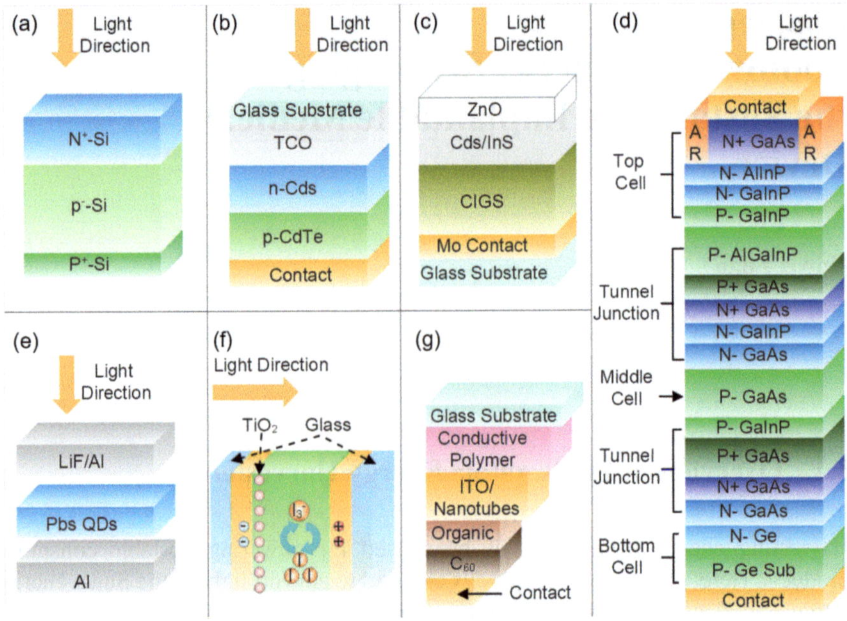

Fig. 5.1 shows the Material and structure of the PV cell. (**a**) shows the PIN structure of crystalline Si PV cell. (**b**) shows the structure of CdTe thin film PV cell. (**c**) shows the structure of CIGS thin film PV cell. (**d**) shows the structure of triple-junction PV cells. (**e**) shows the structure of Quantum Dot PV cells. (**f**) shows the structure of dye-sensitized solar cell. (**g**) shows the structure of organic PV cells

hence causes power dissipation [14]. The efficiency of the crystalline silicon PV cell is up to 26.7%, and the V_{OC}, can be as 0.74 V measured by National Institute of Advanced Industrial Science and Technology **(AIST)** [21]. The other material such as GaAs has the efficiency of 28.89% and the V_{OC} of 1.12 V measured by National Renewable Energy Laboratory (NREL). The increased production and reduced costs in the PV industry led to the new crystallization technologies. The poly-crystalline silicon was highly developed when the price of silicon was over 300 $/kg. However, the efficiency of the poly crystalline Si is lower than that of mono-crystalline Si, which is 15%. The other advantage of poly-crystalline structure rather than mono-crystalline structure is decreasing the flaws at crystal structure and metal contamination [37].

Compared with crystalline, thin film technology is advantageous of lower cost and less hazards to environment during fabrication. Unlike the structure of single crystalline semiconductor, the thin film PV cells are normally deposited in a substrate of glass, which can decrease the deposited layer within 10 micrometers. In this case, the costs of the material and manufacture costs according to throughout deposition process are lowered. The thin film PV cells contains Amorphous Silicon (α-*Si),* Hetero junction structure of Cadmium telluride (CdTe) as well as Copper

Indium Gallium Di-selenide (CIGS). Compared with crystalline Silicon, the α-*Si* is with higher bandgap of 1.7 eV than 1.1 eV, which leads to higher absorption in visible spectrum. The schematic of the CdTe is shown in Fig. 5.1b. The bandgap of CdTe is around 1.45 eV which leads to an efficiency of 15% [21, 22]. While the toxic nature and environmental hazards of fabrication limit the development of the CdTe PV cell. In addition, the material constraints increases the costs of CdTe PV cell. CIGS contains the elements from group I, III and IV in periodic table, which attributes to high absorption coefficient and electrical characteristics. Unlike the p-n junction structure, CIGS PV cells are known as hetero junction structure, which is normally deposited by sputtering, "ink" printing and electroplating [15]. The schematic of CIGS PV cell is shown in Fig. 5.1c.

The Multi-junction PV cell is shown in Fig. 5.1d, which stacks different crystalline layers in order to absorb most of the radiation [37]. As each crystalline layer has different bandgap and absorption coefficient, the layers in the PV cell can be tuned more efficient and suitable for a wider spectrum. The triple junction PV cell can achieve an efficiency of 39% measured by AIST [21]. Originally, the whole crystalline layers are deposited in GaAs substrate because it is matched with GaAs lattice. However, Germanium (Ge) substrate is mostly applied to replace GaAs substrate and reduce the cost. The first cell works as a silicon p-n junction cell, then the GaInP/GaAs as well as GaAs are deposited above to enhance the bandgap, capture optical energy in wider spectrum, then improve the efficiency.

Quantum dot PV cells are one of the efficient with distinctive absorption and emission of spectrum according to different size of particles [37]. The quantum dots are in nano-meter scale and has an adjustable bandgap. The strong coupling of quantum dots is required to enhance the lifetime of excitation and to provide electricity in high voltage. The Schematic of quantum dots PV cells is shown in Fig. 5.1e.

The structure of a dye-sensitized PV cell is shown in Fig. 5.1f. Such structure consists of a semiconductor layer, an electrolytic liquid produced by dissolving a salt in a solvent liquid [37]. The semiconductor and the electrolytic liquid work together to split the electron-hole pairs when the photons are absorbed. A nanomaterial such as Titanium Dioxide (TiO_2) is commonly applied to hold the dye molecules. As the solvent liquid is toxic and hazardous to the environment, the solvent-free dye-sensitized PV cell is developed but its efficiency is low.

The structure of organic PV cells is shown in Fig. 5.1g. The organic PV cell is composed by thin film organic semiconductor within hundreds of nano meters such as polymer and small molecule compounds. Typically, the organic cell contains a glass substrate, a transparent contact of indium tin oxide (ITO), a conducting polymer layer, a photoactive polymer layer and a back metal contact layer, such as Al. As ITO is expensive in the market, the carbon nanotubes are normally used alternatively [46]. The efficiency of Organic PV cells is less than 10% but cost of material and fabrication is very promising [21].

5.2 Fundamental of Photovoltaic Cell

This section is to illustrate the superficial theory of the Photovoltaic Effect including the nature of radiation, the mechanism of the light, the photovoltaic effect, generation and recombination.

5.2.1 *The Theory of Light Radiation*

For implantable power harvesting applications, PV cells cannot harvest *in vivo* bioenergy directly, but it can harvest energy from an ambient light source (natural light or artificial light) [20]. The amount of energy that the sun emits and incident on the earth is much greater than what we see daily. The visible light that humans see from the sun represents a small portion of the electromagnetic spectrum, which is depicted to the right. The electromagnetic spectrum illustrates light as a wave related to a specific wavelength. The electromagnetic spectrum of the solar radiation is shown in Fig. 5.2 [44, 54]. The interaction of incident sunlight with a photovoltaic converter or any other item depends on a number of vital aspects of the incident solar energy. The vital characteristics of the incident solar energy are:

- The spectrum of the input light.
- Solar radiation power density.
  The angle where incident solar light impacts the PV cells.
- the solar radiation received by a specific surface over a specific time.

For an implantable PV cell, the transmittance of the skin layer on top of the PV cell should also be considered.

In addition to spectrum irradiance, photon flux is frequently required for the examination of solar cells. In order to obtain the spectral irradiance from the

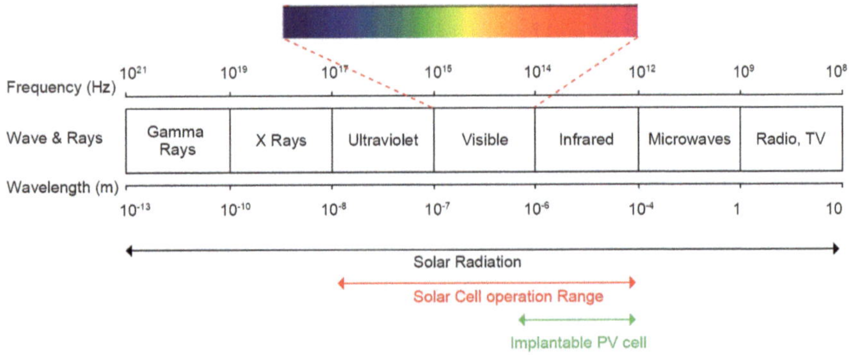

Fig. 5.2 shows the electromagnetic spectrum of solar radiation and also the operation range of normal cell and the operation range of the implantable photovoltaic cell

photon flux, convert the photon flux at a specific wavelength to W/m^2. Atmospheric factors including absorption, reflection, and scattering have numerous impacts on the solar radiation at the Earth's surface. The length of the light's travel through the atmosphere, normalized to its shortest path length, is the Air Mass (AM). The air mass determines the reduction of the radiation from solar light. The majority of PV cells are measured and operated under AM 1.5G. Despite the natural light, artificial light is also applied in powering implantable PV cells to obtain optimal optical transmittance from the skin.

5.2.2 *The Mechanism of the Incident Light*

When the incident light propagates in the materials, part of the light is absorbed, part of the light is reflected, and part of the light is transmitted (refracted and scattered), which is shown in Fig. 5.3a. Besides, Snell's law and Maxwell's equation to analyze light penetration can be indicated by Jones' matrix [35].

To evaluate the optical characteristics in 2D structure, the 2 × 2 Jones Matrix was applied [5], which is also shown in Fig. 5.3b:

$$\begin{bmatrix} E_{Ra} \\ B_{Ra} \end{bmatrix} = \sum_{i=1}^{m} \begin{bmatrix} \cos\varphi_i & \left(\frac{j}{\eta_i}\right)\sin\varphi_i \\ j\eta_i \sin\varphi_i & \cos\varphi_t \end{bmatrix} \begin{bmatrix} 1 \\ \eta_m \end{bmatrix} \tag{5.1}$$

$$M_{total} = M_{0,1} \times M_{1,2} \times M_{2,3} \cdots \times M_{m-1,m} \tag{5.2}$$

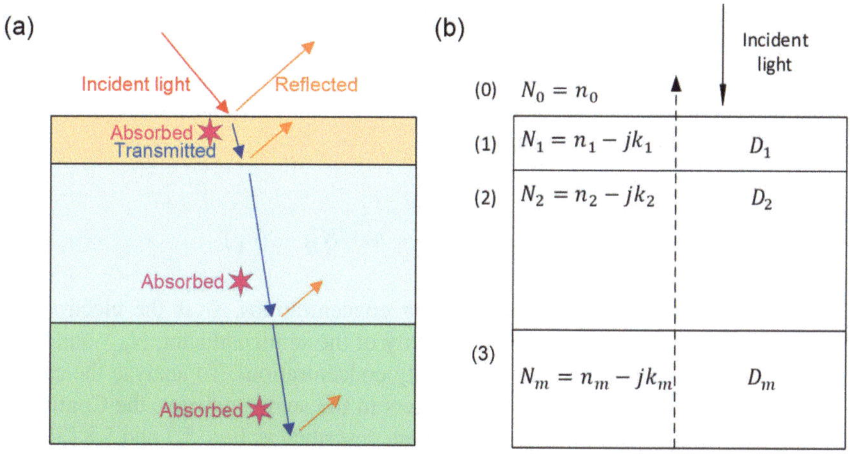

Fig. 5.3 (**a**) shows the light absorption, reflection and transmission in different materials. (**b**) shows the schematic of Jones Matrix

where d_i is the wave phase shift ($d_i = 2\pi N_i d_i cos\ \theta_i/\lambda$) in the ith layer, N_i is the refractive index, d_i the thickness of the ith layer, qi is the pseudo index in ith layer ($\eta i = N_i cos\theta_i$). E_{Ra} or B_{Ra} is the ratio between the electric and magnetic fields of the transmitted light and incident light. M is the total number of layers. The reflectance (fi), absorptance (A) transmittance of light can be evaluated by calculating the components in M_{Total} [5]:

$$R = \frac{4\eta_0 \mathrm{Re}\ \ (\eta_m)}{\left(\eta_0 E_{Ra} + B_{B_n}\right)\ \ (\eta_0 E_{Ra} + B_{Ra})^*} \tag{5.3}$$

$$R = \frac{4\eta_0 \mathrm{Re}\ \ \left(E_{Ra}\, B^*_{Ra} - \eta_m\right)}{\left(\eta_0 E_{Ra} + B_{B_0}\right)\ \ (\eta_0 E_{Ra} + B_{Ra})^*} \tag{5.4}$$

$$T + A + R = 1 \tag{5.5}$$

5.2.3 *The Theory of Photovoltaic Effect*

Most PV cells are fabricated from either monocrystalline or polycrystalline silicon (Si) materials. In its most basic form, a PV cell consists of a pn-junction diode. Typical PV cell efficiencies range from 18% for polycrystalline to 24%. from highly efficient monocrystalline technologies [17]. Most implantable PV cells are made of crystalline silicon due to their high efficiency and non-toxicity [57]. In this case, silicon is mostly applied, and the technology to fabricate silicon semiconductor devices is well developed. Atoms are bound together to form a regular structure in semiconductors. Four of the valence electrons on each silicon atom are shared with the four Si atoms around it to create covalent bonds. The composition of the Silicon semiconductor device is analyzed by the Poisson equation and continuity equation. Poisson's equation determines the electrical potential produced by mass density distribution and given charges [3]:

$$-\nabla \cdot \nabla \psi = \frac{q}{\varepsilon}\ \ \left(p - n + N_D^+ - N_A^-\right) \tag{5.6}$$

Where y and n are the hole and electron concentrations, ψ is the electrostatic potential, q is the charge, ε is the permittivity of the semiconductor, $N_A{}^-$ and $N_D{}^+$ are the ionized acceptor and donor impurity concentrations. To analyze the carrier transport and the gradients of the parameters in the semiconductor, the Continuity Equations with drift and diffusion current are described as Eqs. 5.7 and 5.8 [3]:

$$\frac{\partial p}{\partial t} = \mu_p p \nabla^2 \psi + D_p \nabla^2 p + G - R \tag{5.7}$$

$$\frac{\partial n}{\partial t} = -\mu_n n \nabla^2 \psi + D_n \nabla^2 n + G - R \tag{5.8}$$

Where T is the temperature of the environment, k is the Boltzmann constants μ_p and μ_n, are the hole and electron mobility respectively, Dp and Dn the hole and electron diffusion coefficient ($D_{p,n}=\mu_{p,n}kT/q$), G and R present the Generation rate of electron-hole pairs and relevant Recombination rate. Direct radiative recombination (Eq. 5.9) depends on dopant (p, n), injection factor (g_{eh}) and radiative coefficient (B). Auger recombination (Eq. 5.10) dominates at high dopant densities, and it also depends on the Auger coefficient n and p), injection factor (g_{eeh} and g_{ehh}) and dopant. Shockley-Read-Hall **(SRH)** recombination via states in the forbidden band is shown in Eq. 5.11, which is influenced by recombination lifetime ($\tau_{p,}$). The Recombination rates are as follows [3]:

$$R_{rad} = g_{eh} B \left(pn - n_i^2\right) \tag{5.9}$$

$$R_{Auger} = \left(g_{eeh} C_n n + g_{ehl} C_p p\right) \left(pn - n_i^2\right) \tag{5.10}$$

$$R_{SRH} = \frac{pn - n_i^2}{\tau_p (n + n_1) + \tau_n (p + p_1)} \tag{5.11}$$

The generation rate indicates the generation of electron-hole pairs correlated to the absorption of photons [35]. It is related to the absorption coefficient of silicon $\alpha(\lambda)$, photon flux on the surface $b(\lambda)$, Power flow according to wavelength and the depth into the device inc, *y*, z-direction. The photon flux can be calculated from light intensity of Air Mass 1.5 Global (*AM* 1.5G). The generation rate can be determined by [35]:

$$G(\lambda) = \alpha(\lambda)\, b(\lambda)\, P_s(x, y, z, \lambda) \tag{5.12}$$

The electron-hole pair generation and recombination in the semiconductor is shown in Fig. 5.4.

5.3 Development of Implantable PV Cells

Numerous advancements have been made in terms of materials, PV cell layout, encapsulation, and stability in the development of implantable PV cells. The key

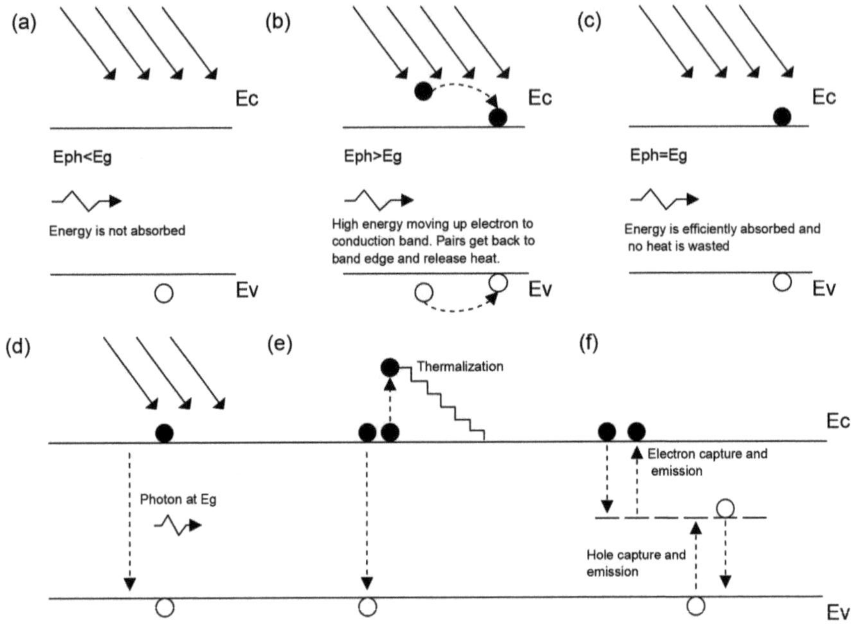

Fig. 5.4 (**a**), (**b**) and (**c**) show the three different conditions of electron-hole pair generation, where (**a**) $E_{ph} < E_g$, (**b**) $E_{ph} >= E_g$ and (**c**) $E_{ph} = E_g$ (**d**), (**e**) and (**f**) show the three types of recombination in the semiconductor, where (**d**) Radiative recombination, (**e**) Auger recombination, and (**f**) SRH recombination

turning points in these processes are depicted in Fig. 5.5. Prior to 2012, implantable PV cells took up more space because of the bulky technology. Implantable PV cell technology is developed between 2010 and 2014 with the help of CMOS technology [6]. The on-chip PV harvesting system which embedded the PV cells with power management unit was invented during this period [6, 33]. The research indicated a transition to lower power density PV cells between 2014 and 2018. This might be explained by a stronger focus on challenges related to better encapsulation and biocompatibility. In implantable applications after 2018, new technologies such organic PV cells were involved in implantable energy harvesting, and it is advantageous in flexibility and biocompatibility [49]. Dan Tchin-iou was one of the first scientists to suggest using PV cells for implantable applications in 1999 [12]. A commercial solar cell with an integrated battery pack was powered by a source of visible light. A synthetic heart was powered by this system. However, this study lacked confirmation of the under-skin device and monitoring of tissue loss. In this case, the light source was limited to visible wavelengths.

Thomas Laube accomplished the first *in vivo* testing of an implanted PV cell in 2004 [32]. The utility of a near-infrared (NIR) light source and the resin-encased PV arrays. The full chip was transplanted into a rabbit, and the PV cells were contained within an ophthalmic microsystem. The system underwent the longest

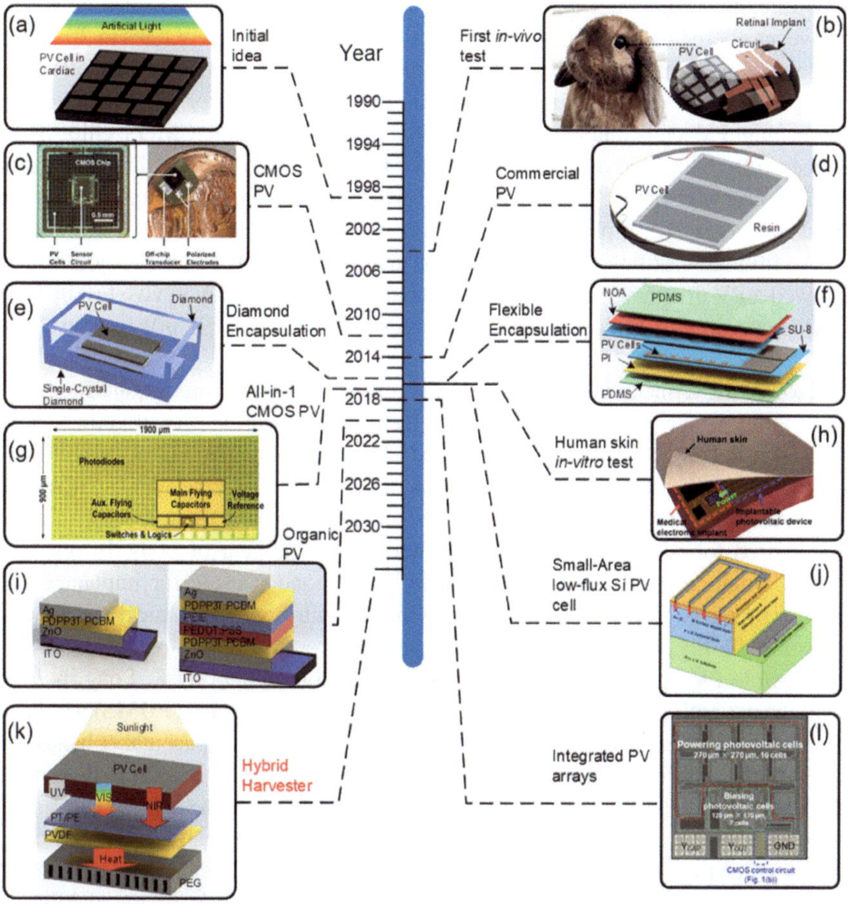

Fig. 5.5 shows the evolution of implantable PV cells between 1999 and 2030. The implanted PV cells have undergone 20 years of development to get from a proposed ideal to a refined approach that may be tested in vivo or in vitro. Also displayed is a 10-year development prospection (2020–2030). The initial design for an implanted PV cell is shown in (**a**). The implantable PV cell's initial in vivo test on an animal is shown in figure (**b**). (**c**) The first CMOS PV cells on-chip. (**d**) Powering pacemakers with commercial PV cells. (**e**) The implanted PV cell is enclosed in a diamond. (**f**) Multilayer flexible encapsulations are used to create flexible PV cells. (**g**) CMOS PV cells that are stacked and have a power management system. (**h**) The first human skin implanted PV cell test. The original organic implantable PV cells (single-junction and tandem Cells). (**j**) Low-flux implantable PV cell with a small surface area. (**k**) Prognostication of hybrid PV with additional energy harvesters. (**1**) Bonding of integrated silicon photovoltaic panels to CMOS circuit

in vivo testing of any implantable device to date, lasting seven months. The full procedure of the clinical trial is well documented.

There were no articles published in the field of implantable PV cells between 2004 and 2012. Implantable PV cells were abandoned as energy harvesters during this time due to their size and the possibility of infection via the connecting cables. However, there has been a resurgence of interest in employing PV cells in 2012 due to developments in integrated CMOS technology. In fact, Sahar Ayazian offered a self-powered and entirely integrated system that included sensor arrays and power-harvesting PV cells integrated onto a 2.5 mm by 2.5 mm CMOS chip. They successfully harvested power from a device implanted 3 mm beneath chicken skin in the microwatt range. They investigated whether a CMOS PN junction might be used as a potential PV power harvester [6].

Additionally, utilizing thin film silicon materials, Haeberlin *et al.* successfully showed a PV-driven pacemaker [23], which underwent a 40-day in vivo test in 2015 [24]. According to Table, the device has a power density of 0.95 mW/cm^2 in Table 5.1.

Expensive semiconductor materials like GaAs and GaInP are required to scavenge more power [50]. In the work [50], the output power density of GaInP/GaAs was reported as 8.46 mW/cm^2 under AM1.5G condition. The GaAs PV cell was reported with output power density of 44 mW/cm^2 by utilizing NIR light source [32]. According to the toxic nature of these materials, a package or a encapsulation is required to isolate the device from internal environment of body. In research [2], A. Ahnood *et al.* actually showed off a diamond capsule for implanted PV cells. The package provides an optical window with high mechanical robustness, biocompatibility and wide transmission spectrum.

Flexible PV cells are employed in the Implantable system by K. Song in 2016 with the patient's comfort in mind [51]. The polymethyl Siloxane (**PDMS**) was used to encase the PV cell. It can successfully supply a commercial pacemaker beneath a rat with the power of 8 mW/cm^2 [51]. A mass spectrometer was used to determine the level of arsenic in order to evaluate the biocompatibility of the device. The results showed a 0.02 μg leakage by applying emerging PDMS encapsulation, which is much lower than the daily intake from air breath (0.60 μg) and water (20.00 μg) by a single adult [50]. Even though there are many studies showing the viability in animals, it is vital to assess the performance of the cells under human skins since the power output of the PV cell greatly depends on the feature of covered skins. In this instance, various locations in 2017 were used to assess the PV cell's performance beneath human skin that was both fresh and fixed. The various results were reported where 2.34 mW/cm^2 is under inner arm, **2.21** mW/cm^2 is under hand dorsum, and 0.96 mW/cm^2 is under forehead with same input optical power [51].

Organic PV cells' sensitivity to NIR light made them suitable for use in retinal applications. In the work [49], the authors showed tandem and single-junction OPV cells built atop a large heterojunction. They used electrical pulses to test the voltage and charging duration of the stimulating electrode in a saltine solution. The open circuit voltage of tandem PV cells (1.31 V) was higher than that of single-junction PV cells (0.67 V). In this case, the tandem PV cell can provide a full charge per pulse window with NIR light source in retinal stimulation application. Moreover, the efficiency of the tandem PV cell was higher (5.6%. in comparison to 5.3% in the single junction). For safety reason, the light intensity in retinal applications was limited from 150 to 600 mW/cm^2. In this situation, the tandem OPV can efficiently tune the charge per pulse with an active area ranging from 2500 to 6250 μm^2 and an electrode diameter of 35 μm, but the electrode of single-junction OPV was 60 μm, which restricts the implantable resolution.

In summary, commercial PV cells are applied in implantable applications in the first decade of twenty-first century. After 2010, CMOS technology was utilized to fabricate the on-chip PV cells as well as the power management circuitry in an effort to decrease the size of solar cells (and increase the power density). Researchers started employing implantable PV cells to power actual devices, such pacemakers, between 2014 and 2017. Since 2017, encapsulation materials were also used to protect the gadget and guarantee its biocompatibility. *in vitro* testing on various human skin types and *in vivo* testing on animals both took place during this time. The emerging technology of PV cells such as organic and thin film was employed in implantable applications since 2018.

5.4 Technologies and Challenges in Implantable Photovoltaic Cells

In this section, the development of implantable PV cells will be demonstrated. Diferent PV cell performance using different materials and technologies will be compared. Next, the fabrication process of PV cells and the encapsulation techniques will be discussed. Here, it is important to emphasize that encapsulation involves protecting the PV cells as well as reducing the adverse side effects to humans. Optical losses due to tissue scattering effects will be analyzed and the optical properties of different skins will also be summarized.

5.4.1 Semiconductor Material

Due to scattering losses from tissue, implantable PV cells are more effective in capturing light in the NIR range [9]. NIR light has problems with heating, though, which could make the wearer uncomfortable. Crystalline and non-crystalline materials can be utilized to fabricate PV cells, which fall into two general categories. In general, crystalline solar cells are less flexible, more expensive, and more efficient than non-crystalline alternatives [57]. In 2015, crystalline silicon materials made up about 93% of all PV cells; 24%. of them were monocrystalline silicon and 69% were multicrystalline silicon [42]. Today, crystalline silicon makes up almost 89% of solar cells, amorphous silicon makes up 10%, and the remaining 5% are made of cadmium telluride, dieseline, copper indium, and Gallium Arsenide [25]. The Table 5.1 in Sect. 5.3 compares the output power, flexibility, and the other features of distinct PV cells manufactured by different materials.

Table 5.1 A comparison of different types of implantable PV cells

Material	Source	P_{out} (mW/cm^2)	Pack.	Flex.	App.	Ref.
c-Si	NIR	9.35^b	Diamond	No	Cochlear	[2]
c-Si	AM1.5G	6.17^a		No	Sensor	[6]
c-Si	Halogen	18.87^b		No		[10]
c-Si	Outdoor	4.94^b	Silicone	No	Cardiac	[23]
c-Si	NIR	2.13^a		No	Temp.	[27]
c-Si	NIR	2.21^b	PLGA	Yes	LED	[36]
GaAs	NIR	44.12^b	Silicone	Yes	Retinal	[32]
GAInP	AM1.5G	8.46^b	PDMS	Yes	Cardiac	[50]
GaAs	LED	42.29^b		No	Neural	[52]
α-Si	Halogen	0.95^b	Silicone	No	Cadiac	[12]
α-Si	Outdoor	0.06^b	Silicone	Yes	Temp.	[56]
Organic	NIR	3.84^b		Yes	Retinal	[49]

Hints: ^{a}On chip PV cell, ^{b}Off Chip PV cell. 'Pack.' is Package, 'Flex.' is Flexibility, 'App.' is Application and 'Ref.' is Reference

Silicon and gallium materials are frequently used in implantable applications due to their advanced technology [10, 50, 39]. For instance, gallium's greater bandgap and lower reverse saturation current make it particularly well suited for NIR light absorption, while a-Si is more adapted to visible light. Additionally, because to the heating problems, Ga has a greater breakdown voltage and is less sensitive to heating 0.14 to 0.15 Ω resistance fluctuation based on the temperature coefficient from 0 to 50 Celsius [8]. For instance, the breakdown voltages of GaP and GaAs at 10^{15} cm^{-3} doping gradient are approximately 800 V and 300 V, which is higher than that of Silicon, which is 200 V [19]. Nevertheless, the high cost and the toxicity nature is the main limitation to apply in implantable medical device.

Biodegradable thin-film silicon PV cells have been researched in addition to wafer-based PV cells [48]. Microelectronics processing methods are required for the manufacture of the thin-film cell in comparison to the wafer-based cells cutting and sawing procedure [48]. Amorphous Silicon (a-Si), Microcrystalline silicon (μ-Si), monocrystalline silicon (c-Si), and amorphous silicon (a-Si) are the most frequently utilized materials for thin-film photovoltaic cells. Crystalline Si thin film cells are created throughout the production process using the transfer printing technique, whereas a-Si and polysilicon thin films are created using the Chemical Vapor Deposition (CVD) technique. OPV cells, on the other hand, have advantages because of their high flexibility, low cost, and simplicity of production [58]. They might make a promising contender for implantable applications if the efficiency of energy conversion can be increased, as well as the cell stability.

Currently, there are two categories of PV cell architectures that are used in implantable applications: On-chip PV cell [6, 9, 10, 27, 28, 32, 36, 38, 39, 50, 51, 52] and off-chip PV cells [2, 12, 23, 56]. On chip PV cells are fabricated by using Complementary metal-oxide-semiconductor **(CMOS)** technology, which applies the junction diode formed by N+ region, P+ region, N well, p well, deep

N well and p substrate [4]. For instance (Shown in Fig. 5.6a), one P+/N-well PV cell and one N+/P-sub PV cell were used as the PMOS source and NMOS source. Each PV cell was able to supply 0.3–0.4 V and different configurations of PV cells can provide different voltage output [4, 10]. To prevent photo-generated current from leaking into the CMOS circuit during operation, a metal layer is frequently used to shade the circuit during design and manufacture. Due to the integration of all the components into a single block, the on-chip PV cell actually takes up less space than an off-chip cell. However, the efficiency of on-chip PV cells is typically lower (209c), and it is challenging to create alternative PV cell configurations due to the negative voltage created by inverted junctions and the P substrate connected to ground [10]. In CMOS fabrication, double-well and triple-well technology is mostly applied in forming the junction of on-chip PV cells, which includes different diffusion regions: P-type substrate/Deep N-type Well (P-sub/DNW), Deep N-type Well/P-type Well (DNW/PW) and P-type Well/Heavy N-type Region **(PW/N+)** [26]. The structure of typical CMOS on-chip PV cells is shown in Figs. 5.6b, c. Depending on their depth, CMOS-fabricated PV cells may produce varied amounts of energy. The performance of three different types of 180 nm **TSMC PV** cells was discussed utilising a halogen light bulb with input power of 1.13 mW/cm^2 in Z. Chen *et al.* 2017 [10]. To be specific, the P-sub/DNW junction diode provides Short-Circuit Current Density, J_{sc}, of 433 pA/ μm^2 as well as an Open-Circuit Voltage, Voc, of 0.55 V. Relatively, the J_{sc} and V_{oc} of the DNW/PW junction diode are 70 pA/μm^2 and 0.53 V respectively, while the J_{sc} and V_{oc}, of the PW/N+junction diode are 74 pA/μm^2 and 0.53 V [10].

The fabrication of on-chip PV cells is based on the stacking of the junction diodes. Figure 5.6b, c demonstrate two stacking typologies to connect three served junction diodes. The topology in Fig. 5.6b is proposed by M. K. Law *et al.* 2011 [33], which applied double-well 0.35 μm **CMOS** technology. The power conversion efficiency of the stacking junction diodes is evaluated by [26]:

$$K > 1 + \alpha$$

$$\eta = \eta_0 \times N \frac{K - 1}{K^N - 1}$$

where K is the area to volume ratio, a is the leakage photocurrent ratio $a = I_{bN}/I_{aN}$ q is the overall efficiency while η_0 is the first-stage efficiency, and N is the number of stages. A high K value affects efficiency; thus it is best to keep it as low as possible. However, K is constrained by $1 + a$. In this situation, minimizing alpha would result in a smaller K. ftp should be reduced to a minimum and I_{aN} should be increased [26, 33]. In the triple-well technology, the current I_{abN} is increased by connecting a DNW/PW (D_{Nb}) and PW/N+ (D_{Na}) diode in parallel. The amount of optics reaching the P-sub/DNW D_{Nc} decreases when the D_{Nb} is inserted deeper, which lowers I_{cN}. The first and second methods have the advantage of boosting PV cell output voltage, which makes it easier to support **MOSFETs** in the energy

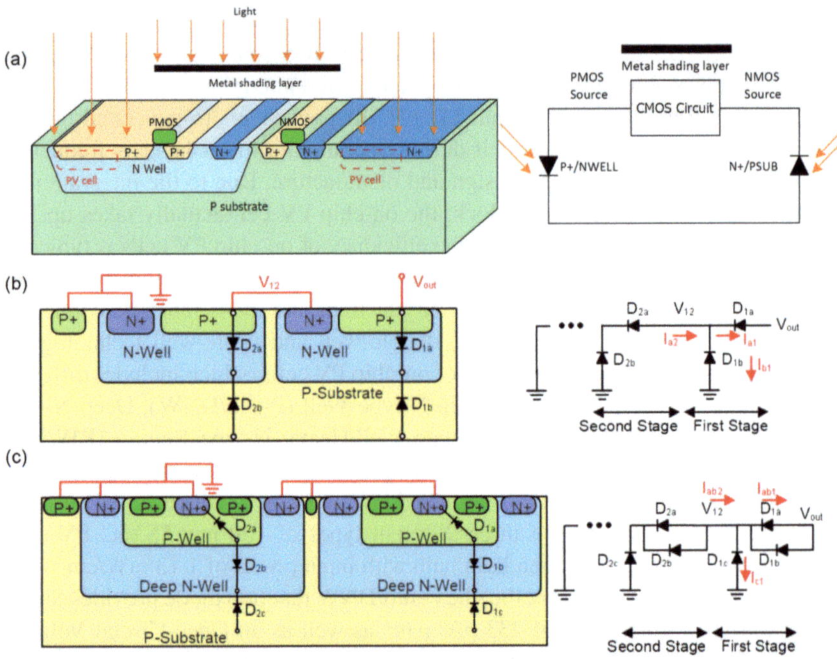

Fig. 5.6 (**a**) The idea of fabricating on-chip PV cells in CMOS technology. (**b**) shows stacking schematic and equivalent circuit of double-well 0.35 pm CMOS technology by M. K. Law et al. 2010 [33]. (**c**) shows stacking schematic and equivalent circuit of triple-well 0.25 pm CMOS technology by G. Hong et al. 2020 [26]

harvesting circuits on the same chip. Additionally, one P-sub/DNW diode in the final stage is bypassed, meaning that not all of the diodes are being utilized [26].

Compared with the conventional use of monocrystalline silicon and the limited stacking topologies of CMOS technology, off-chip PV cells can be based on a variety of materials (inorganic and organic) and stacking topologies. For instance, heterojunction PV cells were previously fabricated using GaInP/GaAs materials in a 2 × 7 array to drive a load [50].

5.4.2 Package and Encapsulation

To prevent the electrical circuitry from corroding, encapsulation is required. The materials used for conventional encapsulation are ceramics, glass, and metals, such titanium and ceramic packing. The majority of the titanium shell in the pacemaker and other implants packaged similarly is accessible to bodily fluids. If all exposed metal surfaces, including tracks, have the same potential, then this is okay. Metal is occasionally used in glass feedthroughs for such implants, however, this is becoming prevalent [24].

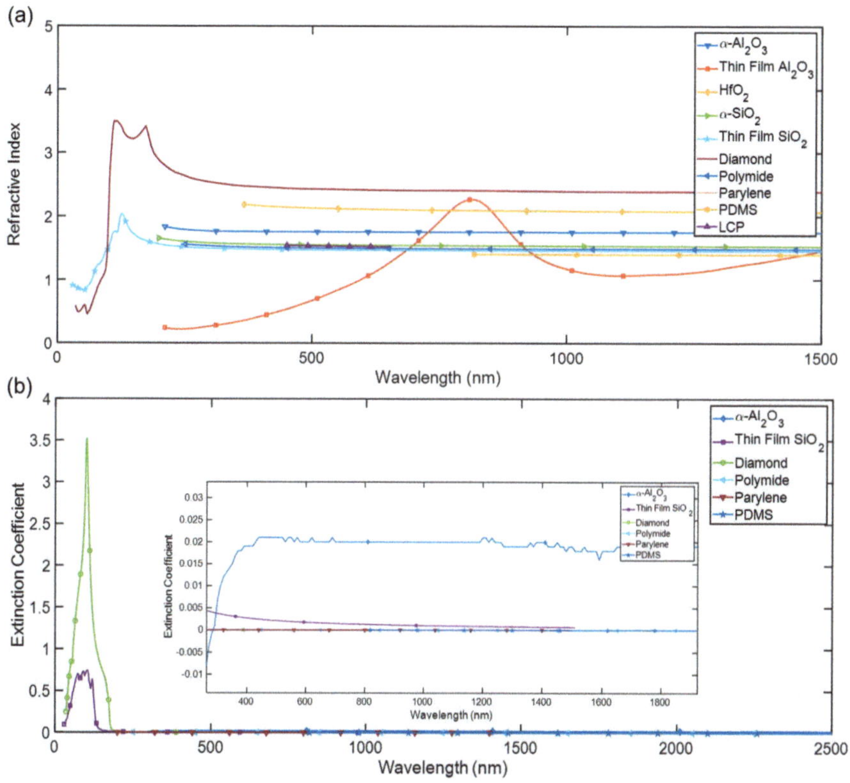

Fig. 5.7 Variation in the encapsulation material properties: (**a**) refractive index as a function of wavelength, (**b**) extinction coefficient as a function of wavelength

Materials that are extremely biocompatible include metals, glass, and ceramics. They are rigid and have a low water vapour permeability [47]. Despite their advantages in harsh environments, many conventional materials are incompatible with the CMOS production process [1, 47]. Consequently, research into new encapsulating materials has been done such as SiO_2 SiC, Al_2O_3, diamond. The other flexible material includes organic silicone polymers, polyimide, PDMS, liquid crystals and SU-8 is also involved [1].

Three considerations, including optical characteristics, biocompatibility, flexibility, and lifetime, can have a big impact on the choice of innovative materials for encasing photovoltaic-driven implanted devices. Figures 5.7a, b and 5.8a, b show the difference in the physical properties of materials which is normally applied in package and encapsulation manufacture. The materials include:

1. **Solid:** Crystalline SiO_2 [18], Diamond [41], Fused Silicon Dioxide (Fused SiO_2) [43], Crystalline Alumina (α-Al_2O_3) [44], Thin-film Alumina [44], thin-film SiO_2[45], Hafnium Oxide (HfO2).[55]

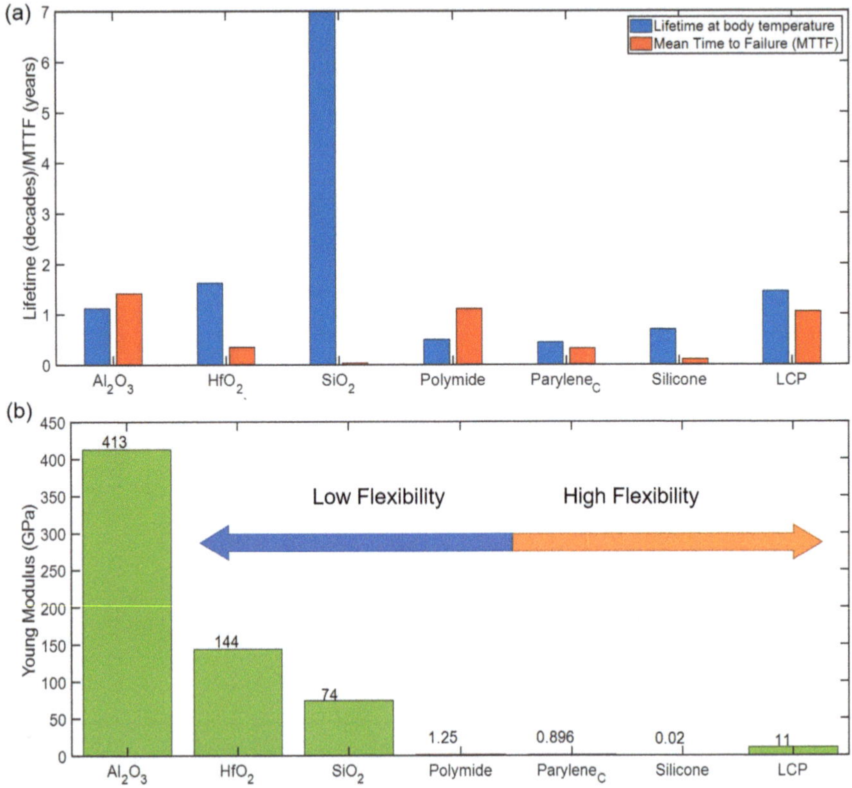

Fig. 5.8 Variation in the encapsulation material properties: (**a**) lifetime in body temperature and mean time to failure, as well as (**b**) Young's modulus

2. **Flexible:** Liquid crystal polymer (LCP) [34], Parylene [40], Silicone (PDMS) [44], Polymide [53].

It is significant to note that these refractive indices in the visible and near-infrared range are very close to 1.5, which is the same as the refractive index of skin [7].

Additionally, the materials' extinction coefficients in the visible and near-infrared ranges are nearly zero, resulting in low light absorption and great optical transparency. These MTTFs of the materials are determined by exposing them to a certain temperature, and the long-term reliability is verified in accelerated aging studies by immersing the samples in a saline solution. The MTTF values (reaction rate in saline solution) and other features of the materials can be used to determine how long these materials will last in the human body [1]. The Young's Modulus variation for various materials is shown in Fig. 5.8b. A high Young's modulus typically indicates that the material is stiff and has little flexibility.

The fabrication methods and physical characteristics of the encapsulating materials are shown in Table 5.2. According to the manufacturing technology, SiO_2

Table 5.2 key factors comparison of different types of encapsulating material

Material	Tech.	Thick (μm)	Thermal[a]	Moisture (%)[b]	Flex.[c]	Failure (nA)[d]
Al_2O_3	ALD + CVD	6.052	4×10^{4} °C^{-1}	4.75[e]	No	1
HfO_2	ALD	0.1	–	–	No	
SiO_2	Thermal	0.1–1	4.5×10^4 °C^{-1}	6.04[e]	No	
Diamond	CVD	300			No	
Polyimide	Spin coating	10	40 ppm°C^{-1}	3	Yes	1000
Parylene C	CVD	6–40	35 ppm°C^{-1}		Yes	1
Silicone	CVD	40	340 ppm°C^{-1}		Yes	
PDMS	Spin coating	150			Yes	

[a]Thermal Expansion Coefficients, [b]Moisture Absorption, [c]Flexibility, [d]Failure Criteria, [e]Moisture Absorption at 30 °C

combines thermal bonding and CVD, whereas $A1_2O_3$ is often deposited by ALD and CVD. However, spin coating and heat bonding are typically used to generate organic encapsulation, such as Polymide, Silicone, PDMS, and LCP (Liquid Crystal Polymer). Despite the presence of Parylene C, complicated surface topography with fissures and sharp edges typically requires the application of CVD.

Alumina (Al_2O_3) is frequently used as the feedthrough with a titanium package because it is regarded as a good encapsulation material with high biocompatibility [1]. Typically, silica is used in anodic bonding and laser welding. However, this encapsulation material has a problem with minimization [1]. A conformal coating of HfO_2 is frequently applied to nanoscale electronics using the atomic layer deposition (ALD) method. It offers excellent mechanical, chemical, and thermal stability [1, 11]. Due to its mechanical rigidity, wear resistance, and chemical resistance [1, 13], diamond is an emerging material for biomedical encapsulation. High pressure and temperature are necessary for the encapsulation process. They are perfect for implantable PV cells due to their optical characteristics [2]. The majority of MEMs procedures can work with polyimide (photosensitive polyimide in photolithography and polymide adhesion in bonding). Since parylene is flexible, inert, and has great optical transparency, it is advantageous in thin conformal layers [1]. Due to its transparency and broad refractive index range (1.38–1.58) [30], silicone is frequently used as an encapsulant for optical devices such as LEDs, photodetectors, and PV cells. In addition, silicone meets the typical requirements of the healthcare and electronics industries due to its low dielectric constant (2.68 at 100k Hz), high thermal conductivity (0.4–1.34) and low resistivity (2.9×10^{14} Ωm) [30] The comfort layer is well protected for encapsulation purposes by silicone rubber, which can also be combined with ceramic material to create an impermeable enclosure. The following are some benefits of silicone:

1. The outer layer is biocompatible, and the sharp layer is covered to protect the host body.
2. The exposed metal electrode is shielded against corrosion.

3. The feedthroughs that convey the signal into and out of the sealed enclosure are insulated with silicone.

5.4.3 Skin Optical Loss

The purpose of skin is to shield the inside organs from light. As a result, it becomes difficult to use implanted PV cells to gather energy from ambient light. Light will therefore be diminished as a result of tissue optical loss. Additionally, different skin types will reflect light in various ways [29]. Skin is a complex medium, and temperature, humidity, and radiation cause its optical properties to change dynamically. Additionally, within tissue borders, absorption, scattering, transmission, and reflection are all necessary for light to go through skin. The interfaces between these boundaries are two mediums with differing optical characteristics. The stratum corneum, epidermis, and dermis are the three layers of skin [16, 31]. The strateum corneum, also known as the non-living epidermis (just dead squamous cells), is the outermost layer and ranges in thickness from 10 to 20 μm. With a 20% lipid, 60% protein, and 20%. water content, the stratum corneum is highly keratinized. Stratum corneum has a refractive index that ranges from 1.47 to 1.51 [7]. Compared to other tissues, the live epidermis is mostly made up of pigments, the majority of which is melanin. From 1.3% (lightly pigmented) to 43%, melanosomes, a particle used to create melanin, are present (Highly pigment) [7]. The amount of pigment varies from person to person, and more pigment might increase melanin absorption. The dermis is known as the main absorber of visible light and is distinguished from the epidermis by a basal lamina. This layer also contains beta-carotene, bilirubin, and blood haemoglobin. The amount of infrared radiation (IR) that is absorbed depends on how much water is present in the dermis [7].

The absorption coefficients of different skin layers are shown in Figure 5.9a, and the absorptions in the visible spectral range are much higher than those in the IR region. For instance, the absorption coefficient of highly pigmented epidermis is around 50 cm^{-1} at 650 nm while it is 10 cm^{-1} at 700 nm. The skin properties are also different from different ethnic groups, which is shown in Fig. 5.9b.

5.5 Conclusion

The implantable PV energy harvesting system is finalized with device fabrication, on-chip power management circuitry, and encapsulations. The polymer encapsulation and hermetic package are applied to protect the PV cell from subcutaneous fluids. The mono-crystalline silicon is mostly used in implantable PV cell fabrication, and the CMOS technology is normally applied in device fabrication. The optimization of using a junction diode is significant to improve the performance of PV cells. CMOS technology is advantageous in minimizing the system size and

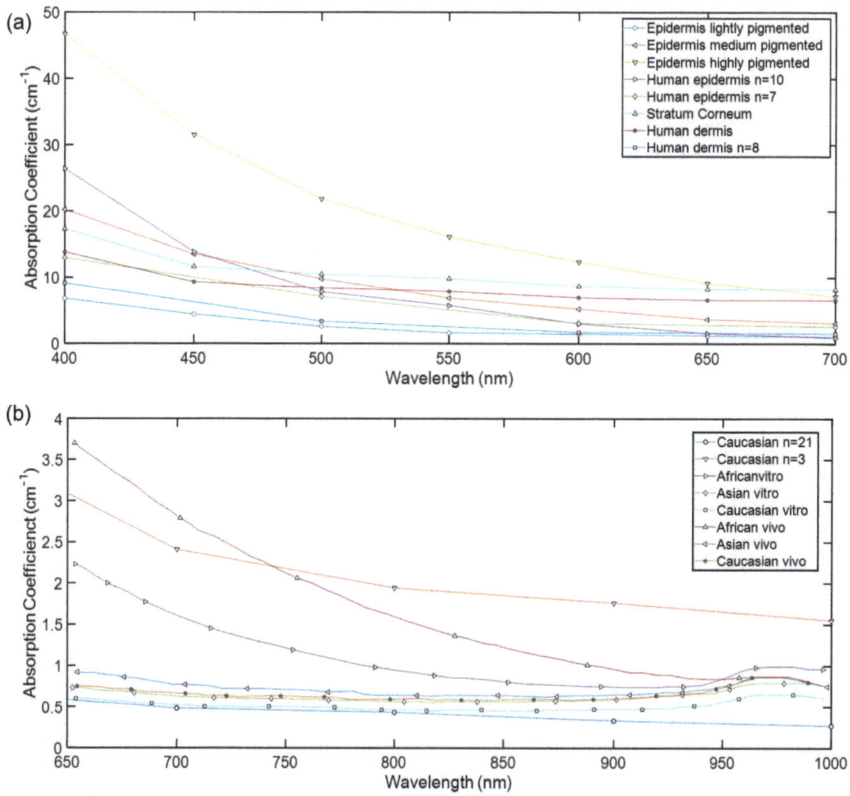

Fig. 5.9 (**a**) demonstrates how human skin absorbs light at various wavelengths according to its absorption coefficient. In A. 2011 Bashkatov et al. [53], Reported on the characteristics of the skin at various depths, from the stratum to the dermis. (**b**) emonstrates the absorption coefficient for various light spectra. The essay analyses the three ethnic groups, which include Caucasians, Asians, and Africans. (The result is from A. Bashkatov et a1. 2011 [53])

ease of embedding PV cells with CMOS circuitry. Turing point to the package and encapsulation, it must be underlined that while the solid hermetic structure offers the best protection for the gadgets, the polymer encapsulation is far more flexible and comfortable. Silicone is one of the best candidates with great optical transparency, flexibility, and advanced manufacture compatibility with optical semiconductor devices.

The performance of a PV cell can also be impacted by additional variables including the light source and optical skin loss. If the heating issue can be resolved, the NIR light is proven as a nice input power source for the implanted PV cells. The human tissue varies with different ethnic groups, geographical locations, ages, and even different body parts in the same person. Additionally, it is demonstrated that PV cells in the hypodermis can provide more stable energy than those in the dermis, which can gather more moving power. Compared to other power harvesting methods, the implanted PV cell has an advantage in providing adequate power in a compact area. The PV cell's circuitry is simpler than those of the other AC sources. Besides, the circuitry is compatible to embed with PV cells in the same chip. Implantable PV cells still have a few limitations, including low light feasibility and intensity, pricey, inflexible materials, a need for further research into long-term performance after implantation, and a requirement for deeper implantation of PV cells into tissue surfaces.

References

1. S.-H. Attn, J. Jeong, S.J. Kim, Emerging encapsulation technologies for long-term reliability of microfabricated implantable devices. Micromachines **10**(8), 508 (2019)
2. A. Ahnood, K.E. Fox, N.V. Apollo, A. Lohrmann, D.J. Garrett, D.A.X. Nayagam, T. Karle, A. Stacey, K.M. Abberton, W.A. Morrison, et al., Diamond encapsulated photovoltaics for transdermal power delivery. Biosens. Bioelectron. **77**, 589–597 (2016)
3. P.P. Altermatt, Models for numerical device simulations of crystalline silicon solar cells – a review. J. Comput. Electron. **10**(3), 314–330 (2011)
4. Y. Arima, M. Ehara, On-chip solar battery structure for CMOS LSI. IEICE Electron. Express **3**(13), 287–291 (2006)
5. D.E. Aspnes, A.A. Studna, Dielectric functions and optical parameters of Si, Ge, GaP, GaAs, GaSb, InP, InAs, and InSb from 1.5 to 6.0 ev. Phys. Rev. B **27**, 985–1009 (1983)
6. S. Ayazian, V.A. Akhavan, E. Soenen, A. Hassibi, A photovoltaic-driven and energy-autonomous CMOS implantable sensor. IEEE Trans. Biomed. Circuits Syst. **6**(4), 336–343 (2012)
7. A.N. Bashkatov, E.A. Genina, V.V. Thchin, Optical properties of skin, subcutaneous, and muscle tissues: a review. J. Innovative Opt. Health Sci. **4**(01), 9–38 (2011)
8. G. Bo, L. Ren, X. Xun, D. Yi, S. Don, Recent progress on liquid metals and their applications. Adv. Phys.: X **3**(1), 1446359 (2018)
9. J.-F. Chen, C.-L. Chun, Y.-J. Hung, Mirror-assisted interdigitated back-contact cmos photovoltaic devices for powering subcutaneous implantable devices, in *2015 International Symposium an Next-Generatian Electronics (ISNE)*, (IEEE, 2015), pp. 1–4
10. Z. Chen, M.-K. Law, P.-I. Mak, R.P. Martins, A single-chip solar energy harvesting IC using integrated photodiodes for biomedical implant applications. IEEE Trans. Biomed. Circuits Syst. **1**(1), 44–53 (2016)
11. J.H. Choi, Y. Mao, J.P. Chang, Development of hafnium based high-k materials–a review. Mater. Sci. Eng.: R: Rep. **72**(6), 97–136 (2011)
12. A.V. Dan Tchin-Iou, B.G. Min, Design of the solar cell system for recharging the external battery of the totally implantable artificial heart. Int. J. Artif. Organs **22**(12), 823–826 (1999)
13. I. Dion, C. Baquey, J.R. Monties, Diamond: the biomaterial of the 21st century? Int. J. Artif. Organs. **16**, 623 (1993)
14. L. El Chaar, L.A. lamont, and N. El Zein., Review of photovoltaic technologies. Renew. Sust. Energ. Rev. **15**(5), 2165–2175 (2011)
15. L. Eldada, F. Adurodija, B. Sang, M. Taylor, A. Lim, I. Taylor, Y. Chang, S. McWilliams, R. Oswald, B.J. Stanbery, Development of hybrid copper indium gallium selenide photovoltaic devices by the fasst printing process. Proc. Euro. Photovolt. Solar Energy Conf. **23**, 2142 (2008)
16. M. Geerligs, *Skin Layer Mechanics* (TU Eindhaven, Eindhoven, 2010)

17. R. Ghannam, P.V. Klaine, M. Imran, Artificial intelligence for photovoltaic systems, in *Solar Photovoltaic Power Plants*, (Springer, 2019), pp. 121–142
18. G. Ghosh, Dispersion-equation coefficients for the refractive index and birefringence of calcite and quartz crystals. Oft. Commun. **163**(1–3), 95–102 (1999)
19. G. Gibbons, S.M. Sze, Avalanche breakdown voltages of abrupt and linearly graded p-n junctions in Ge, Si, GaAs and GaP. Appl. Phys. Lett. **8**, 111–113 (1966)
20. S. Gong, W. Cheng, Toward soft skin-like wearable and implantable energy devices. Adv. Energy Mater. **7**(23), 1700648 (2017)
21. M.A. Green, Y. Hishikawa, E.D. Dunlop, D.H. Levi, J. Hohl-Ebinger, A.W.Y. Ho-Baillie, Solar cell efficiency tables (version 52). Prog. Photovolt.: Res. Appl **26**(7), 427–436 (2018)
22. A.S. Grove, *Physics and Technology of Semiconductor Devices* (Wiley, 1967)
23. A. Haeberlin, A. Zurbuchen, J. Schaerer, J. Wagner, S. Walpen, C. Huber, H. Haeberlin, J. Fuhrer, R. Vogel, Successful pacing using a batteryless sunlight-powered pacemaker. Eurofood **16**(10), 1534–1539 (2014)
24. A. Haeberlin, A. Zurbuchen, S. Walpen, J. Schaerer, T. Niederhauser, C. Huber, H. Tanner, H. Servatius, J. Seiler, H. Haeberlin, et al., The first batteryless, solar-powered cardiac pacemaker. Heart Rhythm. **12**(6), 1317–1323 (2015)
25. M.A. Hannan, S. Mutashar, S.A. Samad, A. Hussain, Energy harvesting for the implantable biomedical devices: issues and challenges. Biomed. Eng. Online **13**(1), 79 (2014)
26. G. Hong, G. Han, Design optimization of photovoltaic cell stacking in a triple-well CMOS process. IEEE Trans. Electron Devices **67**(6), 2381–2385 (2020)
27. Y.-J. Hung, M.-S. Cai, J.-F. Chen, S. Hsiu-Wei, P.-C. Jen, P. Chen, C.-C. Shih, T.-C. Chang, High-voltage backside-illuminated CMOS photovoltaic module for powering implantable temperature sensors. IEEE J. Photovolt. **8**(1), 342–347 (2017)
28. Y.-J. Hung, T.-Y. Chuang, C.-L. Chun, M.-S. Car, S. Hsiu-Wei, S.-L. Lee, CMOS-enabled interdigitated back-contact solar cells for biomedical applications. IEEE Trans. Electron Device **61**(12), 4019–4024 (2014)
29. S.L. Jacques, Optical properties of biological tissues: a review. Phys. Med. Biol **58**(11), R37 (2013)
30. B. Ketola, K.R. McIntosh, A. Norris, M.K. Tomalia, et al., Silicones for photovoltaic encapsulation, in *23rd European Photovoltaic Solar Energy Conference*, (2008), pp. 1–5
31. P.A.J. Kolarsick, M.A. Kolarsick, C. Goodwin, Anatomy and physiology of the skin. J. Dermatol. Nurses' Assoc. **3**(4), 203–213 (2011)
32. T. Laube, C. Brockmann, R. Buß, C. Lau, K. Höck, N. Stawski, T. Stieglitz, H.A. Richter, H. Schilling, Optical energy transfer for intraocular microsystems studied in rabbits. Graefes Arch. Clin. Exp. Ophthalmol. **242**(8), 661–667 (2004)
33. M.K. Law, A. Bermak, High-voltage generation with stacked photodiodes in standard CMOS process. IEEE Electron. Device Lett. **31**(12), 1425–1427 (2010)
34. J. Li, C.-H. Wen, S. Gauza, L. Ruibo, W. Shin-Tson, Refractive indices of liquid crystals for display applications. J. Disp. Technol. **1**(1), 51 (2005)
35. X. Li, N.P. Hylton, V. Giannini, K.-H. Lee, N.I. Ekins-Daukes, S.A. Maier, Multi-dimensional modeling of solar cells with electromagnetic and carrier transport calculations. Prog. Photovolt. Res. Appl. **21**(1), 109–120 (2013)
36. L. Luyao, Z. Yang, K. Meacham, C. Cvetkovic, E.A. Corbin, A. Vázquez-Guardado, M. Xue, L. Yin, J. Boroumand, G. Pakeltis, et al., Biodegradable monocrystalline silicon photovoltaic microcells as power supplies for transient biomedical implants. Adv. Energy Mater. **8**(16), 1703035 (2018)
37. T.K. Manna, S.M. Mahajan, Nanotechnology in the development of photovoltaic cells, in *International Conference on Clean Electrical Power*, (IEEE, 2007), pp. 379–386
38. E. Moon, D. Blaauw, J.D. Phillips, Infrared energy harvesting in millimeter-scale GaAs photovoltaics. IEEE Trans Electron. Device **64**(11), 4554–4560 (2017)
39. E. Moon, D. Blaauw, J.D. Phillips, Subcutaneous photovoltaic infrared energy harvesting for bio-implantable devices. IEEE Trans. Electron Device **64**(5), 2432–2437 (2017)

40. T.P. Otanicar, P.E. Phelan, J.S. Golden, Optical properties of liquids for direct absorption solar thermal energy systems. Sol. Energy **83**(7), 969–977 (2009)
41. H.R. Phillip, E.A. Taft, Kramers-Kronig analysis of reflectance data for diamond. Phys. Rev. **136**(5A), A 1445 (1964)
42. E. Placzek-Popko, Top PV market solar cells 2016. Opto-Electron. Rev. **25**(2), 55–64 (2017)
43. S. Popova, T. Tolstykh, V. Vorobev, Optical characteristics of amorphous quartz in the 1400-200 cm-1 region. Opt. Spectrosc. **33**, 444–445 (1972)
44. M.R. Querry, *Optimal Constants*, Technical report (MISSOURI UNIV-KANSAS CITY, 1985)
45. L.V. Rodriguez-de Marcos, J.I. Larruquert, J.A. Méndez, J.A. Aznàrez, Self-consistent optical constants of SiO_2 and Ta_2O_5 films. Opt. Mater. Express **6**(11), 3622–3637 (2016)
46. M.W. Rowell, M.A. Topinka, M.D. McGehee, H.-J. Prall, G. Dennler, N.S. Sariciftci, H. Liangbing, G. Gruner, Organic solar cells with carbon nanotube network electrodes. Appl. Phys. Lett. **88**(23), 233506 (2006)
47. K. Shen, M.M. Maharbiz, Ceramic packages for acoustically coupled neural implants, in *2019 9th International IEEE EMBS Conference on Neural Engineering (NER)*, (IEEE, 2019), pp. 847–850
48. X. Sheng, S. Wang, L. Yin, Flexible, stretchable, and biodegradable thin-film silicon photovoltaics, in *Advances in Silicon Salar Cells*, (Springer, 2018), pp. 161–175
49. G. Simone, D.D.C. Rasi, X. de Vries, G.H.L. Heintges, S.C.J. Meskers, R.A.J. Janssen, G.H. Gelinck, Near-infrared tandem organic photodiodes for future application in artificial retinal implants. Adv. Mater. **30**(51), 1804678 (2018)
50. K. Song, J.H. Han, T. Lim, N. Kim, S. Shin, J. Kim, H. Choo, S. Jeong, Y.-C. Kim, Z.L. Wang, et al., Subdermal flexible solar cell arrays for powering medical electronic implants. Adv. Healthc. Mater. **5**(13), 1572–1580 (2016)
51. K. Song, J.H. Han, H.C. Yang, K.I. Nam, J. Lee, Generation of electrical power under human skin by subdermal solar cell arrays for implantable bioelectronic devices. Biosens. Bioelectron. **92**, 364–371 (2017)
52. T. Tokuda, T. Ishizu, W. Nattakarn, M. Haruta, T. Noda, K. Sasagawa, M. Sawan, J. Ohta, 1 mm3-sized optical neural stimulator based on cmos integrated photovoltaic power receiver. AIP Adv. **8**(4), 045018 (2018)
53. M.R. Vogt, H. Holst, H. Schulte-Huxel, S. Blankemeyer, R. Witteck, D. Hinken, M. Winter, B. Min, K. Schinke, Ingo Ahrens, et at. Optimal constants of uv transparent eva and the impact on the pv module output power under realistic irradiation. Energy Procedia **92**, 523–530 (2016)
54. M.J. Weber, *Handbook of Optical Materials*, vol 19 (CRC Press, 2002)
55. D.L. Wood, T.Y. Kurt Nassau, Kometani, and DL Nash., Optical properties of cubic hafnia stabilized with yttria. Appl. Opt. **29**(4), 604–607 (1990)
56. T. Wu, J.-M. Redouté, M.R. Yuce, A wireless implantable sensor design with subcutaneous energy harvesting for long-term iot healthcare applications. IEEE Access **6**, 35801–35808 (2018)
57. C. Yan, P.S. Lee, Stretchable energy storage and conversion devices. Small **10**(17), 3443–3460 (2014)

Chapter 6
Ultrasound-Based Wireless Powering Technologies

Milad Zamani, Seyedsina Hosseini, Kjeld Laursen, Amin Rashidi, Saeed Baghaee Ivriq, Yasser Rezaeiyan, and Farshad Moradi

6.1 Wireless Powering Schemes

Biomedical implants, mostly designed in mm to sub-mm scale for less tissue damage, have been introduced for diagnostic and therapeutic purposes in human body. These devices record bio-signals and transmit the data to the external devices [1–7]. Thus, they need to be powered for reliable functioning within body [8, 9]. Batteries, as conventional, cost-saving, available, and reliable method for providing sufficient electrical energy, have been widely used in biomedical implants [10, 11]. Limited lifetime is the main bottleneck for battery, which entails risks as it requires surgery for replacement of the battery every few years. Furthermore, to increase the lifetime, larger sizes of battery would be an option [12]; however, it contradicts the second main criterion for biomedical implants, which is the size. Other side effects are non-eco-friendly and probable leakage of toxic materials into the body.

Another approach is the use of energy harvesters to scavenge different types of energies, as the human body is a powerful source of energy and can serve as an energy carrier. Based on where the source of energy is located, energy harvesters can be categorized into two main categories such as internal and external sources. Among internal sources, energies like thermal, fluid, vibration, kinetic, and chemical can be harvested using a different kind of energy harvesters. Implantable thermoelectric generators (TEGs) provide a solution to reuse the thermal energy within body; however, because of low thermal gradients, their application is limited [13]. Pressure fluctuations in body fluid, e.g., cerebrospinal fluid, are another type of energy that can be scavenged with piezoelectric energy harvesters inside the brain [14]. Vibrations from heartbeat have the potential to be scavenged with piezoelectric

M. Zamani · S. Hosseini · K. Laursen · A. Rashidi · S. B. Ivriq · Y. Rezaeiyan · F. Moradi (✉)
Aarhus University, Aarhus, Denmark
e-mail: moradi@ece.au.dk

R. Das, H. Heidari (eds.), *Wireless Power Technologies for Biomedical Devices*,
Analog Circuits and Signal Processing, https://doi.org/10.1007/978-3-031-52834-7_6

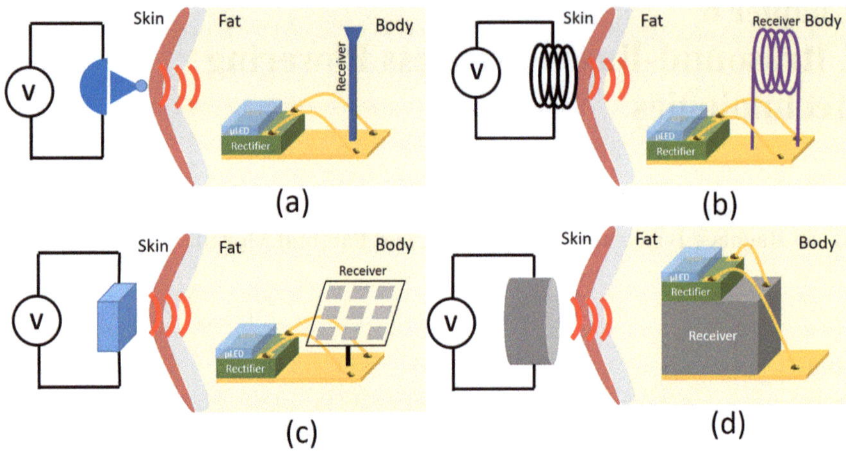

Fig. 6.1 Different powering schemes. (**a**) RF powering with an external and an internal antenna for transmitting and receiving energy, (**b**) inductive powering with two coils as transmitter and receiver, (**c**) optical powering with an LED as transmitter and photocell as receiver, and (**d**) acoustic powering with two acoustic transducers as transmitter and receiver

energy harvesters and can provide electrical energy for implants such as pacemakers [15]. Body motions like diaphragm movement in abdominal, as one of the kinetic sources of energy, can be converted to electrical signals [16]. Glucose, as the most common source of energy in human body, is able to operate fuel cells and provide electrical energy to be consumed by body implants [17].

Given that the internal sources of energy are not sufficient for most of the implants, external sources of energy such as RF radiation, inductive, optical, or ultrasonic are needed.

For RF radiation as a kind of electromagnetic (EM) waves, operating frequency is in the range of 3 kHz–300 GHz. As shown in Fig. 6.1a, transmitting and receiving antennas operate for transferring power in the medium, usually air, in mid- to far-field range. Since the transmitting waves are omnidirectional, power leaks and dissipates in unwanted directions, which causes reduction in efficiency. The main disadvantages of such powering method for bio-implants include reduction in power transfer with distance ($\propto (\frac{1}{distance})^2$) [18], less propagation depth in tissue by increasing the loss, and potential risks for human body such as heat generation in tissue which causes limited output power. On the other hand, the advantages include insensitivity to antenna orientation, longer powering distance relative to the antenna dimension, and no need for coupling between transmitter and receiver antennas [19].

Inductive power transfer (IPT) includes two coils as transmitter and receiver as shown in Fig. 6.1b, operating in megahertz frequency range, mostly 13.56 MHz. Passing an AC signal through the transmitter coil induces a magnetic field around it. The magnetic field inductively generates a voltage inside the receiver coil as a result of the electromagnetic induction, which then a circuit board rectifies the voltage and feeds the direct current (DC) voltage into the biomedical implant after regulation.

This method which is ideal for near-field (<1 cm) is highly dependent on coupling coefficient. The coefficient is directly related to the coil size and indirectly related to misalignment and distance ($\propto (\frac{1}{distance})^3$) [18]. To improve the power transfer efficiency, both coils should be operated in their resonance frequencies [20].

Optical power transfer (OPT) with solar cells, as shown in Fig. 6.1c, has been used for powering biomedical implants less than other schemes due to the limited implant sites which are in eye or under skin. The method, however, is better in terms of loss, scattering, and reflection in near-IR [21].

Acoustic power transfer (APT), as depicted in Fig. 6.1d, usually consists of two piezoelectric transducers, one as transmitter and one as receiver, operating in acoustic (20 Hz–20 kHz) and ultrasound (>20 kHz–MHz range) frequencies. Transmitter is functioning in an indirect effect of piezoelectric material in which electric field across the piezoelectric results in emitting mechanical waves into medium, water, air, tissue, etc. Mechanical waves transfer acoustic energy until they reach the receiver where direct piezoelectric effect induces electric charge across the receiver. Compared to EM with usual wavelength in mm scale, APT offers sub-mm scale wavelength because of the lower acoustic velocity [22]. Therefore, for smaller receivers with longer powering distances, APT outperforms other techniques. Moreover, attenuation coefficient is lower for APT due to the acoustic-induced compression and rarefaction of tissue [23].

As a result of what mentioned earlier and based on the following, APT surpasses the other methods for powering biomedical implants [24].

$$\text{Operating frequency} : \text{RF} > \text{IPT} > \text{APT}$$

$$\text{Implant depth with the same size of receiver} : \text{APT} > \text{RF} > \text{IPT}$$

$$\text{Generated power density with the same size of receiver} : \text{APT} > \text{RF} > \text{IPT}$$

$$\text{FDA-allowed energy intensity} : \text{APT} > \text{EM}$$

6.1.1 *Acoustic Powering*

In acoustic powering method, energy per unit area or acoustic intensity is the main value, which can be calculated from the generated pressure and acoustic impedance in the medium. To find the generated pressure, wave equation needs to be solved. Acoustic impedance defines how easily the acoustic pressure moves into specific medium at the desired frequency that is equal to the product of the density of a material and the speed of sound in the material. Piezoelectric as a common material in APT has acoustic impedance of (30–40 MRayls), and for human tissue this value is (1.58–1.7 MRayls). The difference between these two values results in more required input energy for the transmitter to transfer the desired acoustic intensity into the medium. To efficiently transfer the acoustic power between the transmitter and the receiver, the area and the thickness of both transducers, powering distance, and the operating frequency of transducers should be selected wisely. In APT, maximum continuous, pulse wave acoustic intensity, and mechanical index (to reduce risk of cavitation in body) have been limited by FDA to 7.2 $(\frac{\text{mW}}{\text{mm}^2})$, 1900 $(\frac{\text{mW}}{\text{mm}^2})$, and 1.9, respectively [25].

Piezoelectric Characteristics

Piezoelectrics are a specific kind of crystalline materials without center symmetry, which are highly applicable in electromechanical devices such as sensors, actuators, ultrasonic transducers, etc. As mentioned earlier, they convert mechanical stress into electric polarization (direct effect) or develop mechanical deformation in response to an applied electric field (indirect effect) [26]. By considering a linear relationship between mechanical stress and electric field in piezoelectric materials, the following constitutive equations are valid:

$$S = s^E T + d^t E \tag{6.1}$$

$$D = dT + \epsilon^T E \tag{6.2}$$

where T is stress, S is strain, E is electric field, D is electric displacement, s^E is elastic compliance coefficient, d is piezoelectric charge coefficient, d^t is transpose of d, and ϵ^T is permittivity coefficient [27].

Regarding the different applications, the following characteristics and constants will be introduced for piezoelectric materials:

- Piezoelectric charge coefficient (d): It is the ratio between the generated polarization and mechanical stress T to the material or the mechanical strain S induced by applied electric field.

$$d = \frac{S}{E} \tag{6.3}$$

  It resembles how much the material with specific d_{ij} is appropriate for functioning as an actuator, like acoustic transmitters. Note that the first subscript is the direction of generated polarization in the material or the direction of the applied field. The second subscript is the direction of applied stress or induced strain.
- Piezoelectric voltage coefficient (g): It is the ratio between electric field generated by the material and applied mechanical stress or the ratio between induced mechanical strain in the material and applied electric displacement D.

$$g = \frac{E}{T} = \frac{S}{D} \tag{6.4}$$

By definition, this characteristic indicates the sensing ability of the material like acoustic receivers. In g_{ij} the first subscript shows the direction of the generated electric field or the direction of the applied electric displacement. The second one shows direction of the applied stress or the induced strain.

- Electromechanical coupling factor (k): This parameter is one of the main factors for every piezoelectric material which is defined as the ratio of the mechanical energy in response to an electric field or vice versa.

$$k^2 = \frac{\text{converted energy}}{\text{input energy}} = \frac{\pi}{2}\frac{f_r}{f_a}\cot\left(\frac{\pi}{2}\frac{f_r}{f_a}\right) \tag{6.5}$$

where f_r and f_a are the resonance frequency (point with minimum impedance) and the antiresonance frequency (point with maximum impedance), respectively. Four different coupling coefficients are introduced based on the initial shape of the transducer and aspect ratio ($\text{AR} = \frac{\text{diameter}}{\text{thickness}}$): k_{33} for the case when electric field is applied in direction 3 (parallel to the polarization direction) and longitudinal vibration in the same direction; k_t for thin plates and the case when both electric field and vibration are in direction 3 (high AR, $k_t < k_{33}$); k_{31} for the case when electric field is applied in direction 3 (parallel to the polarization direction) and vibration is in direction 1 (perpendicular to the polarization direction); and k_p for the case when electric field is applied in direction 3 (parallel to the polarization direction) and radial vibrations are in directions 1 and 2 (perpendicular to the polarization direction, high AR). Note that directions 1, 2, and 3 are like the Cartesian coordinate system; however, they will define for each individual material.
- Mechanical quality factor (Q): It corresponds to the power loss over the stored energy in the material which is defined as the ratio between the resonance frequency and the bandwidth.
- Dielectric dissipation factor ($\tan\delta$): It relates to the dielectric loss in the material, and the smaller value, the better for implantable devices as the generated heat in the material will be less.
- Curie temperature (T_c): It is the temperature at which the piezoelectric effect decays due to depolarization. This parameter is important for biomedical implants, especially during the preparation processes like dicing of the material, integration, etc.
- Thickness mode frequency constant (N_t): This parameter is defined by the material supplier and is related to the thickness. It is important for the calculation of the resonance frequency in thickness mode of vibration.

Piezoelectric Materials

Based on the desired application, a wide variety of piezoelectric materials are available; however, for using piezoelectric materials in biomedical implants, biocompatibility is also highly preferable in addition to the high electromechanical coupling coefficient and piezoelectric coefficients. Furthermore, high curie temperature is another important parameter, especially during integration process where different components, e.g., piezoelectric material, microelectronic chip, and modalities for stimulation, recording, and drug delivery, should be assembled. Single-crystal lithium niobate ($LiNbO_3$) and polymer polyvinylidene fluoride (PVDF) have been used for large aperture single-element transducers because of low dielectric permittivity, although $LiNbO_3$ and PVDF are applicable for high-frequency medical imaging and sensing mechanisms, respectively [28]. In addition, PVDF lacks in further applications, especially for powering and energy harvesting, due to the low Curie temperature ($\sim$100 °C). PZT, as a proper piezoelectric material, has been used for decades with wide applications as sensors and actuators because of its high electromechanical coupling coefficient, Curie temperature (200−400 °C based on the types), and sensitivity. Nevertheless, it contains lead more than 60% of its weight. Environmental and health-related concerns have led to make restrictions on lead and lead-included materials in recent years [29]. As a consequence, encapsulation of PZT with biocompatible materials, e.g., polydimethylsiloxane (PDMS) [30] and parylene [31], has been proposed as an approach for avoiding leakage of lead into human tissue. However, challenges with encapsulation like short-term stability, reduction in energy harvesting efficiency, and nonuniformity have led to efforts for replacing PZT with biocompatible materials. Among biocompatible materials, aluminum nitride (AlN) [32], zinc oxide (ZnO) [33], and barium titanate ($BaTiO_3$) [34] have been used at the expense of weaker piezoelectric properties than PZT. As a promising alternative for PZT, potassium sodium niobate (KNN) was discovered in [35] with very similar properties to PZT. KNN has large piezoelectric coefficient, with controllable Curie temperature between 170 and 500 °C which is achievable with addition of lithium. As a result, a lithium doped KNN (LKNN) material is developed with different concentrations of lithium in various thicknesses.

Piezoelectric Crystal Modeling

To simulate piezoelectric crystals with different shapes or materials, studying the effect of their surrounding material on propagation of ultrasonic waves tofrom them, and the exposure of them to different forms of ultrasonic waves, FEM simulation tools such as COMSOL Multiphysics are rather accurate choices. Even though the FEM simulation tools are also good for simulation with simple electrical circuits, their interfaces with chip design tools like CADENCE have not been developed properly. Furthermore, for a preliminary analysis of electromechanical behavior of piezoelectric receiver, a more tangible modeling is required. Electrical lumped element modeling is a popular approach for analysis and simulation of the electromechanical behavior of the ultrasonic links [36–39]. For instance, KLM model [40] is one of the widely employed models, which is illustrated in Fig. 6.2a for a piezoelectric crystal with a thickness of t. This approach suggests to model the mechanical characteristic of the crystal with a transmission line between the front and back mechanical ports and split in half to connect to the electrical port through a

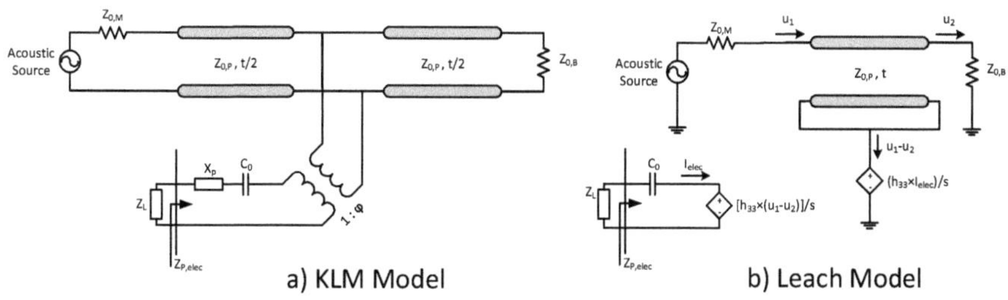

Fig. 6.2 (**a**) KLM model, and (**b**) Leach model of a piezo harvester with thickness of t

modeled transformer with the ratio of φ. In Fig. 6.2a, $Z_{0,M}$, $Z_{0,P}$, and $Z_{0,B}$ are bulk acoustic impedances of media, piezoelectric material, and piezoelectric backing, respectively. From mechanical point of view, the acoustic impedance (Z_0) of a material can be driven out of the below equation:

$$Z_0 = A\rho v \tag{6.6}$$

where A, ρ, and v represent the specific area, density, and the propagation velocity of the material, respectively. Other component values of the KLM model are given by the following equations:

$$\varphi = \frac{Z_{0,p}\omega}{2h_{33}\sin(\frac{\omega}{v}\frac{t}{2})} \tag{6.7}$$

$$X = \frac{h_{33}^2\sin(\frac{\omega}{v}t)}{Z_{0,p}\omega^2} \tag{6.8}$$

where ω is the angular frequency and h_{33} is the piezoelectric coefficient:

$$h_{33} = k_t v\sqrt{\frac{\rho}{\varepsilon_0\varepsilon_r}} \tag{6.9}$$

k_t in the above equation is piezoelectric coupling coefficient, ε_0 is electric constant, and ε_r is piezoelectric relative permittivity. The KLM model is suitable for behavioral analysis of electromechanical properties of piezoelectric crystals. Nevertheless, it is not useful enough for transient simulation in a circuit design CAD tool, since it has a transformer with a frequency-dependent winding ratio which is not compatible with circuit design CAD tools like Orcad PSPICE or CADENCE Virtuoso. Thus, the Leach model is proposed in [41] as illustrated in Fig. 6.2b. In the Leach model, the transformer is omitted, and two dependent voltage sources are added for modeling the mutual influence between mechanical and electrical parts. In Fig. 6.2b, u1 and u2 are the acoustic particle velocity of the acoustic sound on the top and back ports of the piezo. All other parameters are identical to the KLM model and can be driven by the same equations. It is noteworthy that the lumped element models are mainly in thickness mode, where the mechanical aspect ratio (thickness/surface dimensions) is considered to be small. Thus, it is not a good reference for predicting the actual harvestable power out of a crystal. However, it is helpful in prediction of its behavior.

6.1.2 Ultrasonic Transducers

Ultrasonic transducers can be categorized in bulk ultrasonic transducers (BUTs) and micromachined ultrasonic transducers (MUTs) based on the configuration of their active piezoelectric material. In recent years, MEMS technology introduced MUT as a promising alternative for conventional BUTs. Improvements in microelectronics necessitate CMOS compatibility for new ultrasonic transducers which is achievable with MUTs. Moreover, MUTs benefit from the advantages of MEMS technology such as compact size which triggers reduction in noise levels [42], multifunctionality [43], low power consumption [44], and batch production [45]. Increasing demand in mobile phones, tablets, augmented reality, Internet of Things (IoT), etc. has made MUTs more attractive in comparison to other approaches.

Bulk Ultrasonic Transducers

BUTs, as conventional transducers, are single-element piezoelectric transducers with wide applications, just as a few cases, in medical diagnostic, nondestructive testing, sonar applications, material science, acoustics, electrical, and biomedical engineering. Their structure, as shown in Fig. 6.3, includes a thick piezoelectric layer sandwiched between two top and bottom electrodes. Top electrode is connected to the ground in order to reduce noise levels from external sources. When they are driven with an electric signal, ultrasonic waves emit from forward and backward of the piezoelectric layer. Backward waves destroy resolution, for instance, in imaging; thus backing material will shorten backward waves and provide wide bandwidth transducer [46]. Furthermore, to compensate the acoustic mismatch between the transducer and medium, single- or multi-matching layer is

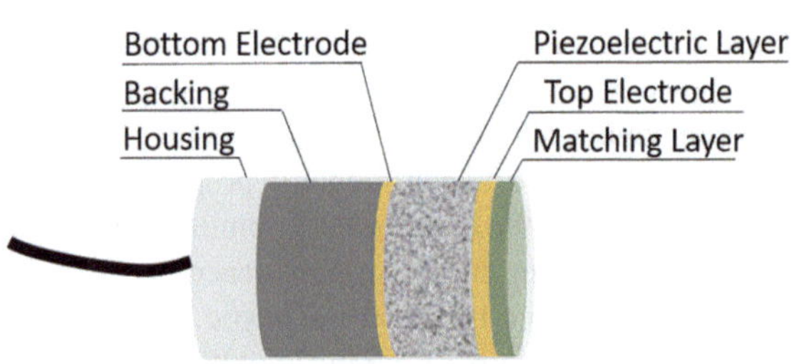

Fig. 6.3 Schematic of the bulk ultrasonic transducer included housing, backing material, bottom and top electrodes, piezoelectric layer, and matching layer

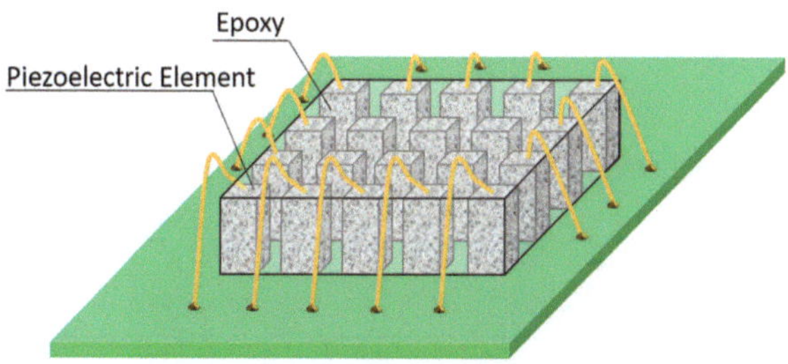

Fig. 6.4 Schematic of the piezocomposite ultrasonic transducer array with piezoelectric elements and epoxy filled in gaps. The main challenge for these transducers is shown as complex wiring, especially for inner elements

located on the front-side of the transducer [47]. The matching layer's thickness is reported to be one-quarter of the wavelength in medium [48]. Ultrasound transmission coefficient that defines the reflected waves at the interface can be found from the following equation:

$$\Gamma = \frac{2Z_m}{Z_p + Z_m} \tag{6.10}$$

where Z_m and Z_p are acoustic impedance of the medium and piezoelectric material, respectively.

To improve the performance of BUTs, another structure is introduced which contains diced piezoelectric material and polymer filled in the gap between piezoelectric and make piezocomposite arrays, like 1–3 composite array. This kind of transducer outperforms conventional BUTs in terms of higher electromechanical coupling factor and lower acoustic impedance [46]. To make piezocomposite arrays, the ratio of piezoelectric material to polymer and piezoelectric thickness are controlling parameters [49]. As depicted in Fig. 6.4, bulk piezoelectric material should be diced and filled the gap with epoxy; therefore their fabrication process is more complex than BUTs. However, they will provide steering ability of the waves; for instance, the 2D phased array of these transducers is able to steer the waves in 3D and make volumetric images, although wiring of each individual element is complex and dicing the elements in smaller dimensions (for high frequency (>10 MHz)) is challenging due to the gap size [50].

Micromachined Ultrasonic Transducers

Demand for small-scale transducer arrays opened up the use of MEMS technology and micromachining for ultrasonic transducers. Capacitive micromachined ultrasonic transducer (CMUT) and piezoelectric micromachined ultrasonic transducer (PMUT) are two MEMS-based devices that are widely researched and used for

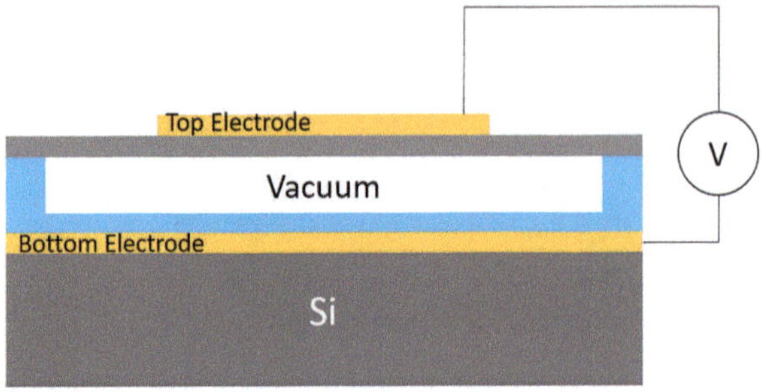

Fig. 6.5 Schematic of cross section of a CMUT with the top and bottom electrodes and the vacuum gap

many applications. The former operates based on electrostatic force between two electrodes, and the latter works based on piezoelectric effect.

As shown in Fig. 6.5, CMUT includes top and bottom electrodes with a vacuum gap in between. When an AC signal is applied between electrodes, a free electrode (membrane) oscillates due to the electrostatic force and emits ultrasonic waves into medium. For efficient operation, CMUT needs to be biased at the pull-in voltage across electrodes, which is the voltage needed to prevent membrane collapse onto the substrate. Based on the pull-in voltages, CMUT operates in three regimes. In the conventional regime, the bias voltage is close to the pull-in voltage [51]. Two other regimes are collapse [52] and collapse-snapback [53] in which the bias voltage is above the pull-in voltage, and it has been reported that the electromechanical coupling factor is larger compared to the conventional regime [54].

The promising characteristics of CMUT are their high bandwidth (larger than 100%) and high electromechanical coupling factor to reach 100% due to the controllable spring constant of two electrodes with a DC bias. On the other hand, the main disadvantages of CMUT are their low ultrasonic intensity due to the limited gap between electrodes, the requirement of high DC bias voltage, and large power consumption.

PMUT, as depicted in Fig. 6.6, contains single- or multi-piezoelectric layer sandwiched between top and bottom electrodes and also passive layers such as silicon (Si) and silicon dioxide (SiO_2). When an AC signal is applied across the electrodes, out-of-plane and in-plane stresses generate into the active piezoelectric layer, located on the Si membrane, and cause perpendicular displacement, hence emission of acoustic waves. Unlike CMUTs, there is not a vacuum gap between electrodes, and therefore membrane displacement is not limited. Compared to CMUTs, PMUTs are also favorable for low-power biomedical implants since CMUTs require an external circuit to measure the change in capacitance under external stresses; however, for measurement in PMUTs, there is no need to supply charge, and the generated charge is accumulated in the passive capacitance. The main drawback of PMUTs is their low electromechanical coupling coefficient (1–6%) compared to CMUTs (~70%) and BUTs (18%) [55]. To improve the coupling

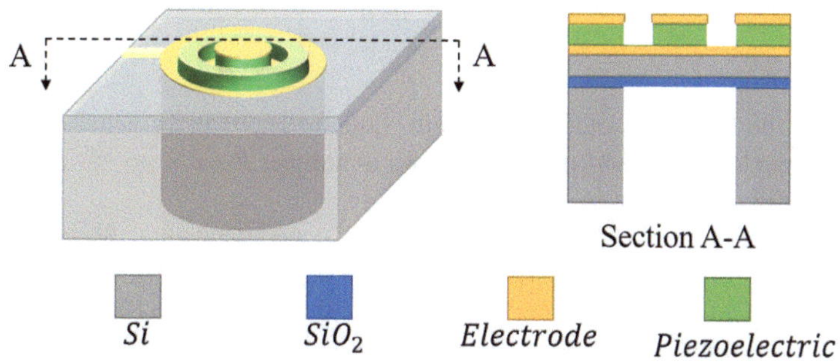

Fig. 6.6 Schematic of the PMUT with the layers as Si, SiO_2, bottom and top electrodes, and piezoelectric layer. Section A-A also shows the cross section of the Si membrane

coefficient of PMUTs, different approaches have been proposed, for example, optimization of the geometry of the PMUT cell like dome-shaped cells [56], reducing the residual stress during fabrication [57], piezoelectric material with higher piezoelectric constant [58], etc. Recently, interesting applications have been suggested for PMUTs such as intravascular ultrasound [59], fingerprint sensor [60], rangefinders in air [61], wireless power supply systems [62], and implantable micro devices [63].

6.1.3 Fundamentals of Transmitters

Acoustic Wave Equation

Acoustic field is the result of a vibrating surface changing with time and position in a medium. Acoustic field can be quantified with a scalar value of pressure P and particle velocity vector of $\vec{u}$. For the linear acoustic waves, the following equations are valid:

$$\rho \frac{\partial \vec{u}}{\partial t} = -\vec{\nabla} p \tag{6.11}$$

$$\frac{\partial \rho\,'}{\partial t} + \rho \vec{\nabla} \cdot \vec{u} = 0 \tag{6.12}$$

$$p = c^2 \rho\,' \tag{6.13}$$

where ρ is the static density and $\rho\,'$ is the acoustic density. By substituting equation of motion, Eq. 6.11, into the equation of continuity, Eq. 6.12, and replacing $\rho\,'$ with the equation of state of the medium, Eq. 6.13, the scalar wave equation for the acoustic pressure is as follows:

$$\nabla^2 P - \frac{1}{c^2}\frac{\partial^2 P}{\partial t^2} = 0 \tag{6.14}$$

where c is the speed of sound in medium. Equation 6.14 in its general form with acoustic wavelength λ, and the acoustic wave number $k = \frac{\omega}{c} = \frac{2\pi}{\lambda}$ is known as Helmholtz equation:

$$\nabla^2 P + k^2 P = 0 \tag{6.15}$$

To solve the Helmholtz equation, different coordinate systems can be chosen based on the initial shape of the transducer. Solutions for the Helmholtz equation may vary based on the region of interest in front of the ultrasonic transducer.

Near- and Far-Field

Acoustic field for ultrasonic transducers can be divided into two main zones in terms of acoustic intensity and beam diameter. Near-field starts from the surface of the transducer where the energy transfers to the medium until it reaches the maximum energy at the focal point and then returns back to the transducer. Therefore, the mass of the medium in near-field increases the mass of the transducer. At the focal point, the beam diameter is half of the transducer's diameter [64]. After that, the beam diverges, and the acoustic intensity gradually decreases in the region called far-field. In far-field, the energy never returns back to the transducer. The separation between the two regions is called Rayleigh distance or focal depth and is determined as follows:

$$d = \frac{d_1^2 f}{4c} \tag{6.16}$$

where d_1 is the transducer's diameter, and f is the frequency. Based on Eq. 6.16, due to the direct relationship between the diameter of the transducer and the focal depth, a larger diameter results in a deeper focus.

Multi-ring Ultrasonic Transducer

The main concept of the MRUT stems from Eq. 6.16, due to the direct relationship between a transmitter's diameter and the focal depth. In other words, a larger diameter of transmitter will result in a deeper focus. To achieve the maximum energy transmission link, both transmitter and receiver should be designed to have the same resonance frequency. Resonance frequency depends on the transducer's size [65]. Here, the dust's size is determined based on the volume limitation in the cortex and the striatum. It was shown that smaller thickness of the dust results in a higher frequency. Given the size limitation of the device, the design of transmitter

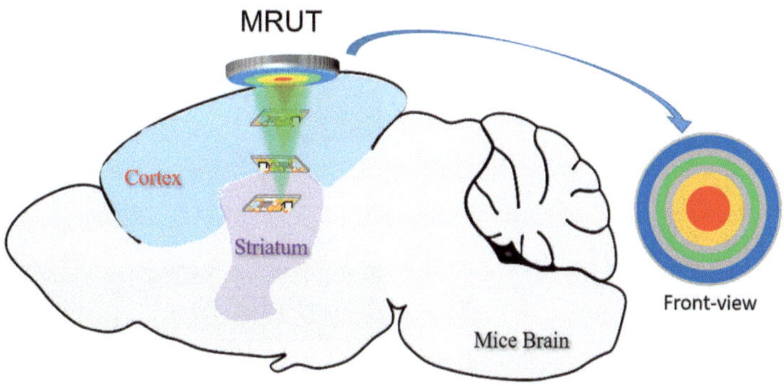

Fig. 6.7 Conceptual diagram of powering dusts in different depths by MRUT

is initiated by taking its diameter and thickness into account. Transmitter's diameter is the main focus for designing the MRUT. MRUT is proposed to focus the waves at different depths. In the literature, focusing the acoustic intensity into a narrow region and improving the lateral resolution have been realized with three methods including external focusing with a lens in front of the piezoelectric material [66], internal focusing with a curved diaphragm [67], and phased array focusing with linear [68] or ring [69] arrays of piezoelectric material, individually controlled by electronics. The last one outperforms the rest due to the versatility in the adjustment of focal depth especially when the location of implants can change due to either the head movements or the use of different implants at different locations (Fig. 6.7).

Furthermore, in single-element conventional transducers or BUTs, there are some side lobes near the main beam with lower acoustic intensity than the central axis. Tissues within these side lobes can be a source of reflections and degrade the lateral resolution as well. However, in MRUT, side lobes can be reduced by exciting each element, separately, using an electric signal [64].

Sectored Multi-ring Ultrasonic Transducer

For focusing light without a lens, FZP theory has been used in the literature [70], which contains concentric rings to diffract waves when they pass through the rings. These waves will constructively interfere with each other, if the distance between the rings is designed accurately. To use this theory for making a new acoustic transducer, two rings with path lengths R_1, R_2, R_3, R_4 and radii a_1, a_3, a_5, a_7 are considered as illustrated in Fig. 6.8. For a full circle, in order to focus the waves at one point, the path length difference between the sources should follow this equation [71]:

$$R_{i+2} - R_i = m\lambda \tag{6.17}$$

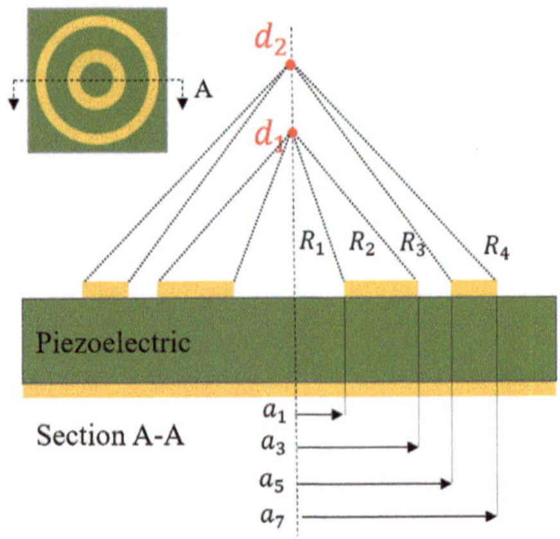

Fig. 6.8 Cross-sectional view of the Fresnel zone plate with multi-symmetrical rings as an acoustic transducer [72]

where i and m are positive integers, and λ is the wavelength in medium. However, for a ring-shape configuration, since the electrodes have a width, the edges of each electrode should fulfill the following criteria to minimize the destructive interference of the waves at the focal point:

$$R_{i+1} - R_i = \lambda/2 \tag{6.18}$$

The S-MRUT is composed of two rings; each of them consists of four equal-length sectors. Let us consider two different focal points d_1 and d_2 for inner and outer rings, respectively. Then, in order to satisfy the conditions in Eqs. 6.17 and 6.18, the radius of electrodes can be calculated using the following equation:

$$a_n = \sqrt{\frac{\alpha\lambda}{2}\left(d + \frac{\alpha\lambda}{8}\right)} \tag{6.19}$$

where d is the focal point, $n = 1, 3, 5, \ldots$, $\alpha = n + 1 + 2\delta$, and δ is the offset value that relates to the discrepancies between the calculated and measured focal points due to the fabrication process with values of $-1 < \delta \leq 0$. The best focusing can be achieved with $\delta = 0$. Therefore, one can design the electrodes to focus the acoustic energy on the axis along the transducer. However, for only one sector of each ring as an acoustic transmitter, the superposition of the waves will be off-axis. The reason for the off-axis focal point is the distance g between acoustic sources and thus the resulting time delay because of the distance.

In order to explain the reason of the off-axis focal point as shown in Fig. 6.9, let us take two acoustic sources a and b as small fractions of inner and outer rings, respectively, into account. Based on the Huygens Principle [64], small sound sources create V-shaped beams. The superposition of the beams will be either constructive or destructive. If they constructively interfere at point x, there is a time delay Δt

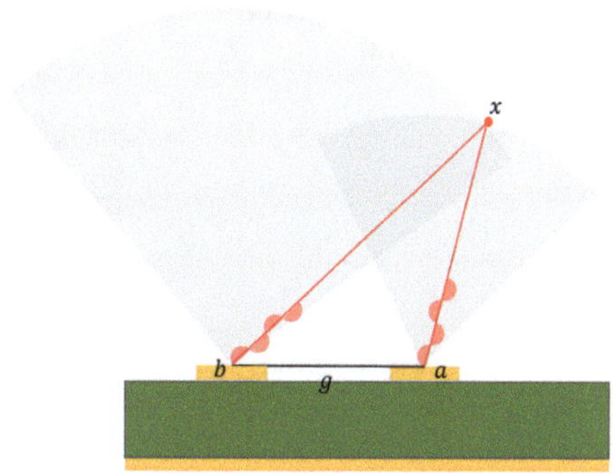

Fig. 6.9 Cross-sectional view of the emitted waves from two fractions of rings [72]

because of the distance g. As a consequence, a phase shift ψ is caused by the time delay that can be calculated with the following equation:

$$\psi = 2\pi f \Delta t = \frac{2\pi}{\lambda} g \tag{6.20}$$

where f is the frequency. For the S-MRUT, the two aforementioned fractions have an equal distance g, and if we integrate their effect over one sector of the ring, the off-axis focal point will be achieved.

Design and Simulation

Design Considerations

For designing a transducer to be able to focus waves in both axial and lateral directions, the main design objectives can be listed as follows:

1. Steering area: To power the dusts in various locations, the transducer should be designed in a way that the maximum acoustic pressure occurs at the desired angle; otherwise the distributed pressure in other directions will reduce the efficiency of the transducer [73]. Therefore, one of the main design objectives is to maximize the steering area A where the maximum pressure takes place; this area is considered as the best place to implant the dust. Let us consider a case where the dusts are implanted on the surface of striatum of mice, which has a spherical shape with a radius of $\approx$1.5 mm [74]. For this case, a large steering area should be targeted for designing the transducer.
2. Ratio of main to side lobe amplitude: In order to reduce the loss of ultrasonic energy in other directions, the amplitude ratio of the main to side lobe needs to be maximized. In other words, a transducer with a smaller main to side lobe amplitude ratio gives higher reflections from the tissues located in regions under the side lobes.

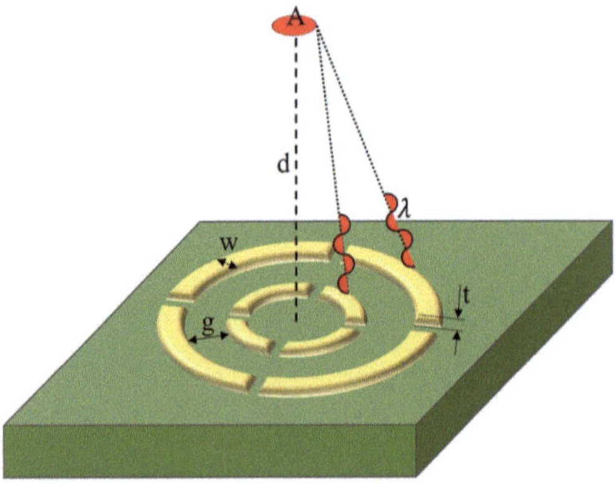

Fig. 6.10 Design parameters of the S-MRUT [72]

3. Main lobe width: To decrease the attenuation of acoustic energy in brain tissue, the beam width of the transducer should be small; otherwise, the sound beam will not be well directed to the target.
4. The focal distance d is also important for powering the dusts with maximum acoustic intensity.

To achieve a proper design, the available design parameters, as shown in Fig. 6.10, i.e., wavelength λ in medium, number of electrodes N, element distance g, element width W, and electrode thickness t, should be selected carefully. Among the aforementioned parameters, the number of electrodes, N, is limited because of the electrical connections on the PCB and the space constraint on the piezoelectric substrate; electrode thickness is fixed at 10 μm. For the distance g and width W, as explained in the previous section, they are calculated when the electrodes' radii are designed using Eq. 6.19 for the desired focal points. Here, we consider a focal distance d of 1 mm–3 mm, which is the distance from the skull's surface of the mice to motor cortex and the striatum. In this work, λ depends on the size of the piezoelectric receiver on the dust, which defines the operating frequency. In other words, the resonance frequency of the piezoelectric receiver (i.e., the proper operating frequency for the dust) and consequently the wavelength of ultrasonic waves in the medium depend on the dimensions of the piezoelectric receiver.

COMSOL Simulation on S-MRUT

Transducer is modeled and simulated in COMSOL 5.4 for characterizing the acoustic pressure and the intensity profiles in the medium and finding the steering area by means of parameters. In all simulations, the transducer and the water domain were modeled in 3D, and PZT-4 is selected as the piezoelectric material. The maximum acoustic intensity for the simulations is considered as 7.2 $(\frac{\text{mW}}{\text{mm}^2})$. To model different sectors on the surface of the piezoelectric substrate, shell module is

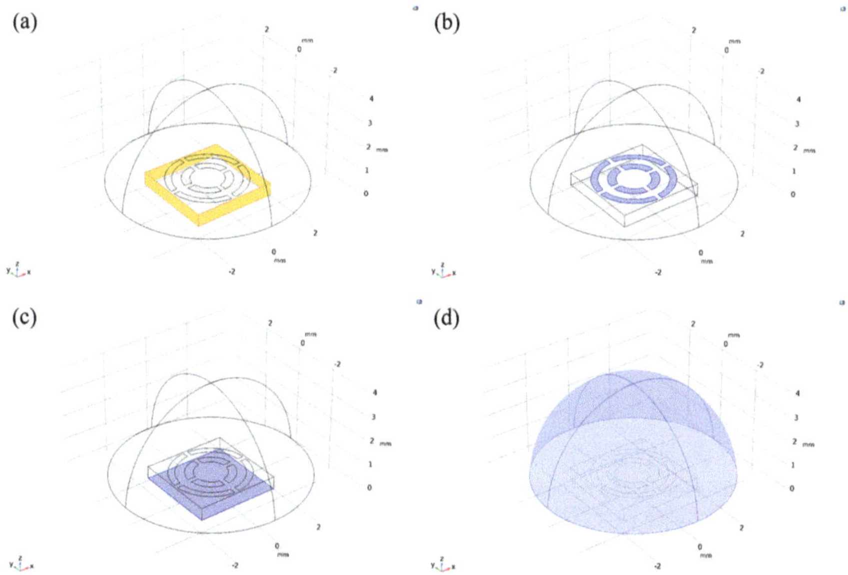

Fig. 6.11 Boundary conditions for (**a**) solid mechanics module with simply supported plate, (**b**) electrostatics with the sectors of top electrode, (**c**) electrostatics with the ground electrode, and (**d**) pressure acoustics with the spherical wave radiation [72]

also implemented in the simulations. Then, the pressure acoustics, solid mechanics, and electrostatics in the frequency domain were solved as physics of the simulation.

Regarding the boundary conditions as shown in Fig. 6.11, the S-MRUT is simply supported in solid mechanics module (Fig. 6.11a). In the electrostatics module, the sectors of top electrodes are considered as the positive electrodes (Fig. 6.11b) and the bottom surface as the ground electrode (Fig. 6.11c). In the pressure acoustics module, the spherical wave radiation node is selected as the radiation boundary condition to remove the reflections (Fig. 6.11d). For coupling the aforementioned physics, the multiphysics node is added to the model with the sub-nodes of piezoelectric effect and acoustic structure boundary. The former is responsible for coupling the solid mechanics and electrostatics and the latter for coupling the pressure acoustics and solid mechanics.

In our simulations, first, the resonance frequency of each cubic size of piezoelectric for the dust is modeled. As mentioned earlier, the resonance frequency and consequently the wavelength were selected based on the resonance frequency of the piezoelectric receiver in the dust. Figure 6.12 shows the results for different cubic sizes of the piezoelectric receivers and the associated wavelength. Since the dusts include a piezoelectric receiver with the dimension of $(0.5\,\text{mm})^3$, the operating frequency of ≈2.9 MHz has been chosen for the S-MRUT. The piezoelectric plate with 0.5 mm thickness is selected for the S-MRUT, the same as the piezoelectric receiver. Before doing the simulations on the acoustic profile of the S-MRUT, it is

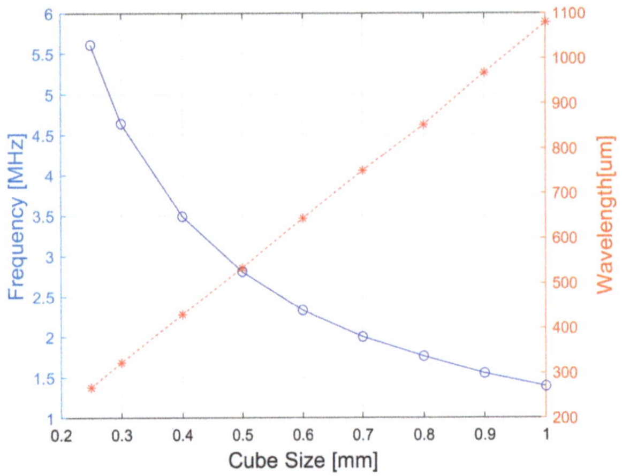

Fig. 6.12 Simulation of the resonance frequency and associated wavelength over the different sizes of piezoelectric receiver [72]

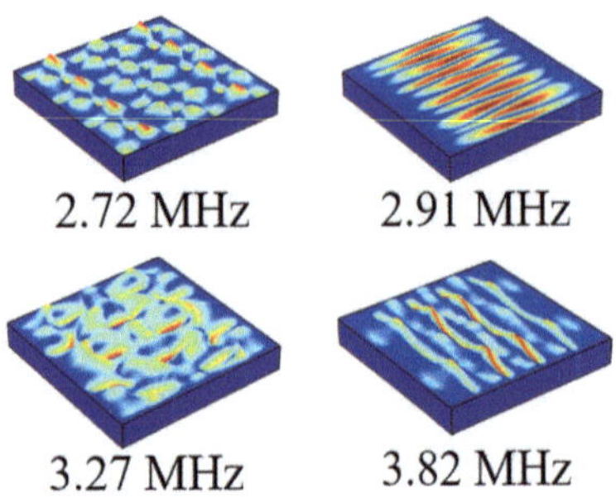

Fig. 6.13 Simulation of the eigenmodes for the piezoelectric plate. Dark blue represents the zero displacement, and dark red is for the maximum one [72]

needed to find the desired mode shape around the selected frequency. Choosing the correct mode shape is one of the important steps in designing the transducer, as each geometry has a specific mode shape with a given profile for the displacement, which relates to the output pressure. The preferable mode shapes are Gaussian and piston-like, as they can symmetrically vibrate with the maximum displacement around the center. Since these two mode shapes occur in the frequency range of kilohertz (in the transducer example), they were not selected. In other words, the frequency range of kilohertz is equal to few millimeter wavelength which is not appropriate in terms of the desired focal distance. To find the desired mode shape, eigenfrequency simulations were carried out as shown in Fig. 6.13. As illustrated in the figure, each mode shape includes vibrations in thickness and flexural modes. However, the desired mode shape is the one with higher vibrations in thickness direction. Based on the figure, the mode shape with frequency of 3.27 MHz has been selected due to the symmetrical vibrations in thickness direction around the center.

Based on the size of the piezoelectric receiver on the dust and the mode shape, the frequency was selected. In the next step, the electrodes' rings on top of the

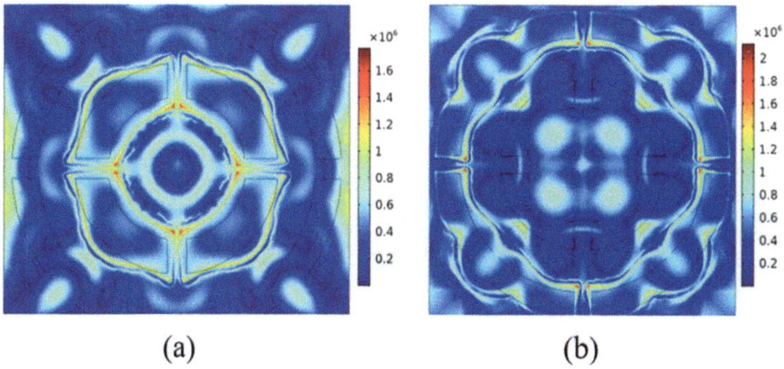

Fig. 6.14 Simulations of the mechanical stress for (**a**) inner active electrodes and (**b**) outer active electrodes [72]

piezoelectric plate should be patterned. Since the active area on the piezoelectric plate is the region with electrodes, only the top electrodes should be designed. However, this depends on the mode shape, and the electrodes should be mainly placed on a region with the highest degree of displacement in eigenfrequency simulation. Thus, the electrodes' radii are designed based on Eq. 6.19 with the desired focal distances (1–3 mm). The electrodes are designed in a way that the inner electrodes have a focal distance at 1 mm, the outer electrodes focus the waves at 2.5 mm, and the whole transducer focuses the waves at 3 mm. As shown in Fig. 6.14a, when the inner electrodes are active, the maximum mechanical stress occurs around the inner electrodes. However, for the outer active electrodes in Fig. 6.14b, the maximum stress and consequently the maximum displacement are around the outer electrodes.

After designing the electrodes with the desired requirements, the measurement of the steering area, the main lobe width, and the ratio of the main to side lobe amplitude are done. In Fig. 6.15, the simulation results for the normalized pressure in a plane normal to the S-MRUT surface and also the polar plot are shown. Figure 6.15a, b illustrates the acoustic profile for the inner electrodes, Fig. 6.15c, d for the outer electrodes, and Fig. 6.15e, f for sectors at the same quarter. As expected, the inner and outer electrodes direct the waves in the axial direction (on-axis) in front of the S-MRUT; however, sectors at the same quarter are able to focus the waves 0.678 mm off-axis. In other words, with the designed electrodes, it is possible to have a steering area of 1.443 mm^2 if only one sector of each ring at the same quarter is driven by an electric signal. The main lobe width for the sectors at the same quarter is 0.46 mm, which is larger than the inner electrodes with 0.4 mm and the outer electrodes with 0.35 mm. The larger main lobe width for the sectors at the same quarter is due to the asymmetrical stimulation of the transducer. The main to side lobe ratio is greater for the inner and outer electrodes when compared to the sectored ones. Increasing the number of electrodes would improve the main to side lobe ratio.

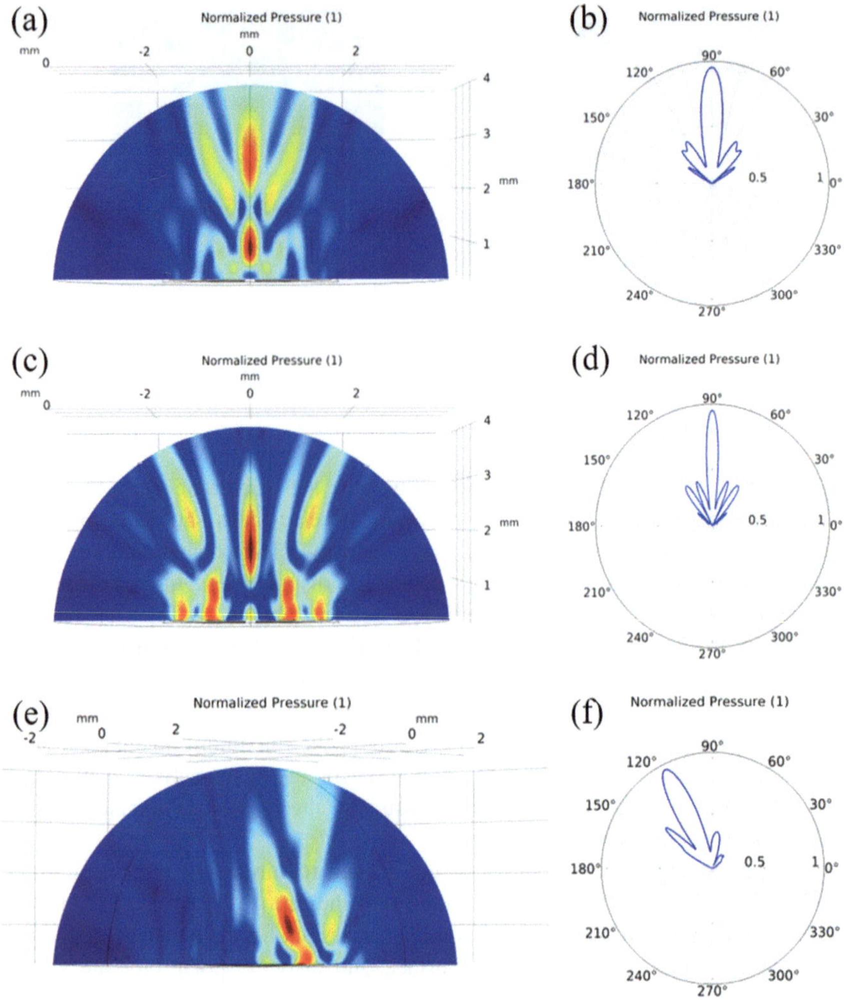

Fig. 6.15 Simulated normalized acoustic pressure for inner electrodes in (**a**) and (**b**), outer electrodes in (**c**) and (**d**), and sectors at the same quarter (**e**) and (**f**). In the plane normal to the S-MRUT surface (**a**), (**c**), and (**e**), dark red shows the maximum pressure, and dark blue is for the minimum pressure [72]

Fabrication Process

After patterning the top electrodes, the PZT plate is diced with a Disco DAD321 with 2000 rpm spindle speed and 1 mm/s feed rate in the size of 3.8 mm × 3.8 mm as shown in Fig. 6.16a. Then, for measurement purposes, the fabricated S-MRUT is mounted on a custom-designed PCB using ESL 1901-S polymer silver conductor paste and cured at 80 °C for 30 minutes, thereby connecting the bottom electrode

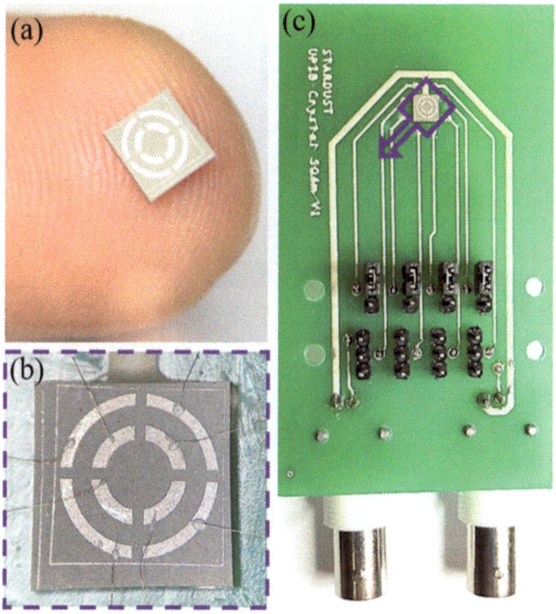

Fig. 6.16 (**a**) Size of the S-MRUT compared to the finger. Magnified view of the S-MRUT in (**b**) on custom-designed PCB in (**c**) [72]

to the PCB. Top electrodes are wire-bonded manually with a 17 μm Au wire to the PCB as shown in Fig. 6.16b. On this PCB, the configuration of the driven sectors is done using some jumpers that can connect each sector to one of the two input BNC connectors, as shown in Fig. 6.16c.

Acoustic Profile Measurement

Acoustic profiles of the S-MRUT for different configurations of the electrodes under excitation are measured using the setup shown in Fig. 6.17. In this setup, an arbitrary signal generator is used to generate the sinusoidal signal and drive the S-MRUT's sectors under test. The acoustic intensity at each relative position to the S-MRUT is measured using a hydrophone followed by a preamplifier, which is connected to oscilloscope. For plotting the S-MRUT's output acoustic profile over a certain volume of water tank, the hydrophone is moved using a micro-positioning system with a resolution of down to 5 μm. The micro-positioning system, signal generator, and oscilloscope are all controlled simultaneously and automatically through MATLAB scripts. The obtained data is retrieved directly from the oscilloscope via Ethernet LAN and saved by MATLAB on a connected computer. For all measurements, the hydrophone is initially located in the center position of the S-MRUT with a 1 mm distance. The initial 1 mm distance is measured based on the time-of-flight of acoustic wave in water (1484 (m/s)). Measurements are conducted in deionized water in a custom-designed plexiglass water tank equipped with wave absorbers to dampen the reflected waves from the tank's walls.

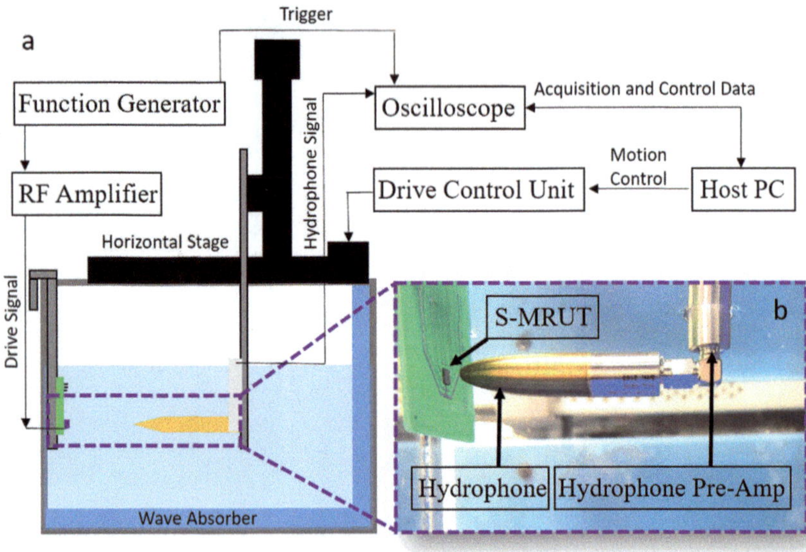

Fig. 6.17 (**a**) Diagram of the measurement setup and (**b**) the real photograph of the scanning region [72]

To characterize different configurations of electrodes, our desirable scanned volume in front of the S-MRUT was a cube of 5 mm × 5 mm in each slice with 5 mm in depth. The resolution of scanning profile in each slice is 200 μm; however, the gap between each slice is 500 μm. In all measurements, all sectors are driven by a sinusoidal signal within the frequency range of 2.5–3.5 MHz with a step of 0.1 MHz and the amplitude of 5 V_{pp}. It should be noted that in each measurement, in total 7436 grid points are scanned.

The acoustic profile of the inner sectors is illustrated in Fig. 6.18a, b. The maximum acoustic intensity occurs at 1 mm distance at 3.5 MHz on the axial direction along the central axis of the S-MRUT, and it starts to decrease after the focal point. The lateral resolution is quantified by calculating the FWHM which is 550 μm. FWHM is a parameter for showing the lateral spatial resolution which it resembles where the acoustic intensity reduces to the half value.

Figure 6.18c, d shows the acoustic profiles of the outer sectors where the maximum acoustic intensity is on the axis along the S-MRUT at 2.5 mm distance at 3.5 MHz. Because of the larger activated area of the S-MRUT, the acoustic profile is distributed more than the inner sectors. The FWHM here is 500 μm.

Figure 6.18e, f shows the acoustic profiles of the S-MRUT when all sectors are driven by the same electric signal. In this case, the acoustic intensity is distributed with higher amplitude and a wider area compared to the situation when only one of the inner or outer sectors is driven. The maximum intensity is obtained at 3 mm on the central axis at the frequency of 3.2 MHz, where the FWHM is 600 μm.

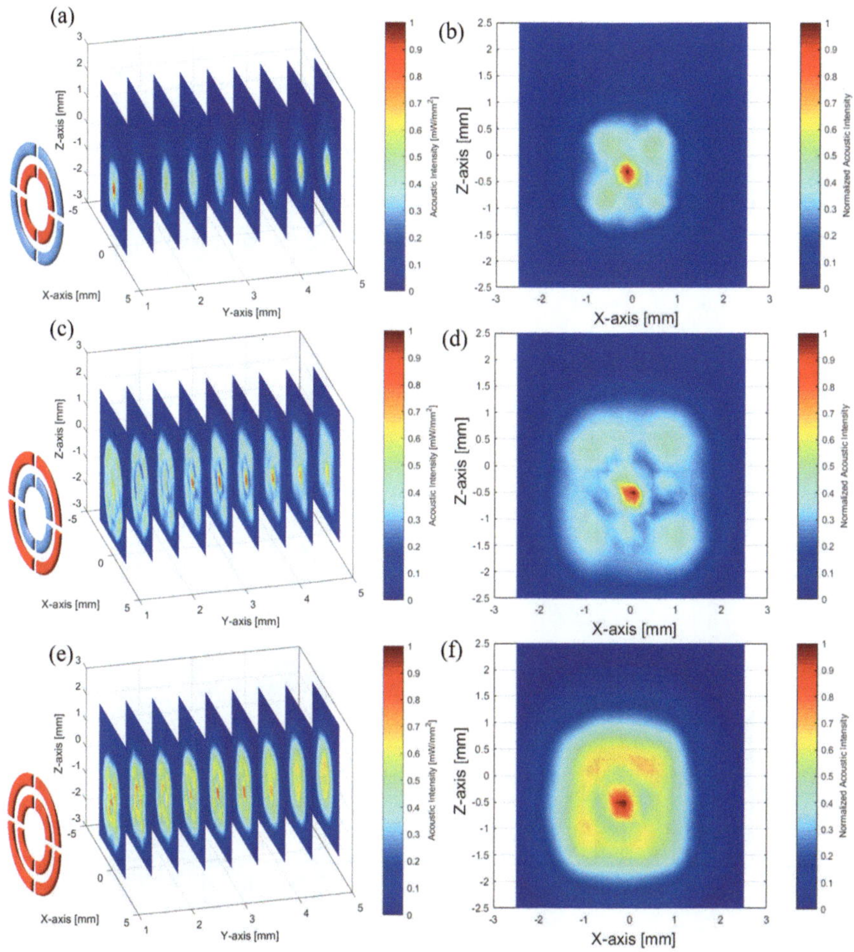

Fig. 6.18 Acoustic profile measurements of (**a**) and (**b**) the inner, (**c**) and (**d**) the outer, and (**e**) and (**f**) all active sectors. It should be noted that in the schematic of the S-MRUT near each measurement, red shows the active sectors, and blue is for the inactive ones [72]

Each sector at the same quarter is driven separately. Figure 6.19a illustrates the acoustic profile of when two sectors at the top left quarter are driven. In this case, the maximum acoustic intensity and FWHM are measured in 700 μm off-axis and 800 μm, respectively. In Fig. 6.19b, the two sectors on the top right quarter are activated, and they have the maximum intensity at 650 μm off-axis and the same FWHM as before. Figure 6.19c shows the two sectors in the bottom left quarter. In this case, the maximum intensity is at 1.1 mm off-axis with the same FWHM. Figure 6.19d shows the two sectors in the bottom right quarter giving the maximum intensity at 1 mm off-axis and the same FWHM; however, the acoustic profile for this setup is different from the others, and the focal point is at 1 mm. The reason for

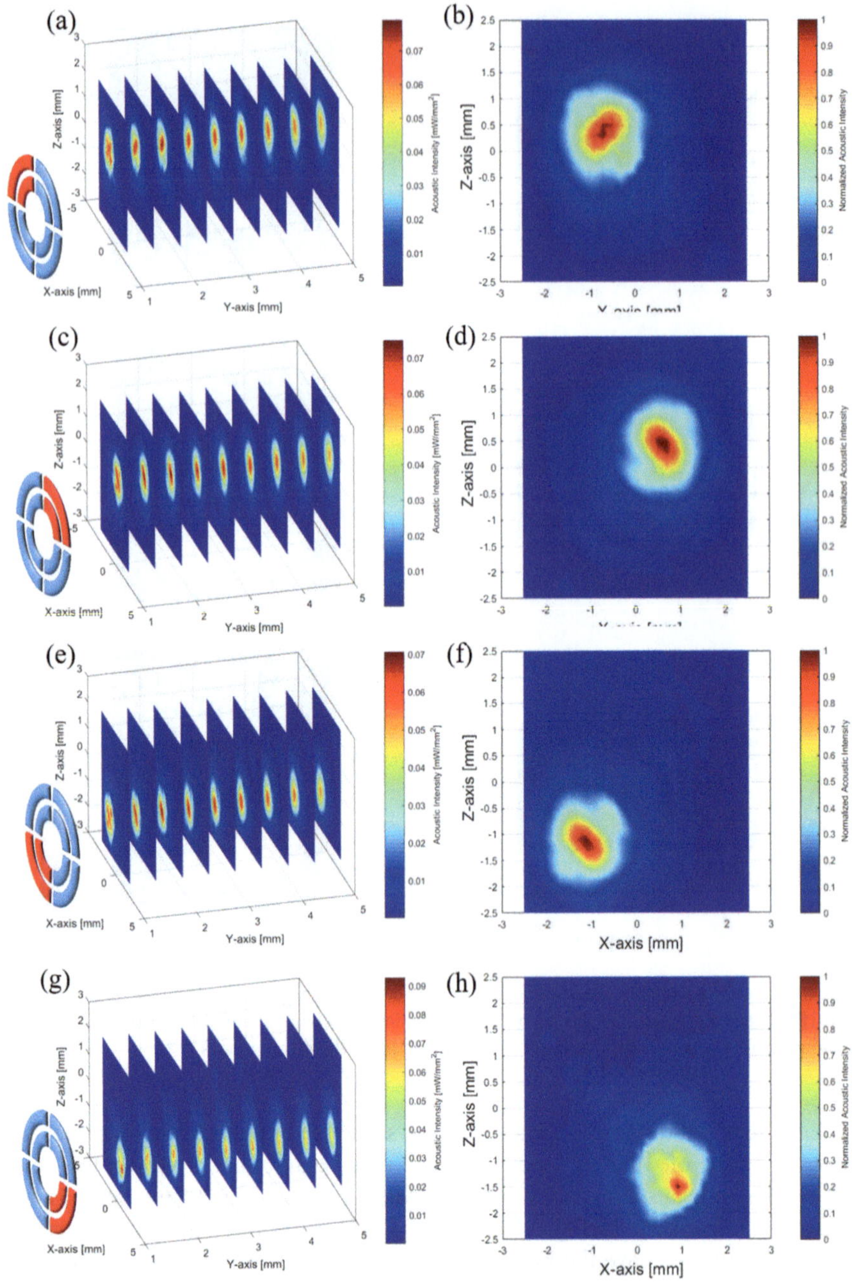

Fig. 6.19 The acoustic profile measurements for two sectors at the same quarter in (**a**) and (**b**) top left, (**c**) and (**d**) top right, (**e**) and (**f**) bottom left, and (**g**) and (**h**) bottom right. It should be noted that in the schematic of the S-MRUT near each measurement, red shows the active sectors, and blue is for the inactive ones [72]

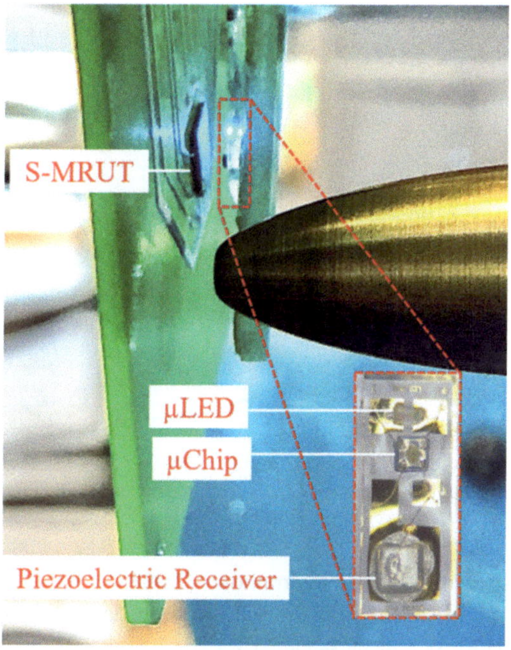

Fig. 6.20 Side view of the measurement setup. The hydrophone, the dust, and the S-MRUT are immersed in the water tank. μLED on the dust is lit up when the S-MRUT is delivering acoustic energy [72]

this discrepancy may come from the variation of the dimension of the sectors due to the etching variability in fabrication process. The maximum intensity for all of these four cases occurs at 3.5 MHz. Experimental results show that for the sectors at the same quarter of the S-MRUT, one can focus the acoustic waves off-axis, thereby increasing the steering area for powering the device.

Verification

To verify the S-MRUT for powering implantable devices, the S-MRUT and an implantable device are immersed in a water tank as shown in Fig. 6.20. The implantable device is glued on the PCB and is located in front of the S-MRUT. The inner sectors of the S-MRUT are driven with the signal generator with amplitude of 220 mV$_{pp}$ and the frequency of 2.935 MHz. The signal is amplified with a 50 dB RF power amplifier. Then, the acoustic intensity is calculated using the following equation:

$$I_{rms} = \frac{V_{pp}^2}{8 \times M_l(f)^2 \times Z_m} \tag{6.21}$$

where V_{pp} is the peak-to-peak value of the voltage of the hydrophone, $M_l(f)$ in $(\frac{V}{Pa})$ is the sensitivity of the hydrophone that can be determined with the

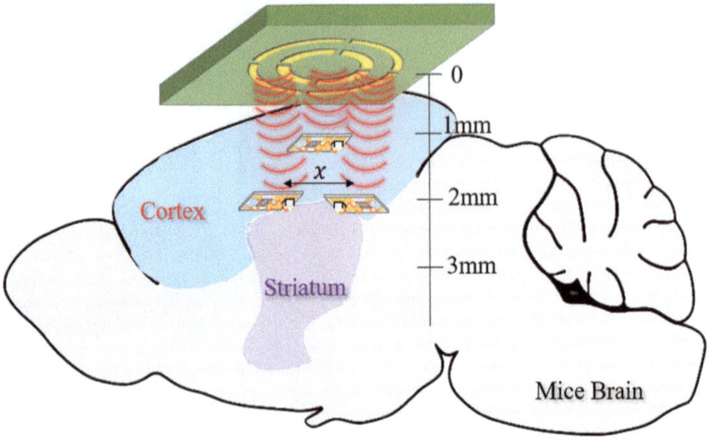

Fig. 6.21 Schematic of cross section of mice brain with only motor cortex and striatum [72]

datasheet of the hydrophone (2.0883 $\times 10^{-}6(\frac{V}{Pa})$ for the given frequency), and Z_m in ($MRayl$) is the acoustic impedance of the medium (1.5 $MRayl$ for water). Thus, the measured acoustic intensity for our S-MRUT is 9.9059 $(\frac{\text{mW}}{\text{mm}^2})$. Note that the FDA-allowed time-averaged acoustic intensity for body implants is 7.2 $(\frac{\text{mW}}{\text{mm}^2})$, although duty cycling the burst signals would allow to send stronger acoustic waves.

As shown in Fig. 6.20, the implantable device is capable of driving a μLED using the harvested ultrasonic power.

As an example of how the S-MRUT works to selectively power the implantable device, let us consider the schematic shown in Fig. 6.21. In mice, the motor cortex starts from 0.5 mm below the skull, and the striatum is at around 3 mm below the skull. Given the S-MRUT with focal distance from 1 mm to 3 mm, transferring ultrasonic power to the device implanted in the cortex or on the surface of the striatum is doable. To vindicate the proposed S-MRUT, three implantable devices are considered as shown in the figure. Two of them are separated by a distance x on the surface of the striatum, and one is located in the cortex. Referring to our simulation and experimental results, each sector in a quarter of the S-MRUT has 0.862 mm off-axis focal point on average. In other words, two implantable devices with maximum 1.725 mm distance can be powered with two excited sectors in different quarters. Furthermore, another implantable device in depth of 1 mm can be excited with the innermost sectors.

Since acoustic intensity in the human brain will be attenuated due to the absorption and reflection of the wave when it passes through the tissues, the S-MRUT is simulated in brain tissue and compared the result with the water domain. Figure 6.22 is given the result of the normalized acoustic intensity on the focal point for the inner excited sectors. As shown in the figure the acoustic intensity is reduced by 15.7% through the brain tissue for the given frequency.

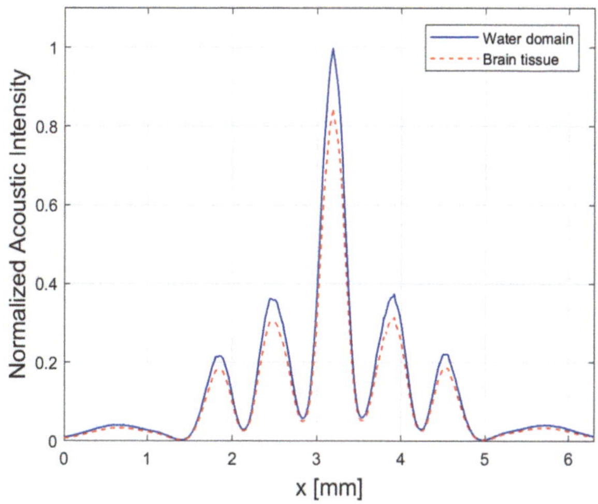

Fig. 6.22 Comparison of acoustic intensity on the focal plane into water domain and brain tissue [72]

6.2 Circuits for Transmitter Side (External Device)

The counterpart to the implantable device is the Ultrasonically Powered Implantable Base or external device, shortly say "UPIB," that is placed outside the body on the surface of the skin.

The external device though is placed outside of the body and thus has more space and volume to work with yet still needs to take the intended host into consideration (Fig. 6.23).

An early rudimentary block diagram only covering the Basic UPIB functionality can be seen in Fig. 6.24, which is a battery-powered device with minimal circuitry for controlling and driving the transducer. In the following, an example UPIB is detailed and described from the early testing of the concept to full-blown integrated solutions that enables significant miniaturization of the UPIB designs.

6.2.1 Transducers

Initially when performing experiments with submerged piezoelectric crystals or implantable prototypes in the water tank setup, industrial transducers driven by a signal amplifier setup serves as the acoustic power source. The industrial transducers are, however, extremely bulky and relatively heavy and not at all suitable for the portable applications. For this reason, a new and significantly more compact solution is needed, one that can be scaled down to a size that can be accommodated on the

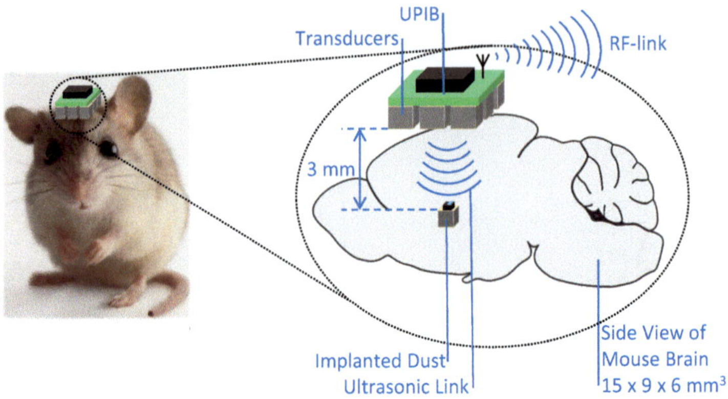

Fig. 6.23 Conceptual sketch of the implanted dust powered by an external device, referred to as an UPIB in the project, that contains piezoelectric crystals used as transducers for sending ultrasonic power to the dust

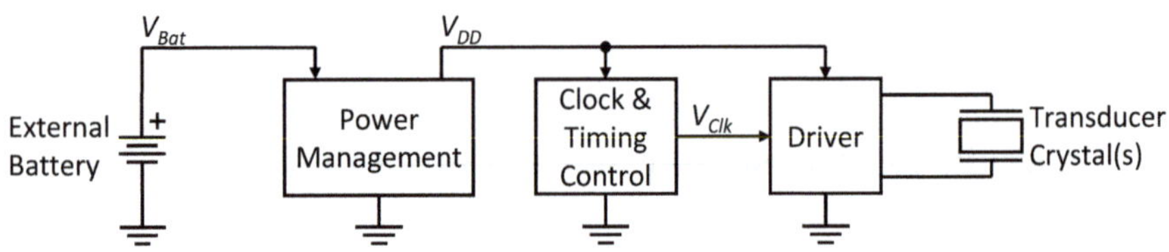

Fig. 6.24 Early basic UPIB block diagram

body (top of the head in brain applications), preferable with the driving electronics integrated into the solution.

6.2.2 *Transducer Prototype for Transmitter Side*

Here, the integrated implantable (dust) prototypes are based on $(500\,\mu m)^3$ crystals, which are responsive in the range of 2.8 MHz–3.0 MHz, and the transducer design needs to match that. A new design taking the aforementioned requirements into account can be seen in Fig. 6.25 which shows a prototype transducer based on a thin flexible PCB with four $(500\,\mu m)^3$ crystals positioned in a two-by-two grid pattern. Including the thickness of the copper layer, the flexible PCB is 170 μm thick and on top sits the crystals with a thickness of 500 μm glued to the PCB with silver epoxy to connect the bottom electrode. The top electrode is bonded with 17 μm in diameter gold bond-wire where the other end is glued to the PCB with silver epoxy as well. The whole structure is covered and protected by a layer of PDMS that must cover the top of the crystal with a minimum of 0.5 mm thick layer, resulting in a recommended minimum total thickness of about 1.2 mm.

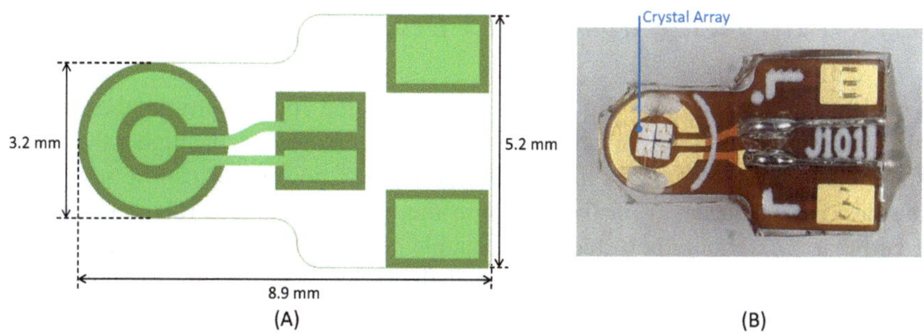

Fig. 6.25 (**a**) Layout of the flexible PCB design with dimensions. (**b**) Picture of the assembled transducer based on four (500 μm)^3 PZT-4 piezoelectric crystals, all protected by a PDMS coating [75]

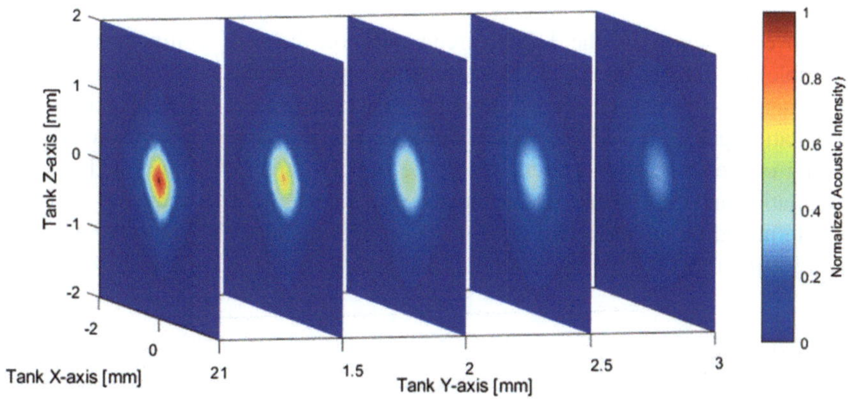

Fig. 6.26 Normalized acoustic intensity output of transducer prototype in a volume, at a distance of 1–3 mm from the surface of the PDMS (Tank Y-axis)

In Fig. 6.26 is shown a scan of the normalized acoustic intensity output of a transducer prototype scanned in a volume at a distance of 1–3 mm, in 0.5 mm steps from the surface of the PDMS. The surface of each scanned slice in the volume is 4 mm by 4 mm, with a grid spacing of 0.2 mm. The transducer was driven by a pure sine wave with an unloaded amplitude of 10 Vpp at a frequency of 2.9 MHz. What the scan shows apart from the expected drop-off in intensity with distance is that there is a line of maximum intensity orthogonal to the plane of the transducer. The exact intensity at a given point is directly dependent on the driving voltage and frequency applied to the transducer, which will be explored in more detail during experiments performed on the UPIB driver circuits detailed in the following subsections. Most importantly, though, it is that the transducer prototype is deemed small enough to be carried by, for example, mice, making the design suitable for that category of animal experiments.

6.2.3 *External Electronic Device Design*

With the initial design of the transducer settled, the focus shifts to the driving electronics. Support for relatively high driver voltage is needed, the higher the better since this directly influences the amount of acoustic power that can be sent from the driven transducer. From these criteria the UPIB driver is a robust electronic board that is externally powered. The example design is built entirely by the use of standard commercially available off-the-shelf devices and is built around simple manual control through trimmers (variable resistors) to control settings such as driving frequency and stimulation timings.

Circuit Design of External Device with Off-the-Shelf Components

The overall design of the UPIB circuit with off-the-shelf components (as an example) is sketched out in the block diagram shown in Fig. 6.27, which includes the connection of the transducer although it is separate from this PCB board. The application guides the design specification regarding how the specification of the timing of the stimulation pattern should be the ON- and OFF-time for the active stimulation and thus when to drive the attached transducer. The frequency however is determined by the choice of crystal used in the implantable prototype, which is the $(500\,\mu m)^3$ PZT-4 (in this example) that puts the resonance frequency at about

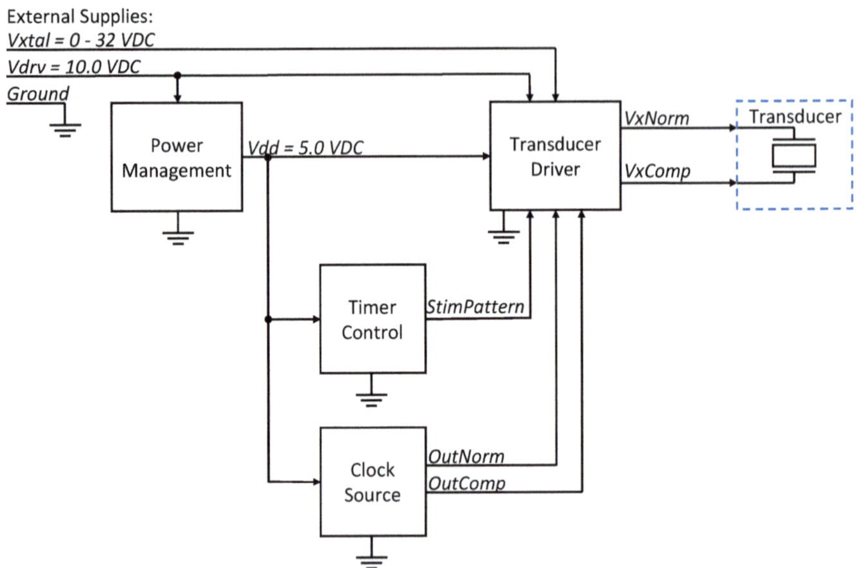

Fig. 6.27 Overall block diagram of the Basic UPIB Driver Part, with the externally connected transducer added for reference to show the context of the system

Fig. 6.28 Picture of Basic UPIB driver PCB, with connectors and variable resistors highlighted

2.8–3.0 MHz, but room for further variance needs to be taken into account in the design.

Figure 6.28 shows a picture of one of the UPIB prototypes with off-the-shelf components, with added info about the connectors and trimmers for tuning the settings of the driver output. The frequency and stimulation pattern timers are however easily reconfigurable by changing a few resistors in the design. One last detail in the final specification is the operating voltage range of the output driver circuit of 0–32 VDC, allowing for great flexibility in the resulting drive strength of the circuit. In the following the individual blocks in the block diagram are detailed further with full schematics and info on how to reprogram the range of any of the variable parameters.

Clock Source

This block controls the clock frequency with which the *Transducer Driver* circuit drives the connected transducer and is designed around the LTC6908-1 (*U101*). The LTC6908-1 is a precision oscillator and has two output clocks with a 50% duty cycle, a normal phase (signal *OutNorm*, phase = 0°), and a complementary phase (signal *OutComp*, phase = 180°). The clock frequency is set according to Eq. (6.22).

$$\text{frequency} = \frac{100 \times 10^9}{R101 + R102} \tag{6.22}$$

With the current component values of *R101* and *R102*, the frequency is adjustable in the range of 2.50 MHz–3.33 MHz theoretically, which is set by the variable resister *R102*, which is found to be very close to the measured results listed later. A schematic of the circuit can be seen in Fig. 6.29.

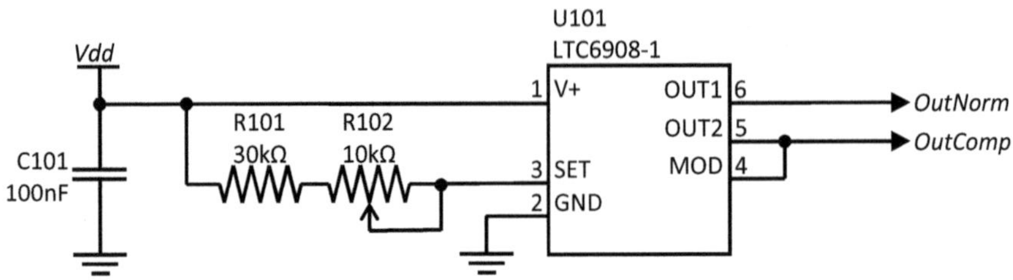

Fig. 6.29 Schematic of the *Clock Source* circuit of the Basic UPIB

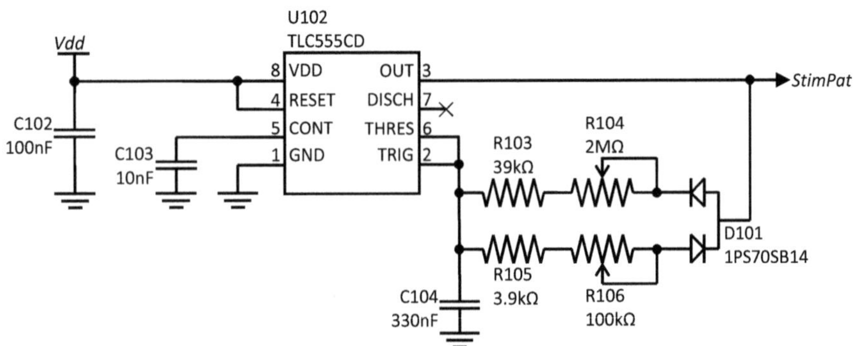

Fig. 6.30 Schematic of the *Timer Control* circuit of the Basic UPIB

Timer Control

The stimulation pattern of the *Transducer Driver* is controlled by separately controlling the drivers' ON- and OFF-time. The circuit is designed around a TLC555CD (*U102*), which is a well-known analog timer controlled by the charge and discharge time of a connected capacitor, here *C104*. Unlike typical setups of this timer, the charge and discharge current paths are separated by the addition of the double diode *D101*, which ensures the ON- and OFF-time can be set independently of each other. A schematic of the circuit can be seen in Fig. 6.30, and the ON- and OFF-time are set according to Eqs. (6.23) and (6.24), respectively.

$$T_{stimON} = (R105 + R106 + 200\Omega) \times C104 \times ln\left(\frac{3 \times V_{D101} - 2 \times Vdd}{3 \times V_{D101} - Vdd}\right) \tag{6.23}$$

$$T_{stimOFF} = (R103 + R104 + 200\Omega) \times C104 \times ln\left(\frac{3 \times V_{D101} - 2 \times Vdd}{3 \times V_{D101} - Vdd}\right) \tag{6.24}$$

V_{D101} is the diode voltage drop which is assumed to be around 0.3 V given the low currents present in this circuit. With the current component values, the ON-time is adjustable in the range of 1.08–27.39 ms theoretically, set by trimmer *R106* which is later found to be very similar to measurement results. Similarly, the OFF-time is adjustable in the range of 10.31–536.5 ms theoretically, set by trimmer *R104*. However, while the minimum OFF-time is very similar to previously mentioned measurement results, the maximum OFF-time differs from practical measurements. The reason for this deviation is believed to come from the fact that when *R104* is set for the full 2 MΩ, then the current through it is around 0.7–1.5 μA theoretically, making the circuit very susceptible to any stray leakage currents such as the reverse bias leakage of the Schottky diodes in *D101*.

Transducer Driver

The main driver circuit is designed around the concept of an H-bridge in order to give the *Transducer Driver* the biggest drive capability from a limited supply voltage, since this kind of configuration effectively drives a load with double the voltage swing of the supplying voltage level. This kind of setup is commonly used for motor drivers, and for this reason the UCC20520 is used, which is a dual-channel high-side/low-side isolated gate driver with a single clock input capable of operating at up to 5 MHz. The four primary switches (*Q201/A,B* and *Q202/A,B*) driven by the gate drivers are dual N-channel metal-oxide-semiconductor field-effect transistors (MOSFETs) with an ON-resistance of 23 mΩ at $V_{GS} = 10$ V, giving the driver a very low output impedance.

The H-bridge consists of two identical sides with the only difference being the clock signal from the Clock Source, where one side is driven by the normal phase signal (*OutNorm*), and the other is driven by the complementary signal (*OutComp*). The overall stimulation pattern is controlled by the signal *StimPattern* from the *Timer Control* block, which is common for both sides of the H-bridge. The outputs of the two sides are the signals *VxNorm* and *VxComp* respectively, and the external transducer is connected between them via the connector *J201*. Figure 6.31 shows the schematic of the full circuit of this block.

Power Management

Overall, the example system needs three different internal supply voltages:

- *Vxtal* = 0–32 VDC, a variable supply for the H-bridge in the *Transducer Driver* that directly sets the circuit drive strength.
- *Vdrv* = 10.0 VDC, a fixed supply needed by the driver circuitry in the *Transducer Driver* block.
- *Vdd* = 5.0 VDC, a fixed common internal supply needed by all controlling integrated circuits (IC) in the different blocks.

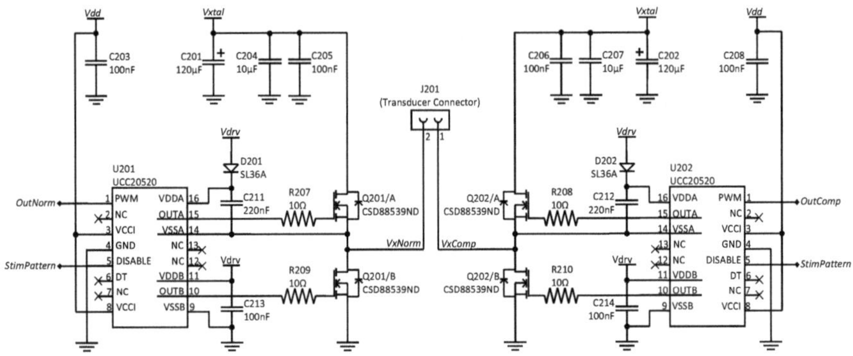

Fig. 6.31 Schematic of the *Transducer Driver* circuit of the Basic UPIB

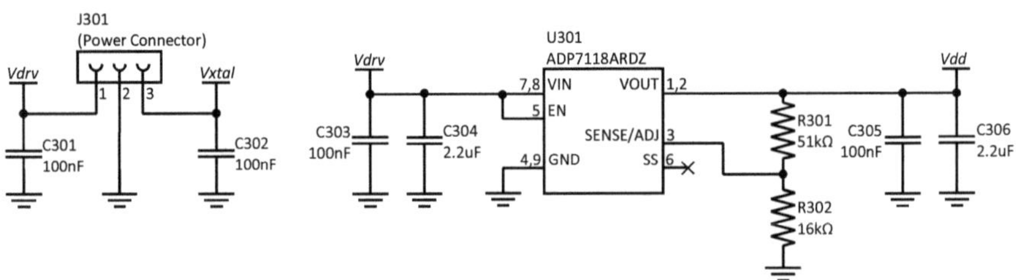

Fig. 6.32 Schematic of the *Power Management* circuit of the Basic UPIB

However, in an effort to reduce the number of needed external DC supply channels, the supply *Vdd* is internally generated from *Vdrv*; thus only two supply channels are needed instead of three. The two external supplies are connected to the circuit via connector *J301* seen schematically in Fig. 6.32, which also shows the circuit based around a Low Drop-Out (LDO) linear regulator that generates the required 5.0 VDC for the internal controller ICs.

Regarding *Vxtal*, the variable voltage of this supply directly governs the drive strength of the H-bridge in the *Transducer Driver* block and thereby indirectly controls the strength of the acoustic output of the connected *Transducer Part*. Furthermore, the individual components interfacing *Vxtal* are all at minimum rated for 50 VDC as a precaution; however, the circuits have only been tested up to 32 VDC (the limit of the external power supply used for these experiments).

6.2.4 Experimental Results

Testing out the UPIB driver with off-the-shelf components capabilities is split into two parts, a simple functional test and a more systematic experiment with multiple transducer prototypes. The functional test consists of connecting up all parts, as

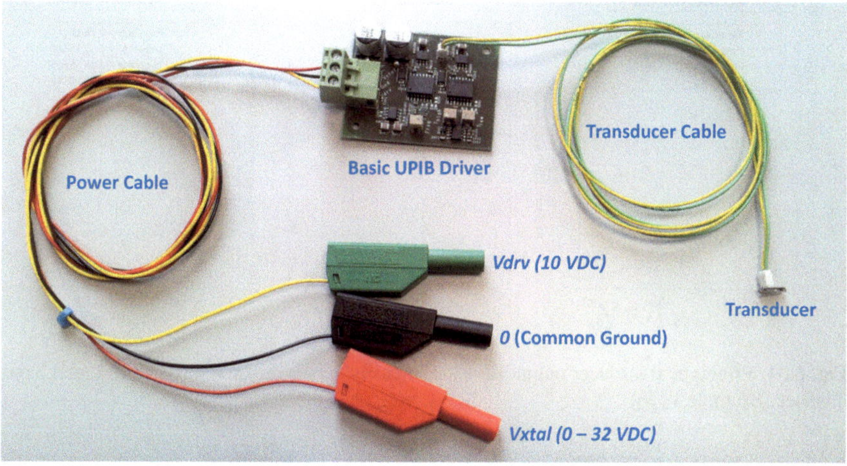

Fig. 6.33 Picture showing the full setup of the Basic UPIB system prototype, with all parts connected by the relevant cables [76]

depicted in Fig. 6.33 where the full UPIB system is prepared and ready to be plugged into a power supply, and then the ranges of the frequency and stimulation timers are measured. The results listed here are based on five assembled prototypes and the resulting average specifications have been measured to be:

- Driver frequency range: 2.51–3.31 MHz
- Stimulation pattern ON-time range: 0.99–24.7 ms
- Stimulation pattern OFF-time range: 9.65–343.6 ms

The second part involves modifying one prototype UPIB driver with off-the-shelf components, where the internal frequency generator and timers are disabled and bypassed, and instead the clock inputs of the *Transducer Driver* circuit are directly controlled by a signal generator. By doing this, it is possible to set up a fully automated test in the water tank for not only testing the drive strength of the UPIB itself under as real operating conditions as possible but also testing and characterizing prototype transducers such as the one shown previously in Fig. 6.25. In Fig. 6.34 is shown an example of such a transducer characterization where the approximate acoustic intensity output versus driver voltage (*Vxtal*) is plotted at 1, 2, and 3 mm distance in demineralized water. The characterizations were performed at four different frequencies, 2.75, 2.80, 2.85, and 2.90 MHz, and the driver voltage swept from 0 to 32 VDC. This characterization of the transducer gives insight into its efficiency for the purpose of helping to target a specific acoustic intensity by adjusting the driver voltage accordingly.

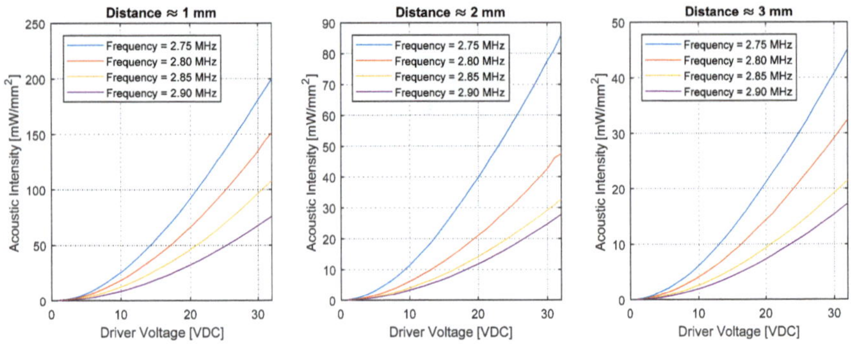

Fig. 6.34 Prototype transducer output acoustic intensity versus driver voltage, at 1, 2, and 3 mm distance. Example 1 [76]

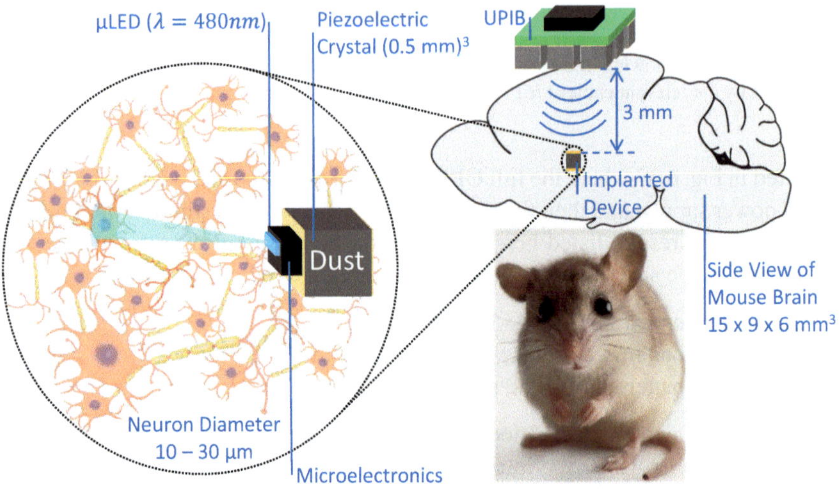

Fig. 6.35 Conceptual sketch of a neuron illuminated by an implanted dust, consisting of a piezoelectric crystal with microelectronics (chip) and LED mounted on top, which in turn is powered by the ultrasonic waves coming from the external host transducer array (UPIB) placed on the surface of the skin [76]

6.3 Implantable Device

6.3.1 Power Conditioning

The implantable device from here on also referred to "dust" since in the application of this case study is brain implants, so it must be small enough to be implantable in the brain of mice (approximately $(500\,\mu\text{m})^3$), as sketched in Fig. 6.35. The functionality of the dust is limited to optogenetics only for the sake of simplicity and showing only the energy harvesting concept [77]. For the dusts to be viable for this, a millimeter- or submillimeter-sized ultrasonically powered integrated deep brain implantable dust, for depths of several centimeters below the skin, is envisioned. The dust itself is built around a piezoelectric crystal for receiving the ultrasonic power, a CMOS chip for power management, and an LED to provide the light for optogenetics.

For practical reasons, the crystal used during testing is a $(1\,\text{mm})^3$ cube, and simulations reflect this, but with an eye on shrinking that down to $(500\,\mu\text{m})^3$ or even as small as $(200\,\mu\text{m})^3$ if an advanced integration is possible. Taking the surface area of a $(500\,\mu\text{m})^3$ crystal as a starting point results in the restriction of the largest allowable size of CMOS chip designs, which has been set to a maximum layout area of 300 μm by 300 μm.

6.3.2 Active Bridge Rectifier

Interfacing and loading the crystal will be a power management block, consisting of a rectifier with storage capacitors and the loads connected on the rectifier's output. Although the load for this dust is simply powering an LED, it is given that future prospects of integrating circuitry into the system will require a DC supply, and thus a rectifier with high efficiency is needed. The simplest form of a bridge rectifier is the commonly known passive diode-based version, but its primary drawback is, however, the relatively high voltage drops across the diodes causing excessive power loss. An alternate and increasingly popular strategy for wireless devices is to replace the diodes with an active circuitry, known as an active bridge rectifier [78–82]. Given the acoustic intensity food and drug administration (FDA) limit of $7.2\,\text{mw/mm}^2$ and an expected crystal size in the range from $(500\,\mu\text{m})^3$ to $(1\,\text{mm})^3$, the total power available will be in the milliwatt range. Also, a high operating frequency range of 1–5 MHz is expected as a direct result of the physical dimensions and material properties of the crystal itself. Finally, the CMOS area of the complete design is restricted to an area of 300 μm by 300 μm on the die, as previously stated, where it will be stacked on top of a crystal together with a μLED. This severely limits the amount of room for on-chip capacitance (to approximately 100 pF in the end for 180 nm CMOS technology), which is all used as the local storage capacitor ($Cstorage$).

In the following subsections, the active bridge rectifier chip design will be detailed. The work and the results based on this chip have also been published in [30].

6.3.3 Integrated CMOS Design

The designed active bridge rectifier follows the standard design [79, 80], with few exceptions described shortly. The full transistor-level schematic of the entire chip

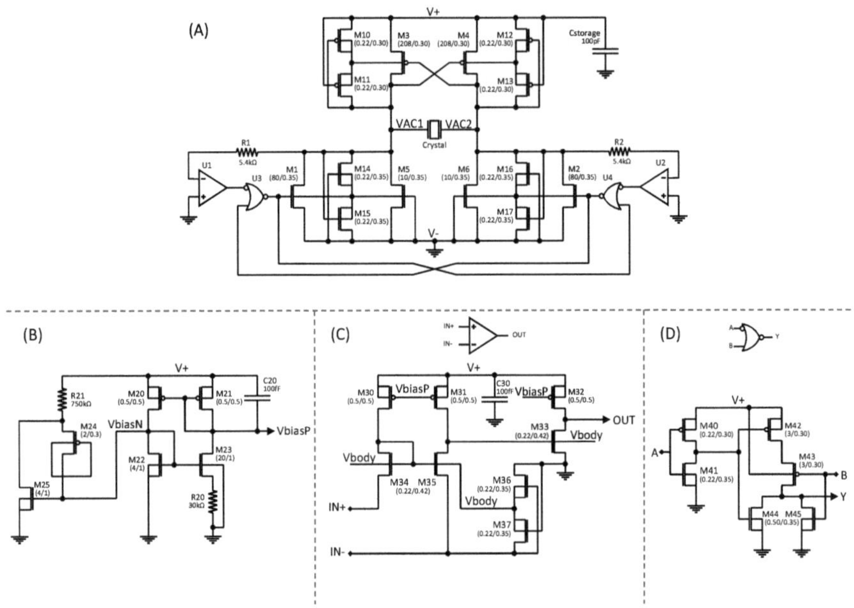

Fig. 6.36 All transistor dimensions are (width/length) in micrometers. (**a**) Active bridge rectifier, including the crystal to illustrate its circuit connections, (**b**) current reference, with start-up circuit, common for the OpAmps, (**c**) common-gate two-stage OpAmp, and (**d**) NOR gate with the A input inverted [30]

design can be seen in Fig. 6.36, and all component references in this subsection refer to this. Only the nominal 3.3 V N-channel metal-oxide-semiconductor (NMOS) and P-channel metal-oxide-semiconductor (PMOS) transistors are used from the standard Taiwan Semiconductor Manufacturing Company (TSMC) 180 nm CMOS technology.

The main current paths in the active bridge design are four transistors, the actively driven *M1* and *M2* and the cross-coupled *M3* and *M4* seen in Fig. 6.36a. *M5* and *M6* are diode coupled and bypass *M1* and *M2* during start-up before the active circuit is up and running, and the eight body-biasing transistors *M10–M17* are there to protect against latch-up. In Fig. 6.36a, the operational amplifiers (OpAmp) *U1* and *U2* are used as static comparators (instead of using dynamic comparators [83]) that monitor the signal levels of the crystal outputs *VAC1* and *VAC2* with respect to common ground and set the inputs of the following logic gates *U3* and *U4* that drive the two main NMOS switches *M1* and *M2*, which all combined forms the two active diodes of the system. The comparators *U1* and *U2* are built around an OpAmp with a common-gate input configuration as the first stage, with a second stage to further boost the gain, as seen in Fig. 6.36c. The individual OpAmps are biased to use a total of 6 μA at 2.5 V supply voltage to keep power usage low, at which point they have a unity gain bandwidth of 1.93 GHz and an open-loop gain of 74.6 dB. While this simple two-stage OpAmp benefits from high bandwidth and gain, it lacks the possibility of making a symmetric layout which leads to uneven parasitic components between the plus and minus inputs. These parasitic components unfortunately give rise to offset errors between the two inputs. Fortunately, they are visible in post-layout simulations and, in this particular case, can be countered sufficiently by adding a series resistor to the minus input, which leads to the addition of *R1* and *R2* as compensation. The OpAmps also share a common resource in the form of a conventional current reference circuit, seen in Fig. 6.36b. The current reference includes a start-up circuit comprised of *M24*, *M25*, and *R21*, which are based on the *Accelerated Leakage Component* concept detailed in [84].

One notable difference from the standard design of active bridges is that the bridge shown in Fig. 6.36a includes two NOR gates with one input inverted, *U3* and *U4*. They were originally two NOT gates in series that acted as gate drivers of the main NMOS switches *M1* and *M2*, respectively. It was discovered during simulation of the circuit, using the Leach model as a power source, that if the bridge output was relatively heavily loaded compared to the input power, then an unfortunate start-up situation could occur, where both main NMOS switches are driven semi-ON. This situation basically shorts the crystal and thereby significantly delays complete start-up of the active bridge. The reason behind this lies in the comparators *U1* and *U2* being unstable until the internal current reference stabilizes, thereby a risk occurs where they both get "stuck" outputting a logic high to subsequent gates. The two NOR gates ameliorate this problem by forming a set-/reset-latch (SR-latch) that is incapable of outputting a logic high on both outputs at the same time and instead simply holds its current output state when both inputs go high. The issue is demonstrated in simulations detailed in the next subsection.

A second notable thing about the design is the lack of any overvoltage protection circuitry between the rectified voltage ($V+$) and common ground ($V-$), thereby negating and saving any power consumption spent by such a circuit. The overvoltage protection is necessary for ultrasonically powered microsystems [85]. The primary reason for this is directly bound to two things: (1) The deliberate choice to only use 3.3 V tolerant thick-gate devices in the standard TSMC 180 nm CMOS technology throughout the entire CMOS design. (2) The intended load of the active bridge is a μLED just like or similar to the one shown and characterized in Fig. 6.37. Based on the current/voltage (IV) curve in Fig. 6.37b, it will take near 19 mA to reach the CMOS limit of 3.3 V, which amounts to about 63 mW that far exceeds any power requirement or results ever achieved during any experiments. The μLED prototype is designed by Tyndall National Institute and emits light at a wavelength of approximately 480 nm.

The final CMOS transistor layout of the active bridge rectifier can be seen in Fig. 6.38, which shows the overall design including the locally placed storage capacitors totaling approximately 100 pF, the four bonding pads, and all of it fitted within the limited area of 300 μm by 300 μm as specified earlier.

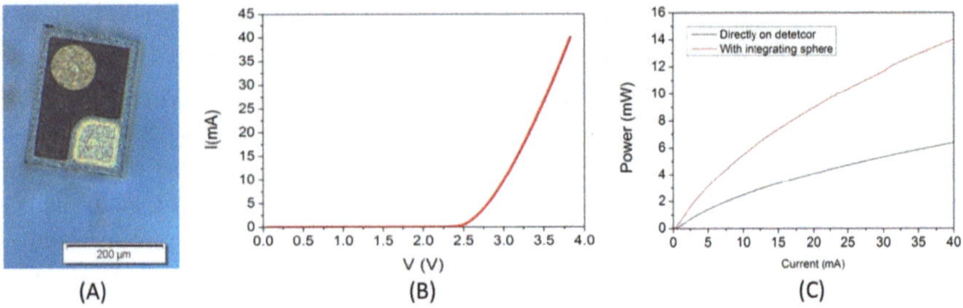

Fig. 6.37 (**a**) Picture of the LED with dimensions of 280 µm by 180 µm and a thickness of 100 µm. (**b**) Current versus voltage of the LED. (**c**) Optical output power versus current [86]

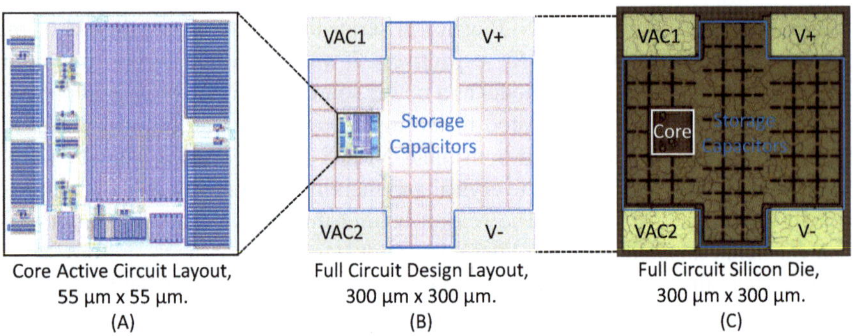

Fig. 6.38 (**a**) Zoom in on the core transistor design of the active bridge rectifier layout. (**b**) Overview of the full active bridge system layout, including bond pads and locally placed storage capacitors totaling approximately 100 pF. (**c**) Close-up picture of the CMOS design implemented on silicon die [30]

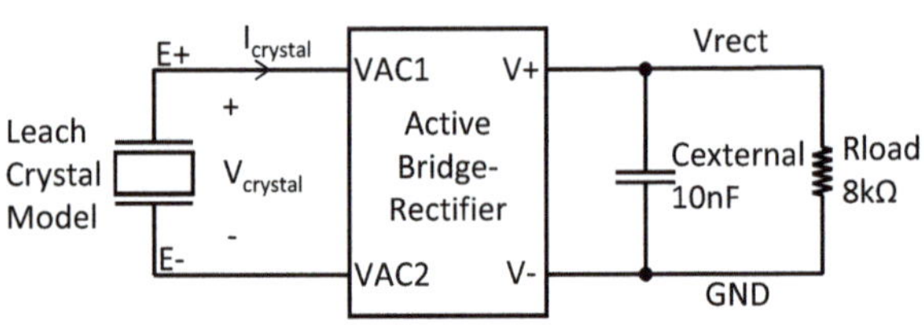

Fig. 6.39 Principle test setup for simulations using the Leach model as the crystal power source for the active bridge rectifier and load [30]

6.3.4 *Simulation Strategy and Results*

During the design phase of the active bridge, the Leach model of the crystal is used as the power source of the system during simulations, as depicted in Fig. 6.39 plus loading of the output. When the model is used in transient mode, as is the case here, it always starts from zero power given the way it is implemented with power initially having to interface through one of the acoustic ports before being

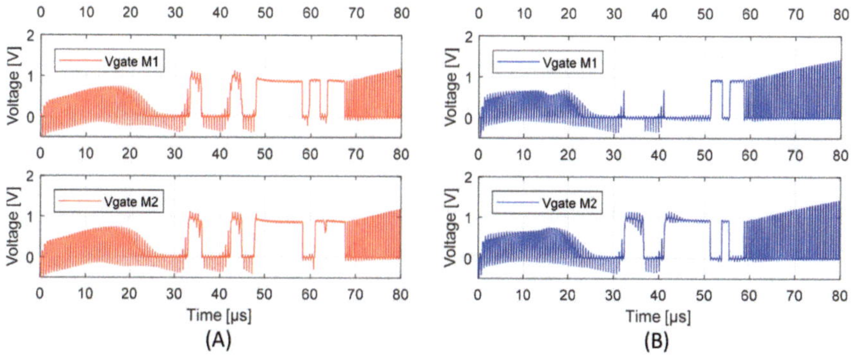

Fig. 6.40 Example simulation result observing the rectified voltage (*Vrect*) and the gate voltage of the two NMOS switches *Vgate M1* and *Vgate M2* during start-up, with respect to common ground. (**a**) Without the NOR gate functionality. (**b**) With the NOR gate functionality implemented [30]

available on the electrical port between $E+$ and $E-$. It was observed during start-up that one or both of the main NMOS switches (*M1* and *M2* in Fig. 6.36) were being driven semi-ON. This state of the switches effectively short-circuits the crystal when it occurs simultaneously to both switches, and this is what leads to the proposed circuit and the addition of the extra gates *U3* and *U4*. Two brief simulation examples of the main NMOS switches of the active bridge getting "stuck" ON at around 1 V during start-up can be seen in Fig. 6.40. Both examples show the internal gate voltages of *M1* and *M2* in the chip along with the rectified voltage (*Vrect*) that also powers the internal transistor circuits with respect to common ground (*GND*). Figure 6.40a shows the response of the design without the functionality of *U3* and *U4* (they function only as buffers here) and how the two main switches are randomly driven ON together most of the time, preventing *Vrect* from rising above 1V and thereby unnecessarily delaying chip start-up. Figure 6.40b shows the response of the design with the functionality of *U3* and *U4* implemented, where the situation is improved by the latching function of the NOR gates and proper start-up happens quicker. In the example described here, the overall circuit is configured as shown in Fig. 6.39, where the crystal is configured and driven at a frequency of 1.7 MHz.

Preference of choosing the Leach model over the KLM model for transient simulations is best illustrated by a comparison of the two under the same test conditions. For this comparison, the overall circuit is still the one shown in Fig. 6.39, with the active bridge circuitry based on the post-layout parasitic extracted model, and both crystal models with the frequency set for 1.7 MHz. This results in both crystal models of having an open circuit voltage and impedance of $V_{OC} = 4.28\,\text{V}$ and $Z_{Crystal} = 2.55\ k\Omega + j4.45\ k\Omega$, respectively. Where the Leach model is left as is, the KLM model needs to be simplified to its Thévenin equivalent model for transient analysis. While the Thévenin KLM model has the same open circuit voltage as the Leach model (V_{OC}), the series impedance is further simplified to only contain the real part of $Z_{Crystal}$, since this has been found to the closest

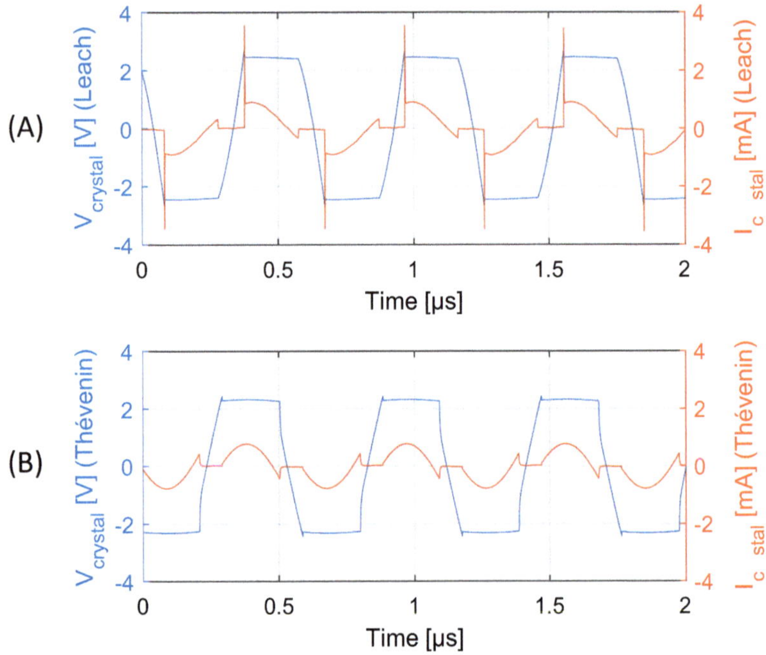

Fig. 6.41 (**a**) Leach model, (**b**) KLM Thévenin model, with simulated voltage (blue) and current (red) curves versus time, of the respective models. Both models are loaded by the same active bridge model created from parasitic extraction of the layout, with the resistor *Rload* and external storage capacitor *Cexternal* on the bridge output, as illustrated in Fig. 6.39 [30]

resemble reality. Yet the two results shown in Fig. 6.41 are starkly different, and later experimental results will confirm that the response of the Leach model (Fig. 6.41a) is strikingly similar to real measurements, confirming the choice.

The average output power of the Leach crystal model in the simulation that Fig. 6.41 is taken from is only 771 μW out of a theoretical maximum of 900 μW. It is believed that this reduction mainly comes from the fact that no matching capacitor has been added in series with the crystal, which potentially could increase the electrical power output of the crystal [87]. However, 712 μW of power is delivered to *Rload*, which in turn gives a chip efficiency of approximately 92% for the simulation, which later will be shown to be very similar to the results of the experimental results in the next subsection.

6.3.5 Experimental Work and Results

Creating a setup for experimentally testing the fabricated chip and the dust as a whole comes with many challenges, not the least of which are reliably measuring

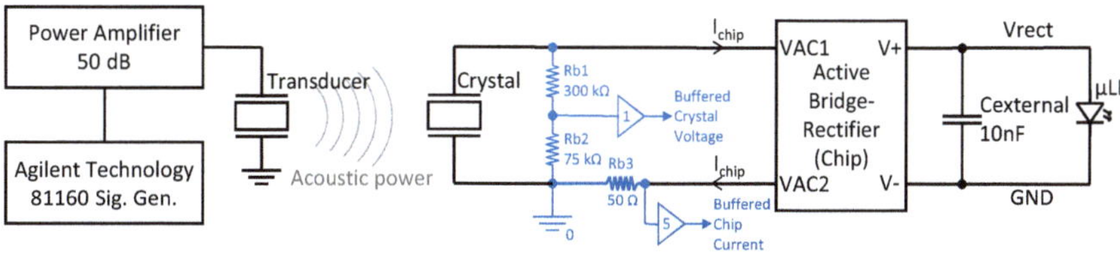

Fig. 6.42 Block diagram of the experimental test setup used for measurements [30]

system power levels before and after the chip. The test setup created is built up around the previously described *Base Board* with a (1 mm)^3 PZT-4 crystal, which functions as the chip power source. On the chip output, the blue μLED shown previously in Fig. 6.37 is used as the load, along with *Cexternal* that adds an additional 10 nF capacitor as external storage. *Cexternal* is the result of a compromise implemented during experiments to suppress the chip output ripple because of adverse problems with measuring the output power reliably. To generate the acoustic power for driving the crystal of the dust system, an industrial V303-SU transducer driven by a waveform generator through a 50 dB power amplifier is used. Figure 6.42 shows the primary components of the experimental setup, which includes a group of additional components shaded in blue that are put in between the crystal and the chip. The blue components serve as locally placed buffer circuits used when measuring the voltage and current of the crystal and the chip. The buffers insulate the sensitive dust circuit from the oscilloscope, as it was found that connecting probes directly to the system's input side caused unwanted loading and disturbed the measurements severely.

The complete experimental test setup is shown in Fig. 6.43a that includes a water tank filled with demineralized water wherein the ultrasonic power transfer happens. Figure 6.43b shows a close-up of the PCB boards that holds the chip under test and the crystal placed in front of the industrial transducer. In the background is also seen the hydrophone with a preamplifier used for calibrating the acoustic intensity. Before testing can commence at a given frequency, the hydrophone is put at the intended position of the crystal, where the acoustic intensity is then tuned to the desired level before swapping the crystal back into place and testing commences. During the experiments described in the following, the acoustic intensity and the frequency are set for 7.2 mW/mm^2 and 1.38 MHz, respectively. The frequency was experimentally found to be the optimum frequency for harvesting power from the (1 mm)^3 PZT-4 crystal.

Measured Input Power

Through the buffers on the input side of the chip, the voltage and currents are measured with the oscilloscope, and thus the power passing from crystal to chip can be determined. Taking the loss in the resistors *Rb1*, *Rb2*, and *Rb3* in the buffer circuit

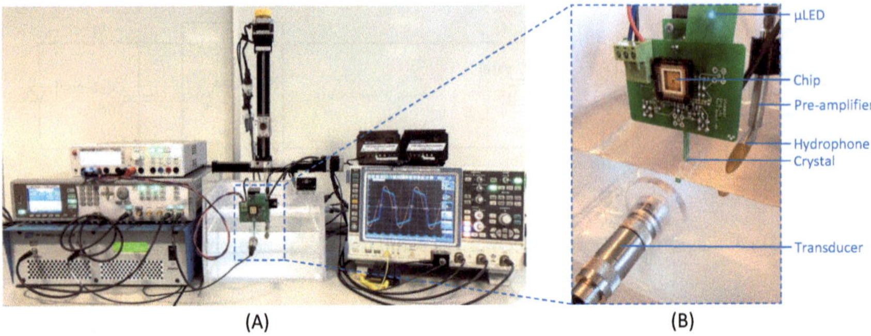

Fig. 6.43 (**a**) Picture of the experimental test setup used for measurements. Test equipment shown is Agilent Technologies 81160A Pulse Function Arbitrary Generator, Electronics & Innovation RF Power Amplifier 50 dB, Rhode & Schwarz RTO 1044 Oscilloscope, Rhode & Schwarz Programmable Power Supply HMP2030. (**b**) Picture of the test boards used for measurements with the active bridge rectifier chip, including the crystal dipped into the demineralized water below. Hanging to the right of the PCB is the HGL-0400 hydrophone and AH-2010 preamplifier from Onda Corporation used for calibrating the acoustic intensity. In the foreground the V303-SU transducer from Olympus can be seen [30]

Table 6.1 Summary of powers in the test system [30]

Ref.	Component(s)	Power	Notes
$P_{crystal}$	Crystal	2.498 mW	Output
P_{loss}	Rb1, Rb2, Rb3	88.71 µW	Loss
P_{chip}	Active bridge	2.409 mW	Input
P_{LED}	μLED	2.198 mW	Load

into account (seen in Fig. 6.42), the power harvested by the crystal and transferred to the chip is all collected in Table 6.1 along with the losses imposed by the buffer circuit resistors. A snapshot of the voltage and current can be seen in Fig. 6.44, which resembles the simulated curves back in Fig. 6.41a extremely well and thus confirms the use of the Leach model in simulation.

Measured Output Power

By measuring the rectified voltage, the chip outputs across the μLED, the output power is determined by holding it against the IV curve of the μLED, and the resulting power is added to Table 6.1. A snapshot of the rectified voltage can be seen in Fig. 6.45, where it can be seen the voltage has settled just above 2.5 V, which from the μLED IV curve gives a current of about 0.9 mA.

With the collected values in Table 6.1 and an incoming acoustic power of 7.2 mW from the crystal, the overall dust system efficiency can be found in (6.25). Similarly, the efficiency of the crystal itself under the load of the chip with the μLED can be found in (6.26). The efficiency of the active bridge rectifier in the chip can be found in (6.27), which is very close to the value found in post-layout simulation.

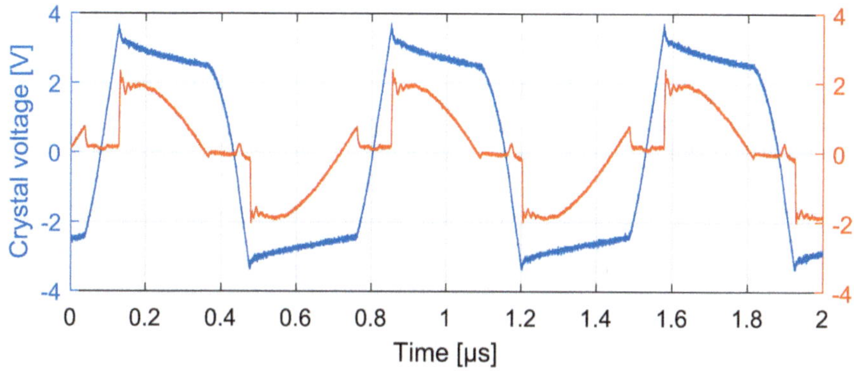

Fig. 6.44 Graph shows the voltage (blue) across the crystal and the current (red) through the chip [30]

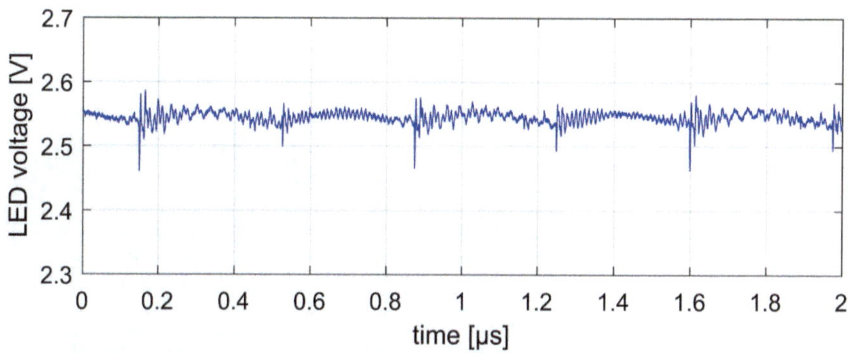

Fig. 6.45 Graph shows the rectified voltage output of the chip measured across the μLED during experiments [30]

$$\eta_{system} = \frac{P_{LED}}{7.2\,\text{mW}} = 30.5\% \tag{6.25}$$

$$\eta_{crystal} = \frac{P_{crystal}}{7.2\,\text{mW}} = 34.7\% \tag{6.26}$$

$$\eta_{chip} = \frac{P_{LED}}{P_{chip}} = 91.2\% \tag{6.27}$$

6.3.6 Integrated Prototype

The promising results of the laboratory experiments show that the dust works. Based on available power in the experiments with the (1 mm)^3 crystal and the fact that it is

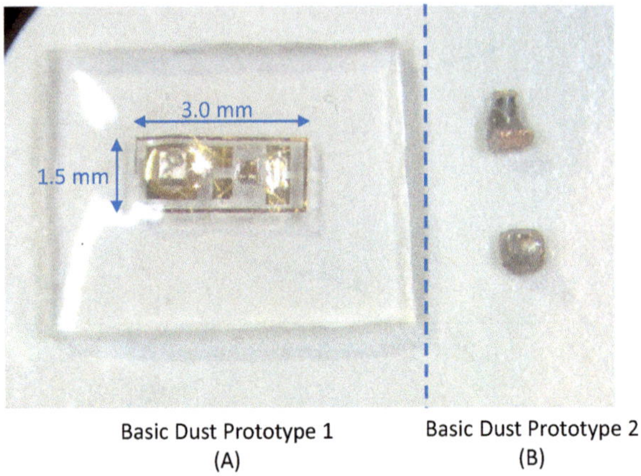

Fig. 6.46 Basic dust prototypes: (**a**) large version with components placed side by side and (**b**) small version with components stacked on top of each other [30]

Fig. 6.47 Picture of the basic dust prototype while being powered by ultrasonic waves that in turn powers the μLED [30]

possible to employ duty cycling of the acoustic power, meaning we can send bursts of much higher intensity while still keeping the average power at the 7.2 mW/mm^2 FDA limit, the crystal dimensions are reduced to only (500 μm)^3. The first compact prototype was constructed by bringing the three main components (crystal, chip, and μLED) close together side by side on a glass substrate, as seen in Fig. 6.46a. To protect the prototypes and ensure biocompatibility, they are all encapsulated in a thin layer of polydimethylsiloxane (PDMS) [30, 88]. With the added PDMS, the volume of the first prototype version is estimated to be 2.85 mm^3. The prototype was tested successfully in the water tank setup shown in Fig. 6.47, where it was submerged and subjected to an incoming acoustic intensity of 7.2 mW/mm^2, but now at a frequency of 3.14 MHz as a direct consequence of the changed crystal dimensions.

References

1. M. Zamani, Y. Rezaeiyan, H.A. Huynh, M. Ronchini, H. Farkhani, F. Moradi, A 2.3-μw capacitively coupled chopper-stabilized neural amplifier with input impedance of 6.7 gΩ. IEEE Solid-State Circuits Lett. **4**, 133–136 (2021)
2. H.A. Huynh, M. Zamani, M. Ronchini, H. Farkhani, F. Moradi, A 53gΩ@dc input impedance multi-channel neural recording amplifier with 0.77 μvrms input-referred noise for deep brain implants, in *2021 IEEE Custom Integrated Circuits Conference (CICC)* (2021), pp. 1–2
3. M. Zamani, Y. Rezaeiyan, O. Shoaei, F. Moradi, A 7.2μw magnitude/phase bio-impedance measurement front-end with PWM output in 0.18μm CMOS, in *2021 IEEE Nordic Circuits and Systems Conference (NorCAS)* (2021), pp. 1–5
4. M. Zamani, Y. Rezaeiyan, O. Shoaei, W.A. Serdijn, A 1.55 μw bio-impedance measurement system for implantable cardiac pacemakers in 0.18 μm CMOS. IEEE Trans. Biomed. Circuits Syst. **12**(1), 211–221 (2018)
5. Y. Rezaeiyan, M. Zamani, O. Shoaei, W.A. Serdijn, Mixed-signal IC with pulse width modulation wireless telemetry for implantable cardiac pacemakers in 0.18-μm CMOS. IEEE Trans. Biomed. Circuits Syst. **12**(3), 589–600 (2018)
6. Y. Rezaeiyan, Y. Koolivand, M. Zamani, O. Shoaei, M. Akbari, F. Moradi, K.-T. Tang, A 4.5 μw miniaturized 3-channel wireless intra-cardiac acquisition system. IEEE Trans. Biomed. Circuits Syst. **17**(5), 1097–1110 (2023)
7. M.N. Christensen, M. Zamani, A. Rashidi, F. Moradi, Ultrasonic backscatter communication for brain implants: mathematical model, simulation, and measurement, in *2021 IEEE Biomedical Circuits and Systems Conference (BioCAS)* (2021), pp. 1–5
8. S. Hosseini, A. Rashidi, K. Laursen, J. Pelloux-Prayer, F. Moradi, Piezoelectric energy harvester with piezo-magnet stack for ultrasonically-powered brain implants, in *2019 IEEE International Ultrasonics Symposium (IUS)* (2019), pp. 201–204
9. A. Rashidi, S. Hosseini, K. Laursen, F. Moradi, Stardust: optogenetics, electrophysiology and pharmacology with an ultrasonically powered dust for Parkinson's disease, in *2019 26th IEEE International Conference on Electronics, Circuits and Systems (ICECS)* (2019), pp. 109–110
10. X. Wei, J. Liu, Power sources and electrical recharging strategies for implantable medical devices. Front. Energy Power Eng. China **2**(1), 1–13 (2008)
11. A. Rashidi, M. Zamani, T. Mondal, S. Hosseini, K. Laursen, B. Corbett, F. Moradi, Ultrasonically powered and controlled microsystem for dual-wavelength optogenetics with a multiload regulation scheme. IEEE Solid-State Circuits Lett. **6**, 33–36 (2023)
12. D.C. Bock, A.C. Marschilok, K.J. Takeuchi, E.S. Takeuchi, Batteries used to power implantable biomedical devices. Electrochim. Acta **84**, 155–164 (2012)
13. Y. Yang, G. Dong Xu, J. Liu, A prototype of an implantable thermoelectric generator for permanent power supply to body inside a medical device. J. Med. Devices **8**(1), 014507 (2014)
14. L. Beker, A. Benet, A.T. Meybodi, B. Eovino, A.P. Pisano, L. Lin, Energy harvesting from cerebrospinal fluid pressure fluctuations for self-powered neural implants. Biomed. Microdevices **19**(2), 32 (2017)
15. M. Amin Karami, D.J. Inman, Powering pacemakers from heartbeat vibrations using linear and nonlinear energy harvesters. Appl. Phys. Lett. **100**(4), 042901 (2012)
16. J. Li, L. Kang, Y. Long, H. Wei, Y. Yu, Y. Wang, C.A. Ferreira, G. Yao, Z. Zhang, C. Carlos et al., Implanted battery-free direct-current micro-power supply from in vivo breath energy harvesting. ACS Appl. Mat. Interfaces **10**(49), 42 030–42 038 (2018)
17. S. Kerzenmacher, J. Ducrée, R. Zengerle, F. Von Stetten, Energy harvesting by implantable abiotically catalyzed glucose fuel cells. J. Power Sources **182**(1), 1–17 (2008)
18. X. Lu, P. Wang, D. Niyato, D.I. Kim, Z. Han, Wireless networks with RF energy harvesting: a contemporary survey. IEEE Commun. Surv. Tutorials **17**(2), 757–789 (2014)
19. Y.-C. Shih, T. Shen, B.P. Otis, A 2.3μw wireless intraocular pressure/temperature monitor. IEEE J. Solid-State Circuits **46**(11), 2592–2601 (2011)

20. K.M. Silay, C. Dehollain, M. Declercq, Inductive power link for a wireless cortical implant with two-body packaging. IEEE Sens. J. **11**(11), 2825–2833 (2011)
21. M. Mujeeb-U-Rahman, D. Adalian, C.-F. Chang, A. Scherer, Optical power transfer and communication methods for wireless implantable sensing platforms. J. Biomed. Opt. **20**(9), 095012 (2015)
22. P.R. Hoskins, K. Martin, A. Thrush, *Diagnostic Ultrasound: Physics and Equipment* (CRC Press, Boca Raton, 2019)
23. T.G. Leighton, What is ultrasound? Prog. Biophys. Mol. Biol. **93**(1–3), 3–83 (2007)
24. H. Basaeri, D.B. Christensen, S. Roundy, A review of acoustic power transfer for bio-medical implants. Smart Mat. Struct. **25**(12), 123001 (2016)
25. U. FDA, Guidance for industry and FDA staff information for manufacturers seeking marketing clearance of diagnostic ultrasound systems and transducers. Rockville MD: FDA (2008)
26. S. Zhang, F. Li, High performance ferroelectric relaxor-pbtio3 single crystals: status and perspective. J. Appl. Phys. 111(3), 2 (2012)
27. C.H. Sherman, J.L. Butler, *Transducers and Arrays for Underwater Sound*, vol. 4 (Springer, Berlin, 2007)
28. K.K. Shung, J. Cannata, Q. Zhou, Piezoelectric materials for high frequency medical imaging applications: a review. J. Electroceram. **19**(1), 141–147 (2007)
29. P.I. Label, P. Label, R. oHS Compliance Engineer, Directive 2002/95/ec of the European parliament and of the council of 27 January 2003 on the restriction of the use of certain hazardous substances in electrical and electronic equipment (2005)
30. K. Laursen, A. Rashidi, S. Hosseini, T. Mondal, B. Corbett, F. Moradi, Ultrasonically powered compact implantable dust for optogenetics. IEEE Trans. Biomed. Circuits Syst. **14**(3), 583–594 (2020)
31. C. Luo, G. Cao, I. Shen, Development of a lead-zirconate-titanate (PZT) thin-film microactuator probe for intracochlear applications. Sens. Actuators A: Phys. **201**, 1–9 (2013)
32. P. Udvardi, J. Radó, A. Straszner, J. Ferencz, Z. Hajnal, S. Soleimani, M. Schneider, U. Schmid, P. Révész, J. Volk, Spiral-shaped piezoelectric MEMS cantilever array for fully implantable hearing systems. Micromachines **8**(10), 311 (2017)
33. M. Haq, Application of piezo transducers in biomedical science for health monitoring and energy harvesting problems. Mat. Res. Exp. **6**(2), 022002 (2018)
34. J. Yan, Y.G. Jeong, High performance flexible piezoelectric nanogenerators based on batio3 nanofibers in different alignment modes. ACS Appl. Mat. Interfaces **8**(24), 15 700–15 709 (2016)
35. Y. Saito, H. Takao, T. Tani, T. Nonoyama, K. Takatori, T. Homma, T. Nagaya, M. Nakamura, Lead-free piezoceramics. Nature **432**(7013), 84–87 (2004)
36. A. Arbabian, T.C. Chang, M.L. Wang, J. Charthad, S. Baltsavias, M. Fallahpour, M.J. Weber, Sound technologies, sound bodies: medical implants with ultrasonic links. IEEE Microwave Mag. **17**(12), 39–54 (2016)
37. J. Charthad, N. Dolatsha, A. Rekhi, A. Arbabian, System-level analysis of far-field radio frequency power delivery for mm-sized sensor nodes. IEEE Trans. Circuits Syst. I Regul. Pap. **63**(2), 300–311 (2016)
38. T.C. Chang, M.J. Weber, M.L. Wang, J. Charthad, B.P.T. Khuri-Yakub, A. Arbabian, Design of tunable ultrasonic receivers for efficient powering of implantable medical devices with reconfigurable power loads. IEEE Trans. Ultrason. Ferroelectr. Freq. Control **63**(10), 1554–1562 (2016)
39. T.C. Chang, M.J. Weber, J. Charthad, S. Baltsavias, A. Arbabian, Scaling of ultrasound-powered receivers for sub-millimeter wireless implants, in *2017 IEEE Biomedical Circuits and Systems Conference (BioCAS)* (2017), pp. 1–4
40. R. Krimholtz, D.A. Leedom, G.L. Matthaei, New equivalent circuits for elementary piezoelectric transducers. Electron. Lett. **6**(13), 398–399 (1970)
41. W.M. Leach, Controlled-source analogous circuits and spice models for piezoelectric transducers. IEEE Trans. Ultrason. Ferroelectr. Freq. Control **41**(1), 60–66 (1994)

42. J. Wu, G.K. Fedder, L.R. Carley, A low-noise low-offset capacitive sensing amplifier for a 50-/spl mu/g//spl radic/Hz monolithic CMOS MEMS accelerometer. IEEE J. Solid-State Circuits **39**(5), 722–730 (2004)
43. M. Hautefeuille, C. O'Mahony, B. O'Flynn, K. Khalfi, F. Peters, A MEMS-based wireless multisensor module for environmental monitoring. Microelectron. Reliab. **48**(6), 906–910 (2008)
44. J. Xie, C. Lee, H. Feng, Design, fabrication, and characterization of CMOS MEMS-based thermoelectric power generators. J. Microelectromech. Syst. **19**(2), 317–324 (2010)
45. J.W. Judy, Microelectromechanical systems (MEMS): fabrication, design and applications. Smart Mat. Struct. **10**(6), 1115 (2001)
46. K.K. Shung, M. Zippuro, Ultrasonic transducers and arrays. IEEE Eng. Med. Biol. Mag. **15**(6), 20–30 (1996)
47. Q. Zhou, S. Lau, D. Wu, K.K. Shung, Piezoelectric films for high frequency ultrasonic transducers in biomedical applications. Prog. Mat. Sci. **56**(2), 139–174 (2011)
48. H. Tohmyoh, Polymer acoustic matching layer for broadband ultrasonic applications. J. Acoust. Soc. Am. **120**(1), 31–34 (2006)
49. A. Abrar, D. Zhang, B. Su, T. Button, K. Kirk, S. Cochran, 1–3 connectivity piezoelectric ceramic–polymer composite transducers made with viscous polymer processing for high frequency ultrasound. Ultrasonics **42**(1–9), 479–484 (2004)
50. W. Lee, S.F. Idriss, P.D. Wolf, S.W. Smith, A miniaturized catheter 2-D array for real-time, 3-D intracardiac echocardiography. IEEE Trans. Ultrason. Ferroelectr. Freq. Control **51**(10), 1334–1346 (2004)
51. A. Caronti, G. Caliano, R. Carotenuto, A. Savoia, M. Pappalardo, E. Cianci, V. Foglietti, Capacitive micromachined ultrasonic transducer (CMUT) arrays for medical imaging. Microelectron. J. **37**(8), 770–777 (2006)
52. O. Oralkan, B. Bayram, G.G. Yaralioglu, A.S. Ergun, M. Kupnik, D.T. Yeh, I.O. Wygant, B.T. Khuri-Yakub, Experimental characterization of collapse-mode CMUT operation. IEEE Trans. Ultrason. Ferroelectr. Freq. Control **53**(8), 1513–1523 (2006)
53. B. Bayram, E. Hæggstrom, G.G. Yaralioglu, B.T. Khuri-Yakub, A new regime for operating capacitive micromachined ultrasonic transducers. IEEE Trans. Ultrason. Ferroelectr. Freq. Control **50**(9), 1184–1190 (2003)
54. K.K. Park, O. Oralkan, B.T. Khuri-Yakub, A comparison between conventional and collapse-mode capacitive micromachined ultrasonic transducers in 10-mhz 1-d arrays. IEEE Trans. Ultrason. Ferroelectr. Freq. Control **60**(6), 1245–1255 (2013)
55. S. Trolier-McKinstry, P. Muralt, Thin film piezoelectrics for MEMS. J. Electroceram. **12**(1–2), 7–17 (2004)
56. A. Hajati, D. Latev, D. Gardner, A. Hajati, D. Imai, M. Torrey, M. Schoeppler, Three-dimensional micro electromechanical system piezoelectric ultrasound transducer. Appl. Phys. Lett. **101**(25), 253101 (2012)
57. T. Wang, C. Lee, Zero-bending piezoelectric micromachined ultrasonic transducer (PMUT) with enhanced transmitting performance. J. Microelectromech. Syst. **24**(6), 2083–2091 (2015)
58. T.R. Shrout, Z.P. Chang, N. Kim, S. Markgraf, Dielectric behavior of single crystals near the (1-X) Pb (Mg1/3Nb2/3) O3-(x) PbTio3 morphotropic phase boundary. Ferroelectr. Lett. Sect. **12**(3), 63–69 (1990)
59. D.E. Dausch, K.H. Gilchrist, J.B. Carlson, S.D. Hall, J.B. Castellucci, O.T. von Ramm, In vivo real-time 3-D intracardiac echo using PMUT arrays. IEEE Trans. Ultrason. Ferroelectr. Freq. Control **61**(10), 1754–1764 (2014)
60. X. Jiang, H.-Y. Tang, Y. Lu, E.J. Ng, J.M. Tsai, B.E. Boser, D.A. Horsley, Ultrasonic fingerprint sensor with transmit beamforming based on a PMUT array bonded to CMOS circuitry. IEEE Trans. Ultrason. Ferroelectr. Freq. Control **64**(9), 1401–1408 (2017)
61. O. Rozen, S.T. Block, X. Mo, W. Bland, P. Hurst, J.M. Tsai, M. Daneman, R. Amirtharajah, D.A. Horsley, Monolithic MEMS-CMOS ultrasonic rangefinder based on dual-electrode PMUTs, in *2016 IEEE 29th International Conference on Micro Electro Mechanical Systems (MEMS)* (IEEE, Piscataway, 2016), pp. 115–118

62. B.E. Eovino, Y. Liang, L. Lin, Concentric PMUT arrays for focused ultrasound and high intensity applications, in *2019 IEEE 32nd International Conference on Micro Electro Mechanical Systems (MEMS)* (IEEE, Piscataway, 2019), pp. 771–774
63. B. Herrera, F. Pop, C. Cassella, M. Rinaldi, AlN PMUT-based ultrasonic power transfer links for implantable electronics, in *2019 20th International Conference on Solid-State Sensors, Actuators and Microsystems & Eurosensors XXXIII (TRANSDUCERS & EUROSENSORS XXXIII)* (IEEE, Piscataway, 2019), pp. 861–864
64. S.K. Edelman, *Understanding Ultrasound Physics* (Baker & Taylor, Charlotte, 2003)
65. S. Hosseini, K. Laursen, A. Rashidi, F. Moradi, Multi-ring ultrasonic transducer on a single piezoelectric disk for powering biomedical implants, in *2019 41st Annual International Conference of the IEEE Engineering in Medicine and Biology Society (EMBC)* (2019), pp. 3827–3830
66. P. Lele, A simple method for production of trackless focal lesions with focused ultrasound: physical factors. J. Physiol. **160**(3), 494–512 (1962)
67. G.-H. Feng, C.C. Sharp, Q. Zhou, W. Pang, E.S. Kim, K. Shung, Fabrication of MEMS ZnO dome-shaped-diaphragm transducers for high-frequency ultrasonic imaging. J. Micromech. Microeng. **15**(3), 586 (2005)
68. M.L. Wang, T.C. Chang, T. Teisberg, M.J. Weber, J. Charthad, A. Arbabian, Closed-loop ultrasonic power and communication with multiple miniaturized active implantable medical devices, in *2017 IEEE International Ultrasonics Symposium (IUS)* (IEEE, Piscataway, 2017), pp. 1–4
69. H. Kim, S. Kim, N.S. Sim, C. Pasquinelli, A. Thielscher, J.H. Lee, H.J. Lee, Miniature ultrasound ring array transducers for transcranial ultrasound neuromodulation of freely-moving small animals. Brain Stimulation **12**(2), 251–255 (2019)
70. D. Tarrazó-Serrano, S. Pérez-López, P. Candelas, A. Uris, C. Rubio, Acoustic focusing enhancement in Fresnel zone plate lenses. Sci. Rep. **9**(1), 1–10 (2019)
71. D. Huang, E. Kim, Micromachined acoustic-wave liquid ejector. J. Microelectromech. Syst. **10**(3), 442–449 (2001)
72. S. Hosseini, K. Laursen, A. Rashidi, T. Mondal, B. Corbett, F. Moradi, S-MRUT: sectored-multiring ultrasonic transducer for selective powering of brain implants. IEEE Trans. Ultrason. Ferroelectr. Freq. Control **68**(1), 191–200 (2021)
73. S.-C. Wooh, Y. Shi, Optimum beam steering of linear phased arrays. Wave Motion **29**(3), 245–265 (1999)
74. N. Kovačević, J. Henderson, E. Chan, N. Lifshitz, J. Bishop, A. Evans, R. Henkelman, X. Chen, A three-dimensional MRI atlas of the mouse brain with estimates of the average and variability. Cereb. Cortex **15**(5), 639–645 (2005)
75. K. Laursen, M. Zamani, Y. Rezaeiyan, S. Hosseini, T. Mondal, B. Corbett, A.M. Ouagazzal, M. Amalric, F. Moradi, An ultrasonically-powered system for 1.06 mm^3 implantable optogenetics and drug delivery dust. IEEE Trans. Circuits Syst. II: Express Briefs **70**(10), 3937–3941 (2023)
76. S. Baghaee Ivriq, K. Laursen, A.M. Jørgensen, T. Mondal, M. Zamani, Y. Rezaeiyan, B. Corbett, B.B. Iversen, F. Moradi, A system-level feasibility study of a lead-free ultrasonically powered light delivery implant for optogenetics. Adv. Intell. Syst. **6**(3), 2300527 (2024)
77. A. Rashidi, K. Laursen, S. Hosseini, F. Moradi, A high-resolution ultrasonically powered and controlled optogenetic stimulator with a novel fully analog time to current converter, in *2020 42nd Annual International Conference of the IEEE Engineering in Medicine & Biology Society (EMBC)* (2020), pp. 3411–3414
78. J. Charthad, T.C. Chang, Z. Liu, A. Sawaby, M.J. Weber, S. Baker, F. Gore, S.A. Felt, A. Arbabian, A mm-sized wireless implantable device for electrical stimulation of peripheral nerves. IEEE Trans. Biomed. Circuits Syst. **12**(2), 257–270 (2018)
79. Y. Lam, W. Ki, C. Tsui, Integrated low-loss CMOS active rectifier for wirelessly powered devices. IEEE Trans. Circuits Syst. II: Express Briefs **53**(12), 1378–1382 (2006)

80. S. Guo, H. Lee, An efficiency-enhanced CMOS rectifier with unbalanced-biased comparators for transcutaneous-powered high-current implants. IEEE J. Solid-State Circuits **44**(6), 1796–1804 (2009)
81. X. Do, H. Nguyen, S. Han, D.S. Ha, S. Lee, A self-powered high-efficiency rectifier with automatic resetting of transducer capacitance in piezoelectric energy harvesting systems. IEEE Trans. Very Large Scale Integr. VLSI Syst. **23**(3), 444–453 (2015)
82. A. Rashidi, K. Laursen, S. Hosseini and F. Moradi, An Ultrasonically Powered Optogenetic Microstimulators with Power-Efficient Active Rectifier and Charge Reuse Capability, 2019 IEEE International Symposium on Circuits and Systems (ISCAS), Sapporo, Japan, pp. 1–5, (2019).
83. M. Hassanpourghadi, M. Zamani, M. Sharifkhani, A low-power low-offset dynamic comparator for analog to digital converters. Microelectron. J. **45**(2), 256–262 (2014)
84. K.A. Makinwa, A. Baschirotto, P. Harpe, *Low-Power Analog Techniques, Sensors for Mobile Devices, and Energy Efficient Amplifiers: Advances in Analog Circuit Design 2018* (Springer, Berlin, 2019)
85. A. Rashidi, K. Laursen, S. Hosseini, F. Moradi, Overvoltage protection circuits for ultrasonically powered implantable microsystems, in *2019 41st Annual International Conference of the IEEE Engineering in Medicine and Biology Society (EMBC)* (2019), pp. 4354–4358
86. A. Rashidi, K. Laursen, S. Hosseini, H.-A. Huynh, F. Moradi, An implantable ultrasonically powered system for optogenetic stimulation with power-efficient active rectifier and charge-reuse capability. IEEE Trans. Biomed. Circuits Syst. **13**(6), 1362–1371 (2019)
87. T.C. Chang, M. Weber, J. Charthad, A. Nikoozadeh, P.T. Khuri-Yakub, A. Arbabian, Design of high-efficiency miniaturized ultrasonic receivers for powering medical implants with reconfigurable power levels, in *2015 IEEE International Ultrasonics Symposium (IUS)* (2015), pp. 1–4
88. M.F. la Cour, M.B. Stuart, M.B. Laursen, S.E. Diederichsen, E.V. Thomsen, J.A. Jensen, Investigation of PDMS as coating on CMUTs for imaging, in *2014 IEEE International Ultrasonics Symposium* (2014), pp. 2584–2587

Chapter 7
Smart Contact Lens

Mengyao Yuan

Electronic wearable devices have gained significant traction and popularity in recent years. Wearable devices are designed to possess data processing for health and fitness [1–3]; biometric tracking, temperature regulation, and posture correction such as smart clothing; immersive experiences of the human-machine interface through Augmented Reality (AR) and Virtual Reality (VR); personal safety and security of emergency assistance or location tracking; and medical and assistive devices such as wearable Electrocardiogram (ECG) monitors, smart prosthetics, and smart contact lens. The emergence of smart contact lenses represents a remarkable advancement in wearable technology, offering numerous potential benefits and transforming the way we interact with the world. These innovative devices, worn directly on the eye, integrate cutting-edge electronics, sensors, and wireless communication capabilities to provide a range of functionalities beyond traditional vision correction. In the future, smart contact lenses are expected to revolutionize various aspects of our lives. They hold immense potential in the fields of healthcare, augmented reality, and personalized computing. One of the key drivers for their development is the need for non-intrusive, unobtrusive health monitoring. Smart contact lenses have the potential to continuously and non-invasively monitor vital signs and biomarkers, enabling early detection of diseases and providing real-time health insights. Figure 7.1 concludes the advanced research in Smart Contact Lens designs including integrated sensors (pressure, strain, temperature, glucose, cortisol, pH, and corneal electrode sensors), WPT technologies (near-field and far-field power transfer strategies), Materials been used, Power generation and power storage plans, which are detailed discussed in the following sections.

M. Yuan (✉)
University of Glasgow, Glasgow, UK

R. Das, H. Heidari (eds.), *Wireless Power Technologies for Biomedical Devices*,
Analog Circuits and Signal Processing, https://doi.org/10.1007/978-3-031-52834-7_7

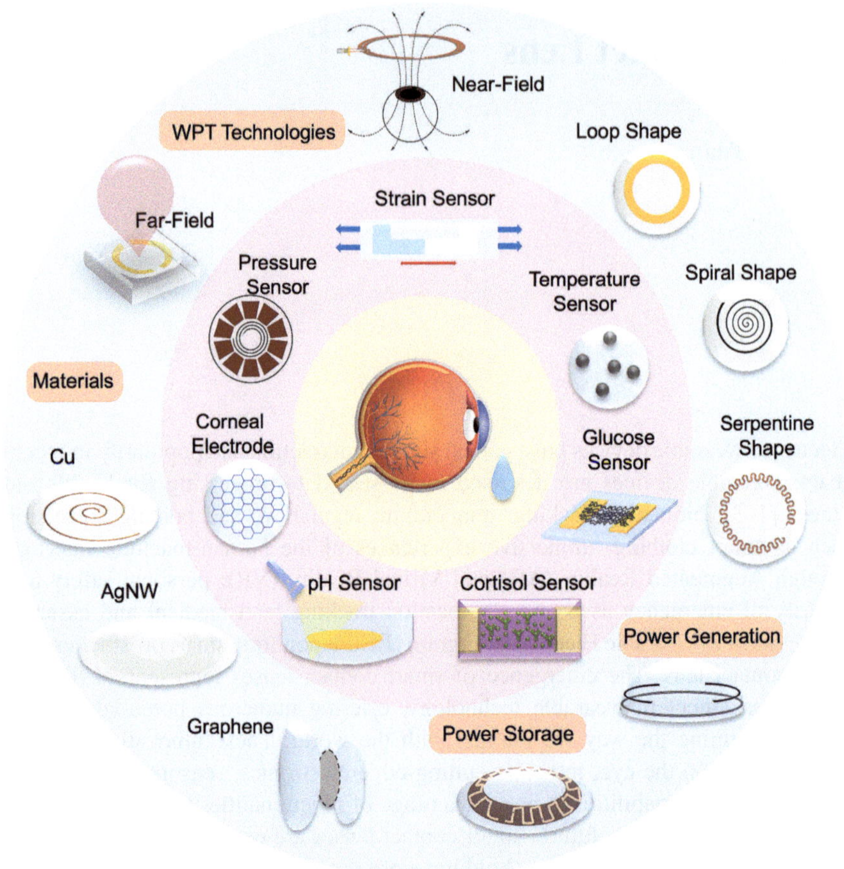

Fig. 7.1 Advancements in Smart Contact Lens designs: Integrated sensors (pressure, strain, temperature, glucose, cortisol, pH, and corneal electrode sensors), WPT technologies (including near-field and far-field power transfer strategies), Materials, Power generation, and power storage plans.

7.1 Ocular Diseases

More than 2.2 billion people worldwide suffer from visual impairment, of whom at least 1 billion have preventable or treatable visual impairment [4]. The leading causes of MSVI contain uncorrected refractive error, Cataracts, Age-related Macular Degeneration (AMD), Glaucoma, and Diabetic Retinopathy (DR), which all have potential risks of blindness [5]. The associated global cost of lost production due to visual impairment due to uncorrected myopia and presbyopia alone is estimated at $244 billion and $25.4 billion annually, respectively [6]. The ocular diseases in human eyes can be divided based on abnormalities of various eye positions:

Cornea-related diseases (e.g. Keratitis, Corneal dystrophy, and Corneal ulcers); Lens-related diseases (e.g. Cataracts and Presbyopia); Retina-related diseases (e.g. AMD, DR, and Retinal detachment); Optic nerve-related diseases (e.g. Glaucoma and Optic neuritis); Conjunctiva-related diseases (e.g. Conjunctivitis); and Eyelid-related diseases (e.g. Blepharitis and Ptosis). The ocular diseases are normally age-related, triggered by other inducements such as diabetes, sclerosis, allergies, and irritants, exacerbations may lead to blurry vision, vision impairment, or blindness.

To diagnose and treat these ocular diseases, various clinical strategies are employed. These include the measurement of visual acuity to assess the clarity of vision; slit lamp examination for a detailed evaluation of different eye structures; ophthalmoscopy to examine the retina, optic nerve, and blood vessels; tonometry to measure intraocular pressure; fluorescein angiography to evaluate blood flow in the retina; genetic testing to identify genetic mutations associated with inherited eye diseases and biopsy with histopathology analysis for suspicious or malignant ocular tumors. However, these clinical methodologies demand extensive facilities, controlled testing environments, and hysteretic result analysis. Nevertheless, some eye conditions like glaucoma, lactose intolerance, and dry eye disease can be diagnosed and monitored through real-time detection of analytes or protein concentrations in tears.

In conclusion, visual impairment is a significant global problem that affects billions of people, but a substantial portion of these cases are preventable or treatable. Continued research and advancements in clinical strategies are crucial to further improve the diagnosis, treatment, and prevention of ocular diseases and to alleviate the burden of visual impairment worldwide.

7.2 Tears

The search for alternative blood samples for non-invasive disease diagnosis has been an area of exploration for many years. Urine and tears have been identified as suitable alternatives and have unique advantages in diagnosing specific diseases [7–9]. Concentrations of analytes in tears and the characteristics of biomedical components in human eyes are limited to normal ranges [10–17], When these values exceed the specific range (see Table 7.1), it can indicate the presence of related diseases [18, 20–23]. Therefore, multiple biomarkers in body fluids are used for disease screening [24]. In general, the tear fluid film is mainly composed of three layers: Mucin, Aqueous, and Lipid [25], which play the role of lubrication and cleaning. Although most of the two body fluids are different, tears share similarities with blood since plasma can penetrate the blood-tears barrier [19], and plasma leakage can be checked in tears instead of blood. The ongoing search for non-invasive disease diagnosis has led to the exploration of alternative samples like tears [26, 27], accurate and sensitive detection of analyte concentrations in tears enables the early diagnosis and monitoring of various ocular diseases and systemic conditions, thereby facilitating timely intervention and personalized treatment plans.

Table 7.1 Analyte concentrations and component ranges in tears

Analyte/Component	Normal range	Related disease	Refs.
K^+	20 – 42 mM	Hyperkalaemia	[12, 13]
Ca^+	0.4 – 1.1 mM	Lacrimal cystic fibrosis	[12, 13]
Na^+	120 – 165 mM	Lacrimal cystic fibrosis	[13–15]
Cl^-	118 – 138 mM	Dry eye disease	[12, 13]
pH	6.4–7.7	Dry eye disease	[16]
HCO_3^-	20 – 42 mM	Dry eye disease	[12, 13]
Zn^+	23.20 – 52.50 ng/ml	Keratitis	[17]
Cu^+	4.60 – 9.75 ng/ml	Hepatopathies	[17]
Lactoferrin	1.01 – 3.43 mg/ml	Keratoconjunctivitis	[18, 19]
Lactate	2 – 5 mM	Lactic acidosis	[12, 13]
Glucose	0.1 – 0.6 mM	Diabetes	[13]
Protein	5 – 11 mM	Inflammation	[20]
IOP	10 – 20 mmHg	Glaucoma	[21]

The analysis of tears can assist in identifying ocular disorders like dry eye disease, glaucoma, diabetic retinopathy, etc.; simultaneously, tears have the potential to provide insights into systemic diseases, as alterations in the concentrations of analytes can indicate conditions such as diabetes, autoimmune disorders, and cardiovascular diseases, etc. The non-invasive nature of tear collection makes it an attractive option for regular patient monitoring, allowing for longitudinal studies and assessing disease progression over time. Moreover, tears can provide real-time and dynamic information, allowing for continuous monitoring of analyte concentrations in response to various stimuli, environmental factors, and therapeutic interventions. This is an advantage when utilizing a smart contact lens, as it offers a non-invasive diagnosis, real-time monitoring, and controllable treatment platform through integrated biosensors.

Advanced contact lens sensors serve various purposes such as pH, temperature, pressure, and chemical sensing, where the powering methods for these sensors can differ. Some contact lens sensors, like those fabricated with photonic crystals, liquid crystals, and phenylboronic acid, can react directly to pressure, temperature, and glucose without needing an external power transfer unit [28–32]. However, the detection of conditions like glaucoma or glucose using electrical and electrochemical sensors may require power generation through wired or wireless power transfer [33–36], which lays the importance in miniaturized battery and wireless power transfer strategy designs. In addition to the electrical considerations associated with sensor implementation and power strategies, there are other critical factors to address. These include the biological evaluations involved in choosing the right materials for encapsulation, which must meet several criteria. These criteria encompass maintaining a transparent appearance to avoid obstructing eyesight, ensuring biocompatibility with the eye to prevent adverse reactions, and allowing sufficient oxygen permeability and wettability to avoid discomfort. Successfully addressing all these considerations represents substantial challenges in the development and integration of advanced sensors within contact lenses [37]. Meeting these challenges is crucial because advanced contact lens sensors have the potential to revolutionize various aspects of healthcare and personalized medicine. However, achieving the seamless integration of advanced sensors into contact lenses requires a multidisciplinary approach, combining expertise in materials science, electronics, power management, biocompatibility, and optics. Continued research and innovations in these areas will drive the development of intelligent contact lenses with enhanced functionalities and improved usability, bringing us closer to a future where wearable technologies seamlessly merge with healthcare.

7.3 SCL Sensors

Unpowered sensors have recently attracted attention in wearable electronic devices for their inherent advantages of flexibility and comfortability, especially when interfacing with wearable layers in contact with the skin without any electrical component [38–41]. Unpowered biosensors on electronic contact lenses are widely used for pressure, strain, glucose, pH, and temperature sensing. In such cases, smartphones and optical microscopes are generally used for data processing.

7.3.1 Intrinsic Ocular Abnormalities Sensors

Typically, microfluidic strain sensing depends on the resistivity change caused by mechanical deformations. Agaoglu et al. presented microfluidic dilatometric sensing for a smart contact lens based on the total volume change, which was composed of a liquid reservoir, an air reservoir, and a liquid-air interface on the sensing channel for strain detection [28]. The proposed microfluidic strain sensor embedded contact lens was tested with a clinical intraocular pressure range of 10–40 mmHg, which presented a comparable pressure test result compared with clinical Tono-pen XL. In order to generate the visible movement, Wu et al. designed a microfluidic intraocular pressure (IOP) sensor with dyed red liquid to make a more precise illustration of the pressure output [30]. The dyed liquid-filled sensing chamber and the liquid flow sensing channels guaranteed the fluid flow direction, and an additional buffer chamber was utilized to avoid liquid leakage. The applied pressure value was obtained from the dyed liquid displacement, with a sensitivity of 0.2832 mm/mmHg under a pressure range of 0–40 mmHg. However, the data obtain process of the accurate pressure values for microfluidic sensors relies on external devices such as optical microscopes, which limits the sensing conditions and significantly reduces real-time practicability.

Other than microfluidic IOP sensing, photonic crystal-based sensors present a quantifiable platform of measurement due to their accurate RGB values. According to Bragg's equation [42], the lattice distance variation in the photonic crystal sensor results in the wavelength shift, reflected by external devices such as optical microscopes or mobile phone cameras. Maeng et al. proposed an IOP sensor fabricated with photonic crystal on top of a circular microhydraulic amplifier system, where the applied pressure was enlarged, with a sensitivity of 0.23 nm/mmHg with a higher testing range up to 110 mmHg [31]. Although pressure sensing with photonic crystal sensors validated a more accurate sensitivity under a higher applied pressure range compared with microfluidic sensing, both methods require external equipment and stationary setups for sensor results validation. The Bragg reflection theory can also be applied to temperature sensing [43]. Yetisen et al. designed the thermochromic liquid crystals (TLC) sensor on a smart contact lens for its rapid and accurate temperature sensing. The proposed sensors were mapped into four separate positions on the central, peripheral, limbal, and paracentral, which reflected the temperature changes from 29 to 40 °C with a deviation of 0.5 °C [44].

Electrical sensors such as piezo-resistive and capacitive types have been popularly researched. Piezo-resistive smart contact lens transforms the imposed force into electrical resistive signals, where the resistivity of the piezo-resistive material changes under the action of a force, thus, the output electrical signal changes upon the application of a force. Leonardi et al. proposed a sandwiched strain gauge platinum-titanium resistive sensor between two polyimide plates, with a sensitivity of 109 μV/mmHg, where the power and data were transferred wirelessly to a portable reader unit [45]. A similar structure was implemented commercially by Sensimed, where the "triggerfish" smart lens sensor was used for continuous pressure measuring in patients with Glaucoma over 24 h [46]. Other than traditional metal materials design, piezo-resistive sensors with organic bi-layer films presents larger resistive changes. Laukhin et al. designed an organic molecular conductor made of nanostructured crystals contact lens for IOP sensing, the resistive changes with the deformation of the conductive layer during cornea curvatures variation caused by pressure, and the proposed sensor responded with a sensitivity of 1.5 Ω/mmHg in an enucleated pig eye [47].

Capacitive sensing is widely used for pressure detection and has been used in applications that include touchscreens, fingerprint sensors, and medical stethoscopes [48–51], where the variations of plate distance, plate cross-section area, and material relative permittivity reflect the external non-capacitive changes through electrical signals. Ali et al. designed a capacitive IOP sensor with Polysilicon/Air/Gold layers, together with four slotted diaphragms on corners, where the capacitance is linearly related to the applied pressure [52]. Chiou et al. proposed a ring-shaped parallel plate capacitive sensor filled with poly-HEMA material, where the variation derived from the gap distance and plate area [53]. The fabricated Ti/Au plate sensor was encapsulated with Parylene-C, which detected pressure under a range of 16–30 mmHg with a sensitivity of 1.2239 pF/mmHg [54]. Apart from using readout circuitry like capacitance-to-digital converter (CDC) chips, inductively coupled coils contribute to an intuitive and more straightforward design for output measurement. The resonant frequency shifts as the capacitance variation through a connected inductive component [55], Chen et al. utilized both capacitive and inductive sensing embodied with a linear relationship of the resonate frequency and applied pressure, which resulted in a response of 8 kHz/mmHg from a pressure range of 5–45 mmHg.

Apart from using readout circuitry like capacitance-to-digital converter (CDC) chips, inductively coupled coils contribute to an intuitive and more straightforward design for output measurement. The resonant frequency shifts as the capacitance variation through a connected inductive component, when the inductive and capacitive components are integrated together into the electronic contact lens, the frequency changes with the implied intraocular pressure. Kim et al. developed a multifunctional contact lens using AgNWs for transparency consideration, and an extra silicon elastomer (Ecoflex) layer put in between two inductive spirals forming the sandwich structure for intraocular pressure sensing [56]. The proposed LC sensor works under a pressure range of 5–50 mmHg and decreased linearly with pressure by the slope of 2.64 MHz $mmHg^{-1}$. Similarly, Yang et al. proposed an LC sensor for IOP sensing under 3.8 GHz in order to distinguish it from the power and data transmission frequency (850 kHz), avoiding the inner interference [57]. The findings indicated that the IOP sensor demonstrated a sensitivity of 1.28 ± 0.09 MHz $mmHg^{-1}$, which can be attributed to the unique cantilever structure where an ultra-soft air film exists between the sensing and reference plates, allowing for mobility. As a result, the cantilever configuration forms variable capacitors that can effectively respond to changes in pressure with a high level of sensitivity. Zhang et al. explored the electrode design for AgSEBs as the component of IOP sensing, where the proposed resonant frequency was at 238.5 MHz with a sensitivity of 0.030 MHz $mmHg^{-1}$ [58].

Smart contact lenses with integrated sensors offer numerous advantages in various applications, particularly in the field of healthcare. One major advantage is their ability to provide real-time monitoring and measurement of vital parameters such as IOP or temperature. These sensors, when integrated into smart contact lenses, enable continuous and non-invasive monitoring, eliminating the need for bulky external devices, which can greatly enhance patient comfort and convenience.

7.3.2 Retinal Sensors: ERG/EOG/VEP Electrodes

Apart from the intrinsic abnormalities of the eyeball itself, retinal-related diseases occupy a crucial portion of the vision impairment causes. Retinal-related diseases refer to a group of conditions that affect the retina, which is the light-sensitive tissue lining the back of the eye, typical retinal diseases contain AMD, DR, retinal detachment, retinitis pigmentosa, retinal vein occlusion, and retinoblastoma, etc. Any damage or impairment to the retina can lead to significant visual problems, including vision loss or blindness, since the retina is responsible for capturing and processing visual information, proper functioning is essential for clear and sharp vision. Diagnosing and treating retinal diseases in their early stages can help preserve vision and prevent further deterioration [59–61]. Regular eye exams and timely treatment are crucial to managing retinal diseases effectively. Despite visual acuity tests and optical coherence tomography, retinal examinations such as electrooculogram, electroretinogram, and visual evoked potential help with the retinal diseases' diagnosis. The typical retinal examination data acquisition equipment comprises three major components: a computer-controlled flash/pattern chamber, a rigid sheet metal electrode for electrophysiological signal acquisition, and an amplifier for signal magnification [62–64]. The principle of electrophysiology acquisition based on the different eye potential under various external stimuli, reflects the abnormalities analysis through various retinal pathways.

Clinical advances in electrophysiology electrode designs can be divided into three main types: fiber [65], contact lens [66], and skin [67] electrodes, among which contact lens electrodes present the highest measured amplitudes and reduced artifacts caused from comfortability [68–70]. Recent studies on contact lens electrodes have taken flexibility, stretchability, and transparency into account. Soft graphene contact lens electrodes (GRACEs) made of graphene with PMMA and Parylene-C encapsulation proposed by Yin et al., improve the transparency and reduces corneal irritation and abnormal, which records stronger signals than ERG-Jet electrode [71]. Kim et al. designed a corneal sensor of gold (Au) and poly(3,4-ethylenedioxythiophene) (PEDOT) in a serpentine shape for stretchable consideration with an encapsulation PDMS layer, which was tested with a significantly higher amplitude than Dawson-Trick-Litzkow (DTL) fiber and ERG-Jet electrode, while the implicit times remained statistically unchanged [72]. For traditional clinical examinations, extra light stimulation equipment is always required. The researcher has proposed the tri-color LED electrode integration to enhance the portability of ERG measurement, although the white light supply was not sufficient with the standard flash [73]. By integrating the flash unit on the smart contact lens with electrode sensing, the need for additional external light stimulation equipment can be eliminated, which allows for a more streamlined and efficient examination process. The flash unit on the contact lens can serve as a self-contained light source, providing the necessary light stimuli directly to the patient's eye. However, the stimulus-sensing integration for physiological detection is still under exploration.

7.3.3 *Tear Sensors: Glucose/Lactate/Cortisol/pH*

Other than the intrinsic disease detected from the human eye and retinal diseases detected from electrophysiological examinations through electrodes, analytes, and biomarkers in tears are vital parameters to evaluate potential diseases and physical conditions.

Badugu et al. proposed the silicone hydrogel (SiHG) contact lenses for pH sensing and individual ionic concentration detection in tears, where the tightly bound ion-sensitive fluorophores on SiHG lenses results in wavelength-ratiometric shift, making it possible for pH value testing and lifetime-based chloride sensing [74]. In a low pH environment, SiHG sensors performed a bright blue emission, while weaker yellow emissions can be detected in higher pH media. Lin et al.

utilized the light scattering phenomenon on smart contact lenses for glucose testing, where the proposed design was made of Phenylboronic acid (PBA)-based Hydroxyethyl methacrylate (HEMA) [75]. The experimental setup contains an image-capturing smartphone, which was placed facing away from a red-light source through the proposed contact lens. Thickness is detected owing to the lens swelling led by the absorption of glucose into PBA material, where the volume enlargement due to the growing glucose concentration provides a promising platform for non-enzyme glucose sensing.

Tear components such as glucose, lactate, and hormones found in the blood can be detected from tears non-invasively through smart contact lenses with electro-chemical sensing, which provides high sensitivity, excellent selectivity, and specific anti-interference ability. The principle of electrochemical sensing relies on chemical enzymatic reactions, where the detected electrons generated from reactions result in voltage or current changes, and immediately reflect the components' concentration. For glucose sensors, hydrogen peroxide is produced through the reaction of glucose and oxygen, and electrons are released through electrochemical oxidation. Current flows led by electron migration can be measured, indicating the concentration of glucose in tears. In the electrochemical sensing system, a three-electrode layout contains a working electrode (WE), a reference electrode (RE), and a counter electrode (CE) [76]. The differential voltage applied on WE and RE terminal guarantees electron movements, which varies from 400 to 700 mV in different case studies depending on electrode materials and chip design. Keum et al. designed a glucose sensor fabricated with chromium (Cr) and Pt as CE and WE terminal, moreover, the Ag/AgCl coated RE terminal on polyethylene terephthalate (PET) substrate was encapsulated with a silicone hydrogel material, consuming 82 μW power including the sensor, integrated chip and drug delivery system [77]. The proposed design presented a linear current increase from 0.41 to 3.12 μA with an increasing glucose concentration from 5 to 50 $mgdl^{-1}$, which also illustrated a good anti-interference of ascorbic acid, lactate, and urea in tears and stability of 63 days. Yao et al. proposed a loop glucose sensor applying the same three-electrode structure, which had a sensitivity of 240 μAcm^{-2} mM^{-1} for a glucose concentration range from 0.1 to 0.6 mM. The glucose electrodes were made from Ti/Pd/Pt mental structure, which was powered by a 3 μW integrated electronic chip [36, 78]. Thomas et al. proposed an enzymatic L-lactate sensor based on similar chemical reactions, where the component concentration can be predicted with the amount of generated hydrogen peroxide and electrons. The reference voltage of 400 mV was applied for optimal signal characteristics, and the lactate sensor showed a high sensitivity of 53 μAcm^{-2} mM^{-1} [79]. Donora et al. set up a series of experiments using spatiotemporal electrochemical sensing on electronic contact lens [80], where Ferrocene methanol (FeMeOH) and Potassium chloride (KCl) concentration variation were measured from four electrode sensors on the contact lens while the tear fluid flowing across the cornea, which successfully detected fluid tracing under a high concentration, laying the foundation of chemical measurement on the eye surface.

The future of tear fluid sensors holds significant potential in the field of healthcare and disease management. As sensor technologies continue to advance, tear fluid sensors are becoming increasingly sophisticated and capable of detecting and analyzing a wide range of components such as glucose, lactate, hormones, and more. These sensors can be integrated into smart contact lenses, enabling continuous and real-time monitoring of biomarkers without the need for invasive procedures.

7.4 WPT Technologies

The WPT technologies applied in smart contact lens design can be divided into two main categories: radiative antennas for simultaneously WPT and signal transmission, and coupled resonators working under lower resonant frequencies (i.e. Near Field Communication (NFC) or lower Megahertz ranges). The main challenges for antennas (or coils) applying on smart contact lenses are the size and thickness limitations under the donut-shaped area (diameter within 8–14 mm) to avoid eyesight impedance while maintaining high power transferring ability. The designed shape varies from loop, spiral, serpentine, and sheet geometries together with materials designed varying from traditional metal (Cu, Au, Ti/Au, etc.) to nanomaterials (AgNW, AgNF, Graphene, etc.). Smart contact lenses with sensors can provide valuable diagnostic and preventive information which can detect and track changes in IOP for glaucoma detection, temperature for early detection of fever or other health conditions, Glucose sensing for diabetic early diagnosis and real-time monitoring, etc. By incorporating a wireless power transfer system, the need for external batteries or power sources can be eliminated, which allows for a more streamlined and unobtrusive design that is comfortable for the wearer where continuous and uninterrupted monitoring can be further ensured. However, challenges remain associated with sensor integration on smart contact lenses, where achieving efficient power transfer to ensure the sensor's performance and the distance between the power transfer unit and the sensors, as well as the efficiency of the power conversion process, need to be carefully optimized. Additionally, the size and weight of the wireless power transfer unit can impact the overall comfort and wearability of the smart contact lens which is required to be compact and lightweight, minimizing any discomfort or interference with the wearer's vision. Table 7.2 summarizes the advanced researches comparison on antenna design on smart contact lenses.

7.4.1 Coupled Resonator

The coupled resonator design on smart contact lenses mainly contains coils resonating at NFC and lower ISM MHz frequency bands [83, 84, 88, 93]. Yang et al. designed a copper-made inductive coil resonate at 850 kHz, which was

Table 7.2 State-of-the-art comparison table on smart contact lens antenna performances

Year	Ref.	Antenna type	Antenna material	Encapsulation material	Operating frequency	Power transferring performance				Efficiency (%)	Test condition	Load	Application
						Bandwidth	S11(dB)	Distance	Power				
2010	[81]	Loop	Au	PET	2.4 GHz	–	–	10 cm	124.9 uW	10	in air	LED	Energy harvesting/ Communication
2013	[82]	Loop	Au/Ti	Parylene C	6 GHz	2.2 GHz	−23	–	2.2 mW	35	in pig eye tissue	Resistor	Energy harvesting
2015	[83]	3 Loops	Cu	Hydrogel-based Contact Lens	13.56 MHz	–	<−10	2 cm	–	49	in saline water	–	Energy harvesting/ Communication
2015	[84]	Octagonal spiral	Ti/Au	Parylene C	13.56 MHz	<1 MHz	−22	20 mm	–	17.5	on pig eye	LED	Energy harvesting
2016	[85]	Monopole	Graphene	Biocompatible soft polymer	10 GHz & 19 GHz	–	−14@10 GHz −19@19 GHz	–	–	40	in air	–	Energy harvesting/ Biosensing
2016	[86]	Loop	Cu	HEMA hydrogel	920 MHz	~90 MHz	−13	1 cm	110 uW	–	on pig eye	IOP sensor	Energy harvesting
2016	[87]	Loop	Au	PET	0.8 –2.0 GHz	–	–	2 cm	117 uW @ 2 GHz	–	on rabbit eye	LED	Energy harvesting
2017	[56]	Spiral	AgNW	Parylene	4.1 GHz	~600 MHz	−26	10 mm	–	–	on bovine eyeball	Glucose sensor/IOP sensor	Glucose/IOP sensing
2018	[88]	Loop	AgNF	Elastic polymer (Elastofilcon A)	50 MHz	–	–	5 mm	–	21.5	on rabbit eye	Glucose sensor/LED	Energy harvesting
2018	[89]	Dipole	Cu	–	850 MHz	~50 MHz	−23.98	18 cm	1 mW	–	in air	–	Energy harvesting/ Communication

(continued)

Table 7.2 (continued)

Year	Ref.	Antenna type	Antenna material	Encapsulation material	Operating frequency	Power transferring performance				Efficiency (%)	Test condition	Load	Application
						Bandwidth	S11(dB)	Distance	Power				
2019	[90]	Loop	Zn	Commercial (Bausch + Lomb)	13.56 MHz	–	−12.6	10 mm	–	17.6	in air	LED	Energy Harvesting/ Power generation
2019	[91]	Spiral	Cu	PDMS	2.4 GHz	90 MHz	−28	–	–	–	in saline water	LED	Energy harvesting
2019	[92]	Spiral	AgNW/AgNF	Elastic polymer (Elastofilcon A)	13.56 MHz	~2 MHz	−12	5 mm	–	–	on human eye	Supercapacitor /LED	Energy harvesting
2020	[93]											NFC chip	
2019	[94]	Loop	Au	PDMS	433 MHz	–	−19.73	1 cm	–	–	in vitro	Glucose	Energy harvesting/ Communication
2021	[95]											Sensor chip	Energy harvesting/ Communication
2020	[77]	Loop	Cu	PDMS	433 MHz	–	–	5 cm	–	~6	on rabbit eye	ASIC chip	Energy harvesting/ Communication/ Drug delivery/ Glucose sensing
2022	[96]												
2022	[58]	Serpentine	AgSEBs/ Silbione/ AgSEBs	PDMS	238.5 MHx	>1 MHz	~16.5	14 mm	–	–	on pig eye	–	IOP sensing
2022	[57]	Spiral	Cu/Au	PDMS	850 kHz	~50 kHz	−30	15 mm	–	–	on rabbit eye		Energy harvesting/ Drug delivery
					3.8 GHz	~100 MHz	−36	6 mm					IOP sensing

determined to be separated significantly from the intraocular sensing resonance, and the optimized design shows a power transfer efficiency of around 48.4% at a testing distance of 6 mm [57]. Kim et al. proposed a wireless system with an octagonal spiral receiver coil on a contact lens for near-field resonance with 17.5% wireless power transfer efficiency and a 20 mm powering distance [84]. The parylene-based coil is proved to be compatible with thin-film process fabrication, which remains stable under moderate device bending, indicating its suitability for daily wearing and simultaneously enabling wireless power transmission. In order to solve the view obstruction issue, Park et al. first designed the transparent smart contact lens coil using AgNFs material by photolithography and a wet etching process for wireless power transfer, which resonates at 50 MHz with a wireless power transfer efficiency of 21.5% [88]. The transparent material provides larger space on the contact lens for wireless power transfer coil design, meanwhile, the vision view is guaranteed through local indices matching, and the load is connected with a graphene-made glucose sensor for real-time tear glucose monitoring. Ku et al. further pushed the design and fabrication boundaries by combining AgNWs with AgNFs as the coil design material, which aims to reduce the locally disconnected AgNFs areas during the photolithographic etching patterning process [93]. The hybrid coil resonates at 13.56 MHz with a matching capacitor and is connected to an NFC chip to realize real-time cortisol monitoring which communicates with a smartphone. In addition, the in-vivo and human pilot experiments in live rabbits have confirmed that the lens has good biocompatibility and reliability as a non-invasive and mobile health monitoring solution.

Apart from reactive near-field antenna design, the magnetic resonance under the Fresnel region is popularly explored together with radio frequency identification (RFID) technology. Chiou et al. designed a WPT system with a loop receiver antenna and a patch transmitter antenna mimicking the eyeglass shape [86]. The proposed system resonated at 920 MHz, which was connected with a silicon sensor chip containing the sensor readout circuit and Gen-2 wireless transceiver, aiming for the intraocular pressure sensing through the micro-capacitive sensor made of top-down parallel plates. The on-lens system enabled the capacitive sensing of the pressure change from 2.25 mmHg to 30 with a powering distance of 1 cm and an incident power level of 26.5 dBm. Jeon et al. have been devoted to the exploration of wireless transmission systems upon smart contact lens wake-up call through a loop antenna, which operates for uplink and downlink at 346 MHz and 433 MHz, respectively [94, 95]. Hahn's group intends to explore the real-time monitoring and therapeutic improvement for intraocular pressure control and diabetic retinopathy treatment [77], where cooper foil was patterned through wet etching photolithography, together with an ASIC for data transmission telemetry. Followed research utilizes a similar loop antenna design to realize the wireless power transfer and data communication under the same frequency with a power transferring distance of 5 cm [96], where the on-demand drug delivery system relying on the external Bluetooth transmission helps with the intraocular pressure control.

The development of smart contact lenses with integrated wireless systems has seen advancements in near-field and magnetic resonance coil designs. However, there are limitations to consider. Near-field coil designs may face view obstruction issues, and while transparent materials mitigate this, they are not entirely eliminated. Magnetic resonance coil designs require specific frequencies for resonance, limiting their versatility. Furthermore, the compact size of contact lenses imposes constraints on the size and power capabilities of the integrated wireless systems. Despite these limitations, the development of near-field and magnetic resonance coils for smart contact lenses shows promise for a range of applications, from wireless power transfer to medical monitoring, providing non-invasive and mobile solutions. With continued research and refinement, these technologies have the potential to revolutionize healthcare and other industries.

7.4.2 Radiative Antenna

While near-field resonance focuses on short-range power transfer with high efficiency, far-field radiation facilitates long-range communication using electromagnetic waves, whereas radiative systems are commonly utilized for long-distance communication, including cellular networks, Wi-Fi, radio, and television broadcasting. Pandey et al. proposed a gold loop antenna on a smart contact lens working at three bandwidths, where signals can be transmitted under 2.4 GHz. With the inherited ASIC die chip design, the system efficiency reaches 10% with a transmission distance of 10 cm [81]. Yuan et al. also explored the Bluetooth and WIFI antenna integrated into a smart contact lens, where a spiral copper-designed antenna guaranteed a bandwidth of 90 MHz, and a reflection coefficient of −28 dB [91]. Lingley et al. depicted gold made loop antenna working for load micro-LED powering, the proposed antenna works under a resonant frequency of 0.8–2.0 GHz with the inherited LED driver circuit, where the wireless power transfer reaches a testing distance of 2 cm and a power consumption of 117 μW at 2 GHz [87]. In order to achieve the communication and power transfer on the same antenna, meanwhile reducing the inner interference, Cheng et al. proposed a rectenna system on the smart contact lens for dual-band sensing and communication under 5.8 GHz [82]. The rectenna was composed of two Au/Ti loop, where the gap in between aims to lower the first resonance mode and for bandwidth increase within the limited design space, which process the wireless power transfer efficiency of 35% tested on the pig eye for 1.5 V/1.2 mA output power level to successful activating 2.4 GHz voltage control oscillator. For transparency and self-powered consideration, Shahini et al. designed a Graphene-based antenna and sensor, which receives and transmits signals at a higher frequency (resonates at 10 and 19 GHz), allowing for sensing in a noisy environment [85]. The Graphene made transistor and the constituted circuit allows efficient frequency multiplication, where the all-graphene wireless sensor can be light-weight, flexible, and optically transparent due to the excellent mechanical and optical characteristics of graphene, additionally, the proposed antenna was also applicable to the application of multilayered transparent ultra-wideband antenna design.

Radiative antennas offer a wider range of communication options and can potentially provide better coverage compared to near-field resonance-based power transfer, and seamless communication over greater distances is beneficial for applications that require reliable and efficient long-range communication, such as transmitting medical data or receiving real-time information. However, there are also disadvantages to consider. One of the main challenges is the limited space available on a smart contact lens for antenna design, where the small size and limited surface area of the lens restrict the complexity of the antenna for integration, which further results in limitations in terms of bandwidth, efficiency, and range. Furthermore, the use of radiative antennas on smart contact lenses may also raise concerns about safety and potential interference. As the antenna emits electromagnetic waves, it is essential to ensure that the radiation levels are within accepted limits and do not pose any health risks to the wearer or interfere with other electronic devices in the vicinity. In summary, sensor-integrated smart contact lenses offer significant advantages in healthcare applications. The incorporation of wireless power transfer units can enhance their functionality and usability. However, careful consideration must be given to factors such as power efficiency, comfort, and interference to ensure the optimal performance of these sensor systems.

7.5 Energy Plan

The energy plan for power generation and power storage components in smart contact lenses is a crucial aspect of their design and functionality, where the lenses require a reliable and sustainable energy source to power the integrated electronic components and sensors. In terms of power generation, various methods can be employed. Energy harvesting techniques, such as solar or ambient light harvesting, can be used to capture and convert light energy into electrical energy [97]. Additionally, other approaches like heat or kinetic energy can also be explored to enhance power generation capabilities [98, 99]. To ensure continuous operation, power storage components are essential, which include supercapacitors or batteries that can be integrated into the lenses to store the generated electrical energy, providing a platform for excess energy storage for use during periods of low or no power generation. However, the power storage capacity should be carefully designed to meet the power requirements of the lens system, ensuring sufficient energy storage for prolonged usage. Overall, a well-designed energy plan is essential for smart contact lenses to ensure a sustainable and reliable power generation and storage system. By utilizing energy harvesting techniques, integrating power storage components, and implementing efficient power management, smart contact lenses can operate autonomously and provide continuous monitoring and functionality without the need for frequent battery replacements or external power sources.

7.5.1 Power Generation

The power generation plan in smart contact lenses is a critical aspect of their design and functionality, these lenses require a sustainable and reliable power source to power the integrated electronic components and sensors, especially the request for self-sufficiency. Smart contact lenses need to operate autonomously without relying on external power sources or frequent battery replacements, which is particularly important as traditional power sources, such as batteries, may be impractical due to their size, weight, and limited capacity in the compact form factor of a contact lens. The power generation plan for smart contact lenses shares similarities with power generation in other wearable devices. For instance, wearable fitness trackers often utilize kinetic energy from body movements through piezoelectric materials [100]. Similarly, smartwatches incorporate solar cells to generate power from ambient light [101, 102]. These power generation techniques enable wearables to operate for extended periods without interruption.

To address this challenge, Takamatsu et al. proposed the DC power generation by a Zn loop anode and a bilirubin oxidase (BOD) biocathode in an artificial tear, together with the wireless power transfer coils for hybrid power source [90]. When the AC voltage is powered off, a DC offset of 1.5 V is generated by connecting the metallic anodes to BOD biocathodes, which generate the maximum power 74 μW at 0.7 V, and with AC power supplied, the output voltage was boosted. The designed hybrid power source can potentially realize wearable electronics in body fluids and promises versatile sources for many applications such as displays and iontophoretic drug-delivery systems. Other than powering from metallic electrodes, Guang et al. designed a snowflake-shaped piezoelectric contact lens generating energy from eye blinks and intend for corneal injury repair application [103]. The flexible blink-driven piezoelectric contact lens (BPCL) was composed of a piezoelectric polypropylene electret film generator (PEG) with aluminum (Al) electrodes for electric pulse generation, a concentric circular indium tin oxide (ITO)/polyethylene terephthalate (PET) film electrode, and a micro rectifier as the intermediate connection. The optimized snowflake geometry with 12 dendrites generated a maximum 1.3 V output voltage with an increased pressure from 1 to 20 kPa. The in-vivo experiment reveals the effective corneal injury repair through the stable vertical electric field strength of $\sim$100 mV/mm at 5 mm deep inside. Overall, these power generation techniques offer promising prospects for the future development of smart contact lenses. With further advancements, these lenses could become self-sufficient power sources for various applications in the field of wearable technology.

7.5.2 Power Storage

Unlike batteries, which store energy chemically, supercapacitors store energy electrostatically. This allows them to charge and discharge rapidly, making them ideal for applications requiring high-power bursts. The design of supercapacitors for smart contact lenses involves incorporating thin layers of electrode materials into the lens. These electrodes are typically made of conductive polymers or carbon nanotubes, which provide a large surface area for energy storage. The supercapacitor is then integrated with other electronic components, such as sensors and wireless communication modules, to create a fully functional smart contact lens. Supercapacitors show great potential as energy storage components for smart contact lens systems due to their high-power density, extended cycling stability, and minimal material toxicity [104, 105]. Chiou et al. once proposed a thin-film supercapacitor design on a smart contact lens, forming a semi-passive system with the 3D packaging technology [106]. The micro-electrode was composed of Au/Ti and graphene materials, which were dropped by PVA-H_3PO_4 gel electrolyte on the top, and parylene-C was deposited for electrolyte insulation and whole encapsulation. The proposed supercapacitor was measured to process a good electrical double-layer capacitor characteristic of 9.53 F/cm^3 at 10 mV/s and 6.60 F/cm^3 at 800 mV/s in saline solution, which has great potential to be extended to more wearable electronic applications. Park et al. designed a monolithically integrated solid-state supercapacitor (MIS-supercapacitor) for power storage, integrated with a glucose sensor and wireless power transfer system [92]. The supercapacitor has an arc-shaped form factor that aims to be compatible with securing maximum energy density within the limited area of this contact lens, which allows certain deformability under the stretching of this soft lens. In addition, the design of the MIS-supercapacitor includes an in-plane electrode configuration to minimize the potential for the failure of the internal short circuit during abrupt conditions. The advantages of using supercapacitors in smart contact lenses are numerous. They offer quick charging times, longer lifespan, and higher power density compared to traditional batteries. Additionally, supercapacitors are more flexible and can be seamlessly integrated into the lens without compromising comfort or vision. In conclusion, the integration of supercapacitors into smart contact lenses holds great potential for the advancement of wearable technology. This innovative design allows for efficient energy storage and delivery, paving the way for more comfortable and functional smart contact lenses in the future.

References

1. X.P. Liang et al., Fusion of wearable and contactless sensors for intelligent gesture recognition. Adv. Intell. Syst. **1**(7), 1900088 (2019)
2. Y.J. Gao et al., Wearable microfluidic diaphragm pressure sensor for health and tactile touch monitoring. Adv. Mater. **29**(39) (2017). https://doi.org/10.1002/adma.201701985

3. W. Gao et al., Fully integrated wearable sensor arrays for multiplexed in situ perspiration analysis. Nature **529**(7587), 509-+ (2016)
4. T.R. Fricke et al., Global prevalence of presbyopia and vision impairment from uncorrected presbyopia: systematic review, meta-analysis, and modelling. Ophthalmology **125**(10), 1492–1499 (2018)
5. J.D. Steinmetz et al., Causes of blindness and vision impairment in 2020 and trends over 30 years, and prevalence of avoidable blindness in relation to VISION 2020: The right to sight: an analysis for the global burden of disease study. Lancet Glob. Health **9**(2), e144–e160 (2021)
6. K.D. Frick et al., The global burden of potential productivity loss from uncorrected presbyopia. Ophthalmology **122**(8), 1706–1710 (2015)
7. R.R. Holman et al., 10-year follow-up of intensive glucose control in type 2 diabetes. N. Engl. J. Med. **359**(15), 1577–1589 (2008)
8. S.H. Lim et al., Rapid diagnosis of tuberculosis from analysis of urine volatile organic compounds. ACS Sens **1**(7), 852–856 (2016)
9. C. Lin et al., Toward the development of a glucose dehydrogenase-based saliva glucose sensor without the need for sample preparation. J. Diabetes Sci. Technol. **12**(1), 83–89 (2018)
10. D. Harvey, N.W. Hayes, B. Tighe, Fibre optics sensors in tear electrolyte analysis: towards a novel point of care potassium sensor. Cont. Lens Anterior Eye **35**(3), 137–144 (2012)
11. K. Mitsubayashi, T. Arakawa, Cavitas sensors: contact lens type sensors & mouthguard sensors. Electroanalysis **28**(6), 1170–1187 (2016)
12. C.R. Taormina et al., Analysis of tear glucose concentration with electrospray ionization mass spectrometry. J. Am. Soc. Mass Spectrom. **18**(2), 332–336 (2007)
13. L.G. Carney, T.F. Mauger, R.M. Hill, Buffering in human tears: pH responses to acid and base challenge. Invest. Ophthalmol. Vis. Sci. **30**(4), 747–754 (1989)
14. A. Cancarini et al., Trace elements and diabetes: assessment of levels in tears and serum. Exp. Eye Res. **154**, 47–52 (2017)
15. A. Kijlstra, S.H. Jeurissen, K.M. Koning, Lactoferrin levels in normal human tears. Br. J. Ophthalmol. **67**(3), 199–202 (1983)
16. V. Ng, P. Cho, The relationship between total tear protein concentrations determined by different methods and standards. Graefes Arch. Clin. Exp. Ophthalmol. **238**(7), 571–576 (2000)
17. H. Hashemi et al., Distribution of intraocular pressure in healthy Iranian individuals: the Tehran eye study. Br. J. Ophthalmol. **89**(6), 652–657 (2005)
18. A. Rastegar, M. Soleimani, Hypokalaemia and hyperkalaemia. Postgrad. Med. J. **77**(914), 759–764 (2001)
19. N.M. Farandos et al., Contact lens sensors in ocular diagnostics. Adv. Healthc. Mater. **4**(6), 792–810 (2015)
20. A. Alghadyan et al., Lacrimal glands in cystic fibrosis. Saudi J Ophthalmol **27**(2), 113–116 (2013)
21. L.S. Feiler et al., Herpetic keratitis in zinc-deficient rabbits. Invest. Ophthalmol. Vis. Sci. **22**(6), 788–795 (1982)
22. Y. Danjo et al., Ocular surface damage and tear lactoferrin in dry eye syndrome. Acta Ophthalmol. **72**(4), 433–437 (1994)
23. C. Terres-Martos et al., Serum zinc and copper concentrations and cu/Zn ratios in patients with hepatopathies or diabetes. J. Trace Elem. Med. Biol. **12**(1), 44–49 (1998)
24. N.V.U. Hohenstein-Blaul, S. Funke, F.H. Grus, Tears as a source of biomarkers for ocular and systemic diseases. Exp. Eye Res. **117**, 126–137 (2013)
25. J. Murube, Basal, reflex, and psycho-emotional tears. Ocul. Surf. **7**(2), 60–66 (2009)
26. A. Mann, B. Tighe, Contact lens interactions with the tear film. Exp. Eye Res. **117**, 88–98 (2013)
27. N.J. Van Haeringen, Clinical biochemistry of tears. Surv. Ophthalmol. **26**(2), 84–96 (1981)
28. S. Agaoglu et al., Ultra-sensitive microfluidic wearable strain sensor for intraocular pressure monitoring. Lab Chip **18**(22), 3471–3483 (2018)

29. E. Sollier et al., Rapid prototyping polymers for microfluidic devices and high pressure injections. Lab Chip **11**(22), 3752–3765 (2011)
30. H.B. An et al., Microfluidic contact lenses for unpowered, continuous and non-invasive intraocular pressure monitoring. Sens. Actuat. A. Phys. **295**, 177–187 (2019)
31. B. Maeng, H.K. Chang, J. Park, Photonic crystal-based smart contact lens for continuous intraocular pressure monitoring. Lab Chip **20**(10), 1740–1750 (2020)
32. P.J. Chen et al., Unpowered spiral-tube parylene pressure sensor for intraocular pressure sensing. Sens. Actuat. A. Phys. **127**(2), 276–282 (2006)
33. M. Miyashita et al., Development of urine glucose meter based on micro-planer amperometric biosensor and its clinical application for self-monitoring of urine glucose. Biosens. Bioelectron. **24**(5), 1336–1340 (2009)
34. M.X. Chu et al., Soft contact lens biosensor for in situ monitoring of tear glucose as non-invasive blood sugar assessment. Talanta **83**(3), 960–965 (2011)
35. H. Yao et al., A contact lens with integrated telecommunication circuit and sensors for wireless and continuous tear glucose monitoring. J. Micromech. Microeng. **22**(7), 075007 (2012)
36. H. Yao et al., A contact lens with embedded sensor for monitoring tear glucose level. Biosens. Bioelectron. **26**(7), 3290–3296 (2011)
37. A. Kunzmann et al., Toxicology of engineered nanomaterials: focus on biocompatibility, biodistribution and biodegradation. Biochim. Biophys. Acta **1810**(3), 361–373 (2011)
38. S. Halldorsson et al., Advantages and challenges of microfluidic cell culture in polydimethylsiloxane devices. Biosens. Bioelectron. **63**, 218–231 (2015)
39. S. Li et al., Advanced wearable microfluidic sensors for healthcare monitoring. Small **16**(9), e1903822 (2020)
40. R.V. Nair, R. Vijaya, Photonic crystal sensors: an overview. Prog. Quantum Electron. **34**(3), 89–134 (2010)
41. X.F. Wang, J. Engel, C. Liu, Liquid crystal polymer (LCP) for MEMS: processes and applications. J. Micromech. Microeng. **13**(5), 628–633 (2003)
42. I. Sage, Thermochromic liquid crystals. Liq. Cryst. **38**(11–12), 1551–1561 (2011)
43. R. Moreddu et al., Contact lenses for continuous corneal temperature monitoring. RSC Adv. **9**(20), 11433–11442 (2019)
44. Y.R. Lin et al., Noninvasive glucose monitoring with a contact lens and smartphone. Sensors **18**(10) (2018)
45. M. Leonardi et al., Wireless contact lens sensor for intraocular pressure monitoring: assessment on enucleated pig eyes. Acta Ophthalmol. **87**(4), 433–437 (2009)
46. G.E. Dunbar, B.Y. Shen, A.A. Aref, The Sensimed Triggerfish contact lens sensor: efficacy, safety, and patient perspectives. Clin. Ophthalmol. **11**, 875–882 (2017)
47. V. Laukhin et al., Non-invasive intraocular pressure monitoring with a contact lens engineered with a nanostructured polymeric sensing film. Sens. Actuat. A Phys. **170**(1–2), 36–43 (2011)
48. D. Dias, J.P.S. Cunha, Wearable health devices-vital sign monitoring, systems and technologies. Sensors (Basel) **18**(8) (2018)
49. A.C. Kevat et al., A systematic review of novel technology for monitoring infant and newborn heart rate. Acta Paediatr. **106**(5), 710–720 (2017)
50. G. Walker, A review of technologies for sensing contact location on the surface of a display. J. Soc. Inf. Disp. **20**(8), 413–440 (2012)
51. B.Y. Won, H.G. Park, A touchscreen as a biomolecule detection platform. Angew. Chem., Int. Ed. **51**(3), 748–751 (2012)
52. M. Ali et al., Optimization of MEMS intraocular capacitive pressure sensor, in *2016 IEEE International Conference on Semiconductor Electronics (ICSE) Proceedings*, (2016), pp. 173–176
53. J.C. Chiou et al., Toward a wirelessly powered on-lens intraocular pressure monitoring system. IEEE J. Biomed. Health Inform. **20**(5), 1216–1224 (2016)

54. J.C. Chiou, Y.C. Huang, G.T. Yeh, A capacitor-based sensor and a contact lens sensing system for intraocular pressure monitoring. J. Micromech. Microeng. **26**(1), 015001 (2016)
55. G.Z. Chen et al., Soft wearable contact lens sensor for continuous intraocular pressure monitoring. Med. Eng. Phys. **36**(9), 1134–1139 (2014)
56. J. Kim et al., Wearable smart sensor systems integrated on soft contact lenses for wireless ocular diagnostics. Nat. Commun. **8**, 14997 (2017)
57. C. Yang et al., Intelligent wireless theranostic contact lens for electrical sensing and regulation of intraocular pressure. Nat. Commun. **13**(1), 2556 (2022)
58. J. Zhang et al., Smart soft contact lenses for continuous 24-hour monitoring of intraocular pressure in glaucoma care. Nat. Commun. **13**(1), 5518 (2022)
59. D.C. Hood et al., Assessment of local retinal function in patients with retinitis pigmentosa using the multi-focal ERG technique. Vis. Res. **38**(1), 163–179 (1998)
60. L.J. Wilsey, B. Fortune, Electroretinography in glaucoma diagnosis. Curr. Opin. Ophthalmol. **27**(2), 118–124 (2016)
61. H.P. Scholl, E. Zrenner, Electrophysiology in the investigation of acquired retinal disorders. Surv. Ophthalmol. **45**(1), 29–47 (2000)
62. J.V. Odom et al., ISCEV standard for clinical visual evoked potentials: (2016 update). Doc. Ophthalmol. **133**(1), 1–9 (2016)
63. A.G. Robson et al., ISCEV standard for full-field clinical electroretinography (2022 update). Doc. Ophthalmol. **144**(3), 165–177 (2022)
64. P.A. Constable et al., ISCEV standard for clinical electro-oculography (2017 update). Doc. Ophthalmol. **134**(1), 1–9 (2017)
65. G.B. Arden et al., A gold foil electrode: extending the horizons for clinical electroretinography. Invest. Ophthalmol. Vis. Sci. **18**(4), 421–426 (1979)
66. H.M. Burian, L. Allen, A speculum contact lens electrode for electroretinography. Electroencephalogr. Clin. Neurophysiol. **6**(3), 509–511 (1954)
67. A.G. Fernandes et al., Full-field electroretinogram recorded with skin electrodes in normal adults. Arq. Bras. Oftalmol. **79**(6), 390–394 (2016)
68. L. Esakowitz, A. Kriss, F. Shawkat, A comparison of flash electroretinograms recorded from Burian Allen, JET, C-glide, gold foil, DTL and skin electrodes. Eye (Lond) **7**(Pt 1), 169–171 (1993)
69. M.P. Hennessy, Vaegan, Amplitude scaling relationships of Burian-Allen, gold foil and Dawson, Trick and Litzkow electrodes. Doc. Ophthalmol. **89**(3), 235–248 (1995)
70. A.H. Brouwer et al., Effects of DTL electrode position on the amplitude and implicit time of the electroretinogram. Doc. Ophthalmol. **140**(3), 201–209 (2020)
71. R. Yin et al., Soft transparent graphene contact lens electrodes for conformal full-cornea recording of electroretinogram. Nat. Commun. **9**(1), 2334 (2018)
72. K. Kim et al., All-printed stretchable corneal sensor on soft contact lenses for noninvasive and painless ocular electrodiagnosis. Nat. Commun. **12**(1), 1544 (2021)
73. S. Mizunoya et al., Electroretinogram contact lens electrode with tri-color light-emitting diode. Acta Ophthalmol. Scand. **79**(5), 497–500 (2001)
74. R. Badugu et al., Contact lens to measure individual ion concentrations in tears and applications to dry eye disease. Anal. Biochem. **542**, 84–94 (2018)
75. Y.R. Lin et al., Noninvasive glucose monitoring with a contact lens and smartphone. Sensors (Basel) **18**(10) (2018)
76. P.G. Boswell, P. Buhlmann, Fluorous bulk membranes for potentiometric sensors with wide selectivity ranges: observation of exceptionally strong ion pair formation. J. Am. Chem. Soc. **127**(25), 8958–8959 (2005)
77. D.H. Keum et al., Wireless smart contact lens for diabetic diagnosis and therapy. Sci. Adv. **6**(17), eaba3252 (2020)
78. Y.T. Liao et al., A 3-mu W CMOS glucose sensor for wireless contact-lens tear glucose monitoring. IEEE J. Solid State Circuits **47**(1), 335–344 (2012)
79. N. Thomas, I. Lahdesmaki, B.A. Parviz, A contact lens with an integrated lactate sensor. Sens. Actuat. B Chem. **162**(1), 128–134 (2012)

80. M. Donora et al., Spatiotemporal electrochemical sensing in a smart contact lens. Sens. Actuat. B Chem. **303**, 303 (2020)
81. J. Pandey et al., A fully integrated RF-powered contact lens with a single element display. IEEE Trans. Biomed. Circuits Syst. **4**(6), 454–461 (2010)
82. H.W. Cheng et al., The rectenna design on contact lens for wireless powering of the active intraocular pressure monitoring system. Annu. Int. Conf. IEEE Eng. Med. Biol. Soc. **2013**, 3447–3450 (2013)
83. L.Y. Chen, G. Shaker, S. Safavi-Naeini, Energy harvesting system integrated on wearable contact lens, in *IEEE International Symposium on Antennas and Propagation & Usnc/Ursi National Radio Science Meeting*, vol. 2015, (2015), pp. 358–359
84. Y.J. Kim, J.M. Maeng, P.P. Irazoqui, Eyeglasses-powered, contact lens-like platform with high power transfer efficiency. Biomed. Microdevices **17**(4), 75 (2015)
85. A. Shahini et al., Self-powered and transparent all-graphene biosensor, in *2016 IEEE Sensors*, (2016)
86. J.C. Chiou et al., A wirelessly powered smart contact lens with reconfigurable wide range and tunable sensitivity sensor readout circuitry. Sensors (Basel) **17**(1) (2017)
87. A.R. Lingley et al., A single-pixel wireless contact lens display. J. Micromech. Microeng. **21**(12), 125014 (2011)
88. J. Park et al., Soft, smart contact lenses with integrations of wireless circuits, glucose sensors, and displays. Sci. Adv. **4**(1), eaap9841 (2018)
89. L.Y. Chen et al., Cellular wireless energy harvesting for smart contact lens applications. IEEE Antennas Propag. Mag. **60**(5), 108–124 (2018)
90. T. Takamatsu et al., Multifunctional high-power sources for smart contact lenses. Adv. Funct. Mater. **30**(29) (2020)
91. M.Y. Yuan et al., Electronic contact lens: a platform for wireless health monitoring applications. Adv. Intell. Syst. **2**(4) (2020)
92. J. Park et al., Printing of wirelessly rechargeable solid-state supercapacitors for soft, smart contact lenses with continuous operations. Sci. Adv. **5**(12), eaay0764 (2019)
93. M. Ku et al., Smart, soft contact lens for wireless immunosensing of cortisol. Sci. Adv. **6**(28), eabb2891 (2020)
94. C. Jeon et al., A smart contact lens controller IC supporting dual-mode telemetry with wireless-powered backscattering LSK and EM-radiated RF transmission using a single-loop antenna. IEEE J. Solid State Circuits **55**(4), 856–867 (2020)
95. C. Jeon, J.Y. Sim, A 2.5-nW radio platform with an internal wake-up receiver for smart contact lens using a single loop antenna. IEEE J. Solid State Circuits **56**(9), 2668–2679 (2021)
96. T.Y. Kim et al., Wireless theranostic smart contact lens for monitoring and control of intraocular pressure in glaucoma. Nat. Commun. **13**(1), 6801 (2022)
97. J. Zhao et al., Self-powered implantable medical devices: photovoltaic energy harvesting review. Adv. Healthc. Mater. **9**(17), e2000779 (2020)
98. M. Thielen et al., Human body heat for powering wearable devices: from thermal energy to application. Energy Convers. Manag. **131**, 44–54 (2017)
99. Y.M. Choi, M.G. Lee, Y. Jeon, Wearable Biomechanical Energy Harvesting Technologies. Energies **10(10)**, 1483 (2017)
100. Y. Liu et al., Piezoelectric energy harvesting for self-powered wearable upper limb applications. Nano Select **2**, 1459–1479 (2021)
101. J.Q. Zhao et al., A fully integrated and self-powered smartwatch for continuous sweat glucose monitoring. Acs Sensors **4**(7), 1925–1933 (2019)
102. J.Q. Zhao et al., Recent advances in wearable self-powered energy systems based on flexible energy storage devices integrated with flexible solar cells. J. Mater. Chem. A **9**(35), 18887–18905 (2021)
103. G. Yao et al., Snowflake-inspired and blink-driven flexible piezoelectric contact lenses for effective corneal injury repair. Nature Communications **14**(1) (2023)

104. X. Wang et al., Flexible energy-storage devices: design consideration and recent progress. Adv. Mater. **26**(28), 4763–4782 (2014)
105. B. Song, K.S. Moon, C.P. Wong, Recent developments in design and fabrication of graphene-based interdigital micro-supercapacitors for miniaturized energy storage devices. IEEE Trans. IEEE Trans. Compon. Packag. Manuf. Technol **6**(12), 1752–1765 (2016)
106. J.E. Chiou et al., The methodology to make smart contact lens become a semi-passive system, in *2019 IEEE 32nd International Conference on Micro Electro Mechanical Systems (Mems)*, (2019), pp. 1006–1009

Chapter 8
Next-Generation SIMO Converter: A Step Beyond SISOs with Novel Approach

Kaung Oo Htet

8.1 Research Gap for Single Input, Single-Output (SISO) State-of-the-Art

Nowadays, biomedical devices with integrated microelectronics are incorporated into modern healthcare systems. Advances in CMOS fabrication processes are further miniaturising these devices. These devices will be integrated on Implantable neural probes [1], pacemakers [1, 2]and many more. Thus, enhancing the quality of day-to-day healthcare. Such devices can be divided into several building blocks, see Fig. 8.1, such as sensors [3, 4], ADC [5], MEM devices [6], memory [7] and so on. These components are implemented to serve the purpose of recording, stimulation, and communication. Each of these not only has different functionalities also require different operating voltages and load currents powered by a battery or energy harvester through DC-DC converters. These converters are based on the charge pump designs. In search of reliable and efficient charge pump converter quality can be examined by performance and reliability, depicted in Fig. 8.2. Performance of the converter is defined by the power efficiency and voltage conversion ratio. On the other hand, reliability is defined upon converter's ability to handle variation in output side (Load Regulation) and input side (Line Regulation).

Well-known charge pumps such as Dickson [8, 9] and cross-coupled topologies [10] are widely used for boost conversion purposes and often used as energy harvesters as their input source [11]. However, only the integer gains can be configured in charge pump design, making them inconvenient for applications with varying inputs or outputs. Whenever there is a change in input voltage, the circuit may need to be re-adjusted to keep up with the high conversion ratios.

K. O. Htet (✉)
University of Hertfordshire, Specter building, College lane Campus Hatfield, Glasgow, UK
e-mail: k.oo-htet@herts.ac.uk

R. Das, H. Heidari (eds.), *Wireless Power Technologies for Biomedical Devices*, Analog Circuits and Signal Processing, https://doi.org/10.1007/978-3-031-52834-7_8

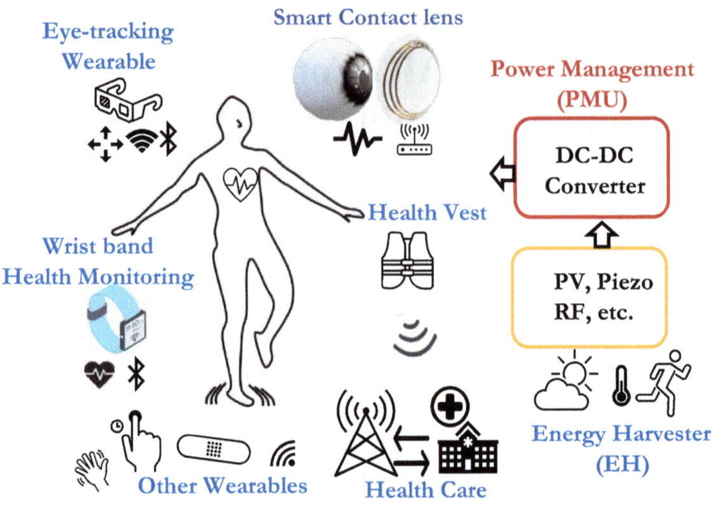

Fig. 8.1 Overview of Power management system in Biomedical devices

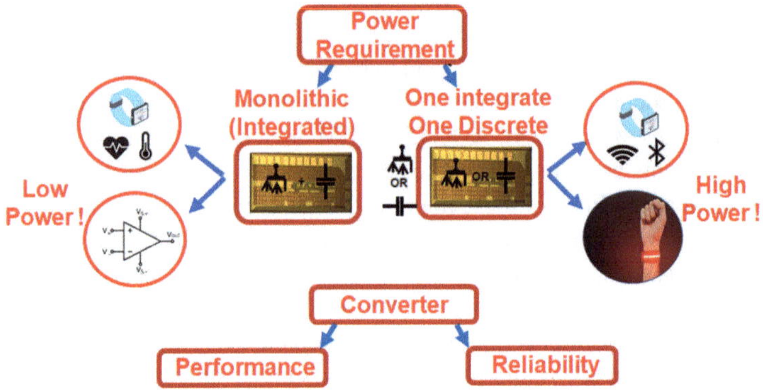

Fig. 8.2 Classification of two types of power management systems and validation based on performance and reliability

Therefore, series-parallel converters [12] that can configure non-integer gain values and reconfigure DC gains [13, 14] are becoming more popular. Present series-parallel converters are reconfigurable. Sometimes, a specific load operating at the highest voltage requires maximum power from the converter for an extended period. Thus, the converter drives at its optimum gain almost all the time to provide sufficient voltage to all the present loads connected to a single output line. Moreover, since all the loads are connected to a single output line, small loads with low operating voltages require further drop down using additional Linear Drop Out (LDO) circuits. It leads to power losses across different components.

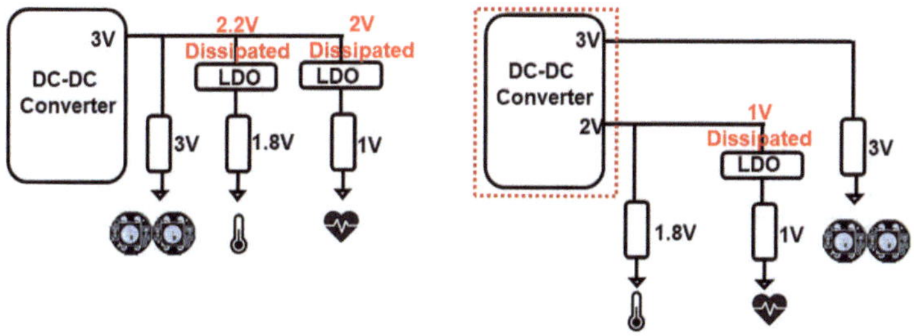

Fig. 8.3 Example of the energy-efficient delivering method compared to (**a**) SISO design (**b**) SIDO design

In search for the solution to this challenge, a multi-output converter can greatly benefit applications consisting of electronic modules with different operational voltages, see Fig. 8.3.

8.2 State-of-the-Art Single-Input-Multi-outputs Converters

Single-input-multi-output designs are becoming popular due to their ability to supply an independent range of voltages to the load. Triple output voltage switch capacitor regulator is proposed in [15]. By using a binary reconfigurable SC and two Dickson's charge pumps in parallel, controlled by independent frequency modulations to produce two of 3/1 and 1/2 outputs for wide range accessible load currents. However, stacking three regulators resulted in the large area of overhead.

Next, the implementation of a two-rung SC ladder circuit topology for several discrete power modes for a power-aware application is presented in [16]. Three different voltage regions are generated spontaneously by two uneven flying capacitors in each cell sharing charge with an identical subcell, which operates in anti-phase, across odd number switches in two-phase mode, Fig. 8.4a. The ripple performance of all three sub voltages outputs, super-threshold (V_{dd}), near-threshold ($2V_{dd}/3$) and sub-threshold($V_{dd}/3$), are further enhanced by N-times interleaving, Fig. 4b. By which, increase the subcells to N-times to reduce the output ripples, Fig. 4c, and subsequently compensate for the increase in the chip area by reducing the size of the capacitors of each network by 1/N factors. The large capacitance capacitor is used for those at the large load region (V_{dd}) twice the bottom region ($V_{dd}/3$). Further optimisation is done using unequal flying capacitors instead of conventional equal capacitances across the network.

However, although the total area remains the same, the complexity of design for multiple interleaving grows as the number of sub-cells increases. The demand for considerable change in physical mechanics rather than optimising through system design is not ideal for implant power management. Secondly, a change in the size of the charge pump capacitor due to expanding the sub-cells will require the system

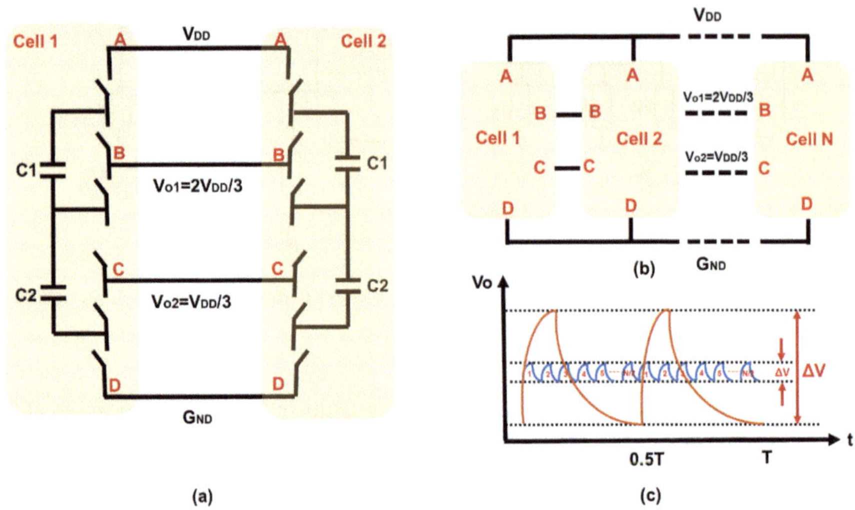

Fig. 8.4 (**a**) a two-rung SC ladder topology produces three sub-voltage regions (**b**) multiphase N time interleaving with identical sub-cells (**c**) Output ripple performances comparison of N-interleaving (small) and non-interleaving (large) [16]

to re-adjust the operating frequencies to discharge/charge and install an additional phase controller fully. Thirdly, the conversion efficiency will drop once the input voltage changes due to the fixed ratio.

The parallel SC converter, which uses four-channel time multiplexing to produce two simultaneous outputs, is proposed in [17]. The proposed SC converter in 65 nm is reconfigurable, generating gains of 1 and 1/2, respectively, during two-phase operations. Although the converter is reconfigurable, no step-up conversion is available. Therefore, it is not suitable for energy harvester applications.

A dual output converter integrated design producing ($2V_{in}$ and $-V_{in}$) for liquid crystal display (LCD) application is presented in [18], and it is fabricated in 0.13 μm technology. This work uses dual-side-dual-output (DSDO), enabling simultaneous step-up and step-down conversions. There are always two-phases charging and the other two phases for energy delivery at the output. This design's downside is that the fixed gain dual outputs and two independent voltage inverters and voltage doublers are used, resulting in area overhead.

Similarly, the capacitive DC-DC step-down converter governed by a traditional hysteretic controller using a simultaneous dual output design is fabricated in 90 nm CMOS technology and presented in [19], Fig. 8.5a–e. From the 1.2 V input battery, using the four charge pump capacitors and switches, two spontaneous dual outputs of 0.755 V and 0.32 V are gathered and displayed efficiencies between 68.6–75.3%. However, whenever the input voltages change due to the degraded battery life, the conversion efficiency will drop from the fixed gain ratio design.

The problem with this design is only down conversions are available at the output, and efficiency is relatively low, considering there are no step-up conversions.

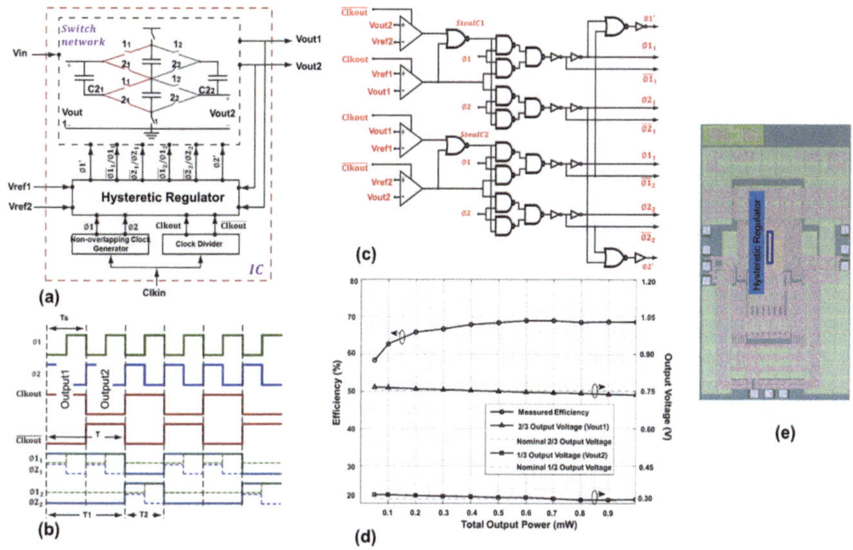

Fig. 8.5 (**a**) Ladder topology converter controlled by Hysteretic Regulator to produce dual output (**b**) Outputs (Out1, Out2) timing diagram, (**c**) schematic of Hysteretic regulator (**d**) Efficiency and Output voltage over the total output power (**e**) die photograph [19]

Though spontaneous voltages are available at the production, the design technique involves a lot of fast switching whenever the output voltage falls below reference voltages. Moreover, charge sharing to the second network across the switch and configuring the amplitude of the second network by switching pulses required additional control methods.

The best example of cascaded charge sharing is discussed in insect-scale pico-aerial vehicles (Fig. 8.6) [20]. This application employed the switched-capacitor 4:1 Voltage Regulator integrated power design described in [21]. Conversion of 4:1 is achieved by cascading 2:1 conversion twice and generating two simultaneous outputs from each network. The first network converts 3.7 V battery input into 1.8 V half conversions. It then feeds into the second network to yield 0.8 V. Both networks have an equal number of switches and charge pump capacitors. However, each network operates at different operating frequencies, different transistor sizes and independent feedback loops. The additional voltage-controlled oscillator was employed for the second network to optimise the dual output and manage the contact with the battery differently from the first network charging time to enable better start-up conditions. Furthermore, the second network is driven by a 4-time slower frequency than the first network as higher output voltages which require frequent energy delivery, are carried out by higher frequency switching. Whenever fast switching frequency in switch capacitor network for low output voltage will charge settling time in each cycle, and low load usually does not require fast energy delivery.

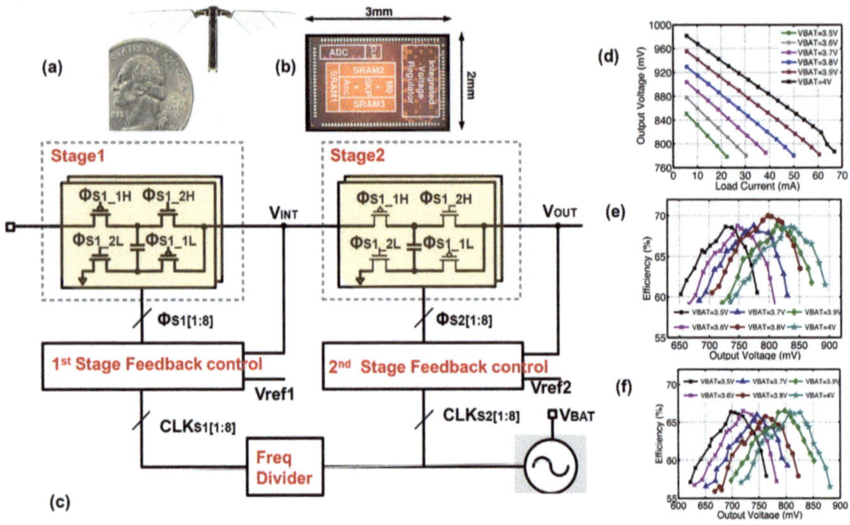

Fig. 8.6 (**a**) Size comparison of inset-scale pico-vehicle, (**b**) die photograph, (**c**) schematic of 2:1 and 2:1 cascaded dual output converter, (**d**) output voltage over load current and efficiency change over (**e**) line regulation and (**f**) load regulation [21]

In the SIDO SC converter design presented in [22], two gain pairs voltages of (×3, ×2 or ×2, ×2) can be configured at two different times. In each gain pair, 180-phase shifts, two different voltages amplitude are gathered at the V_{ol} and V_{oh} outputs. The operation can be best understood by following the green arrow line described in Fig. 8.7a. Vc_{f1} is charged to V_{dd} via switches S_{11} and S_{13}, while V_{cf2} is discharged to V_{oh} through S_{22} and S_{24}. Since V_{cf2} hold the value of $2V_{dd}$ from the previous charging phase, discharging to V_{oh} in series with the input source yields $3V_{dd}$ voltage at the output. The operation of Vc_{f1} and Vc_{f2} are swapped in the next phase. During the discharged phase of Vc_{f1}. The input source provides $2V_{dd}$ at the V_{ol}, and since it is in parallel with C_{f2}, the second capacitor, Vc_{f2}, is charged to $2V_{dd}$. Similarly, for Fig. 8.7b, simultaneous voltages are generated by charging two capacitors, Vc_{f1} and Vc_{f1}, in one phase to V_{dd} and discharging them independently in series with the input source to get $2V_{dd}$ at the outputs.

This work not only provides two different dual outputs by using only two capacitors, but it also allows for reconfiguring into different gain pairs. It is the only recent state-of-the-art SIDO converter in which conversion ratios are not fixed. However, the only downside is, in case 1, the converter does not simultaneously supply two outputs. Likewise, in case 2, although two spontaneous outputs are gathered in one phase, the charging phase is required for the discharged capacitor to reset. Therefore, this limitation is overcome by introducing this topology to our novel 4-phase rotations scheme and enhancing into simultaneous dual outputs with no idle condition for all four phases.

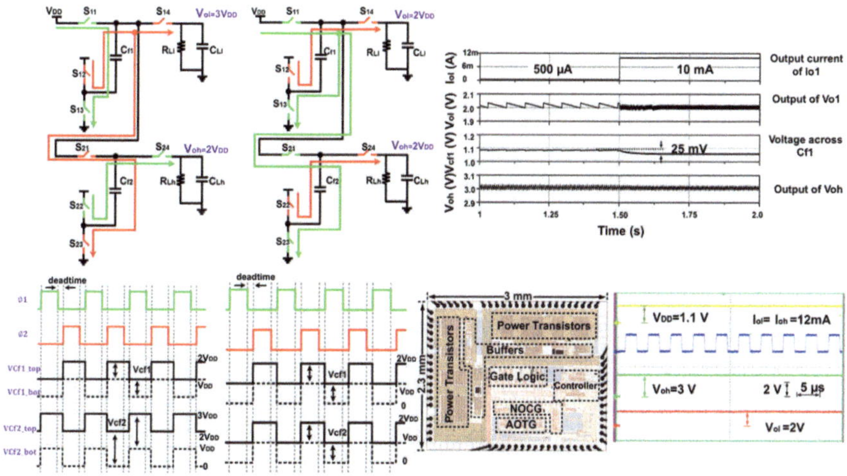

Fig. 8.7 Dual-output SC converter (**a**) (×3, ×2) gains (**b**) (×2, ×2) gains (**c**) cross-regulation of simulated load step change at Vol under Load regulation of $V_{DD} = 1.5$ V and (**d**) measurement output waveforms under $V_{DD} = 1.1$ V [22]

8.3 Abstract of Proposed Single-Input-Dual-Output Converter

This section presents a novel inductor-less switched capacitor (SC) DC-DC converter, which generates simultaneous dual-output voltages for implantable electronic devices. Present dual output converters are limited to fixed ratio gain, which degrades conversion efficiency when the input voltage changes. The proposed power converter offers step-up and step-down conversion with 4-phase reconfigurable logic. With an input voltage of 1 V provided by photovoltaic (PV) cells, the proposed converter achieves step-up, step-down and synchronised voltage conversions in four gain modes. These are 1.5 V and 0.5 V for Normal mode, 2 V and 1 V for High mode, 2 V for Double Boost mode, and 3 V and 2 V for Super Boost mode with a ripple variation of 14–59 mV. The converter circuit has been simulated in standard 0.18 μm CMOS technology, and the results agree with state-of-the-art SC converters. However, this proposed monolithically integrated PV-powered circuit achieves a conversion efficiency of 85.26%. It provides extra flexibility in terms of gain, which is advantageous for future microelectronics applications with various inputs.

8.4 Contribution of 4-Phase Rotation Dual Output Converter

This work introduces a miniaturised power management DC-DC converter design, an integrated on-chip power system for wearable devices. An inductor-less,

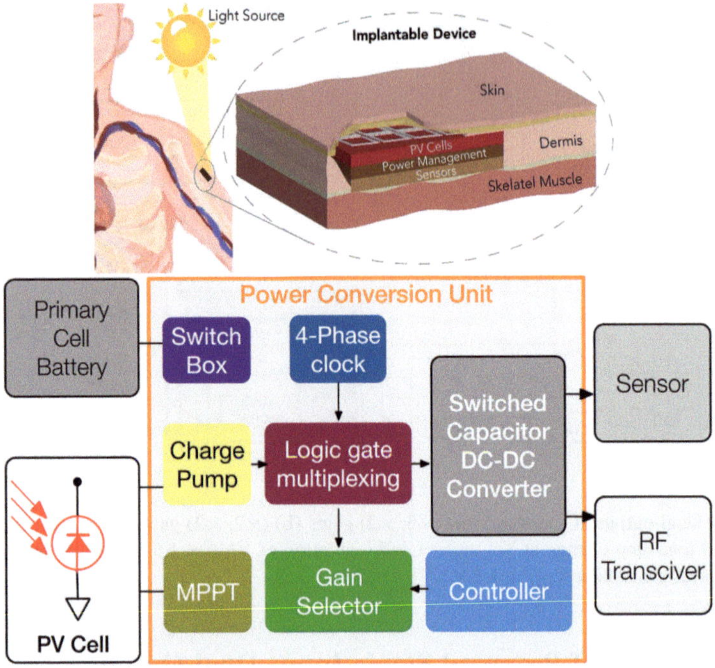

Fig. 8.8 Block Diagram of power conversion for energy harvesting using photovoltaic cells in Implantable applications

switched-capacitor, DC-DC converter design is proposed to remove the involvement of bulky inductors. Moreover, power management is obtained using series-parallel connections of flying capacitors via MOSFET power switches. These are all digitally controlled using digital modules, as shown in Fig. 8.8. The PV cell and its energy storage device are primary energy inputs. Depending on the power conversion unit's load requirements, the proposed system performs step-up and step-down voltage conversions simultaneously using the novel 4-phases rotation scheme with four flying capacitors. The appropriate gain pairs are chosen accordingly to produce a different gain of step-up conversion simultaneously. These various VCR gain pairs can be categorised into four modes (Normal, High, Double Boost, and Super Boost). This novel topology, which uses a single-input-dual-output (SIDO) design, is inspired by the single-input-single-output (SISO) 3-phase rotation topology presented in [23], whereby one idle capacitor in every charging phase is always present. Two outputs need to be produced independently to achieve the same objective. It will require either doubling the SISO converters [24], which can only generate one gain at a time or stacking the capacitors as a capacitor bank [21].

8.5 System Description

Using a switch-mode regulation, this converter illustrated in Fig. 8.9a, b has achieved the reconfigurable dual outputs (Vout1, Vout2), producing regulated voltage pairs of both boosting or boosting and buck at the same time. Four different gain modes represent a set of voltage gains generated from two internal conceptual

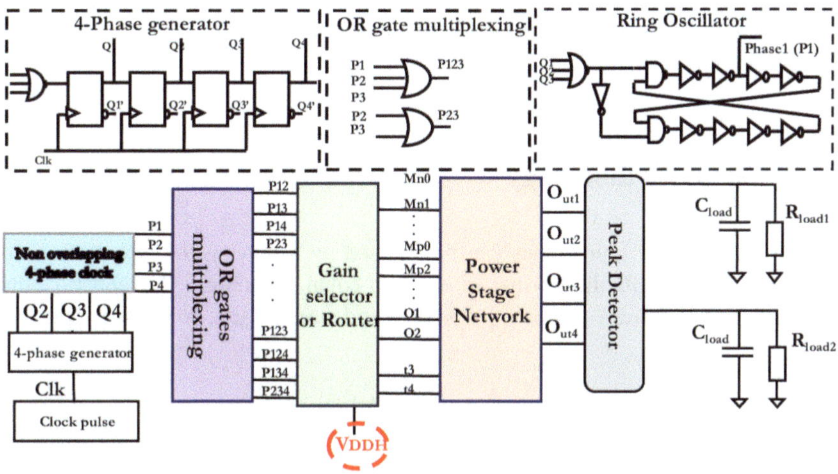

Modes	Normal	High	Double	Super
V_{out1}	$\frac{3V_{in}}{2}$	$2V_{in}$	$2V_{in}$	$2V_{in}$
V_{out2}	$\frac{V_{in}}{2}$	$\leq V_{in}$	$2V_{in}$	$3V_{in}$

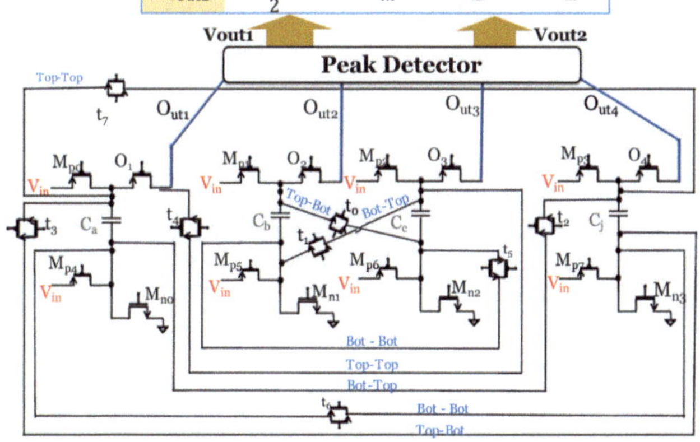

Fig. 8.9 (**a**) Overall schematic diagram of the proposed converter and the crucial blocks are highlighted in colours. (**b**) Circuit level diagram of the power stage network in which a non-overlapping 4-phases clock digitally controls the operation

networks (Net0, Net1). As the methodology suggests, various DC gains are configured in each network by constructing series-parallel configurations of capacitors and switches. Since its operation changes follow each capacitor's position in every phase, the corresponding internal outputs (Out1–Out4) also change accordingly. Henceforth, these internal outputs shall be referred to as "discontinuous" outputs (Out1-Out4), whereas "continuous" output refers to two outputs (Vout1, Vout2) connected to external loads. Four DC gain modes can be configured in (Vout1 and Vout2), each consisting of two different voltages. Therefore, discontinuous-to-continuous switches or rectifier circuits (peak detector) can be employed to correspond to the connection of internal outputs (Out1–Out4) to external loads (Vout1 and Vout2) at different phases.

8.6 System Methodology

The operation of all four modes can be realised by following the arrow rotations described in Fig. 8.10. In Normal mode, two equal capacitors, "C_b and C_c," are being charged in series with the input source and stored $V_{in}/2$. Assuming a present

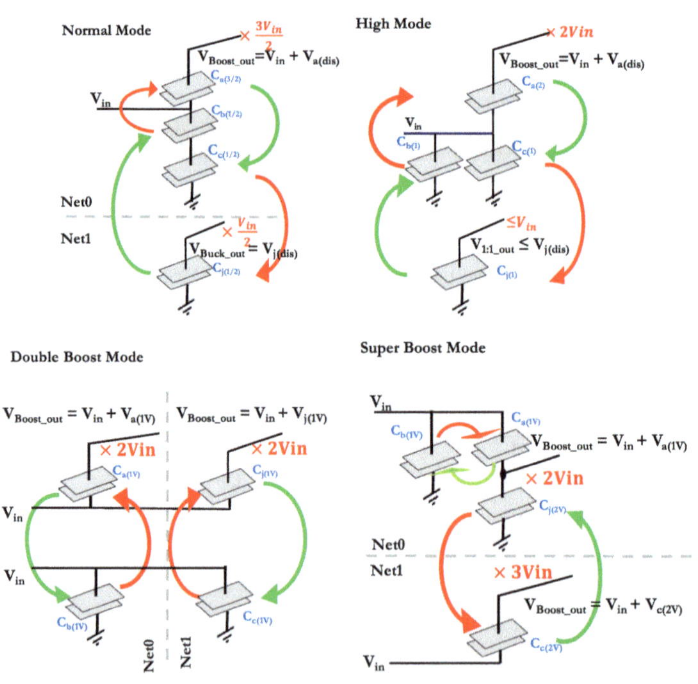

Fig. 8.10 The methodology of (**a**) Normal Mode, (**b**) High Mode, (**c**) Double Boost Mode, and (**d**) Super Boost Mode configurations

set-up is in a steady state, a step-up gain of Net0 output is obtained $3V_{in}/2$ by discharging capacitor "C_a" in series with the input source. Meanwhile, a capacitor, "C_j", is independently discharged to output at Net1 with the step-down gain of $V_{in}/2$. In the next phase, two charged capacitors are swapped with two discharged capacitors, as shown in Fig. 8.10a. Instead of constructing two charging capacitors in series, parallel connections are made so that each capacitor is charged to Vin values. Thus, the output of Net1 achieves up to Vin gain. Since this proposed converter not only provides fixed gain outputs but rearranges series, also-parallel connections of the power stage network can configure higher gain than Normal mode, as shown in Fig. 8.10b.

The reconfigurable dual-output SC converter with the two flying capacitors was presented in [22], whereas two different step-up gain pairs are discretely output to loads in two separate phases. These configurations are further enhanced by interleaving with a novel 4-phase clock topology to enable no idle charging time for all phases. As a result, continuous outputs of 2Vin are obtained for Double Boost and 3Vin and 2Vin for Super Boost in every cycle. Figure 8.10c, d show the proposed methodology.

8.7 Output Voltages and Output Power Losses

The derivation of output voltages for Normal and High modes has been described in [25]. By taking the total amount of charge for steady-state charging and discharging voltage-time ΔQ_{Mode} for Normal mode, High mode, Double Boost, and Super Boost are $\frac{I_O}{2\,f_s}$, $\frac{I_O}{3\,f_s}$, $\frac{I_O}{4\,f_s}$ and $\frac{I_O}{2\,f_s}$, respectively. The gain ratio multiplier (k) for different modes and the output voltage for individual modes can be generalised as:

$$V_{boost,buckMode} = kV_{in} - \frac{\Delta Q_{Mode}}{\alpha\, f_s C_p}\left(\frac{1}{1-e^{-\frac{t_d}{T_d}}} + \frac{1}{1-e^{-\frac{t_c}{T_c}}} - 1\right) \quad (8.1)$$

where t_d represents discharging time and t_c represents the charging time, the RC time constant (T_d and T_c) are subjected to on-resistance along the path of charging-discharging R_{on}, α is the charge multiplier varies over different modes and flying capacitor C_p which has the most dominant capacitance in the power stage. The conduction power loss can be expressed from the second term of (8.1). Multiplying by the output current I_o gives (8.2):

$$P_{conduct\_Mode} = \frac{I^2_{oboost,buck}}{\alpha\, f_s C_p}\left(\frac{1}{1-e^{-\frac{t_d}{T_d}}} + \frac{1}{1-e^{-\frac{t_c}{T_c}}} - 1\right) \quad (8.2)$$

$$P_{sw\_Mode} = f_s \sum_i C_{gsi} V^2_{gsi} \cong f_s C_{OX} \sum_i W_i L_i V^2_{gsi} \quad (8.3)$$

Equation (8.3) represents switching power loss (P_{SW}), which is influenced by charge-discharge mechanisms of a gate-source parasitic capacitance of MOSFET transistors (C_{gs}), the frequency of switching (f_s) and the gate-source voltages (V_{gs}) of individual switches along the path [26]. The other variables are the width (W_i) and length (L_i) of individual transistors along the path of operations and the transistor gate-oxide capacitor (C_{ox}), which has the approximate value of 8.31 fF/μm^2 for 0.18 μm technology. Due to the symmetric nature of output waveforms in every mode, redistribution loss is considered by taking the KQL of a one-phase operation for a steady state. Whereas present phase discharge capacitor (C_d), switching time (T_s) and the number of parallel connections with C_d (n) are described in (8.4):

$$P_{redis\_Mode} = \frac{I^2_{o\_boost,buck}\,T_s}{4\,C_d} + \frac{I^2_{o\_boost,buck}\,T_s}{8\left(nC_d + C_{load1,2}\right)} \quad (8.4)$$

8.8 Design and Implementation

This design is implemented in a standard 0.18 μm CMOS technology, and Fig. 8.11 represents the system diagram of the proposed converter controlled by digital modules. A series of four D-flip flops generates the 90-degree phase shift four-phase clock on every rising edge of the input clock. The ring oscillator circuits are used to construct a 0.1 ns non-overlapped region to benefit full charge settling and prevent reversion loss by taking phase generator outputs. OR gate-multiplexing accommodates all combinations of 4-phase pairs as clock signal inputs in the next router module. Depending on the topology selection, the same transistor can be active for all 4-phases. Its output voltage, Vout1 or Vout2, supplies the V_{DDH} of the router, depending on the highest output at chosen gain mode. V_{DDH} can be directly connected to the highest output Vout at a time.

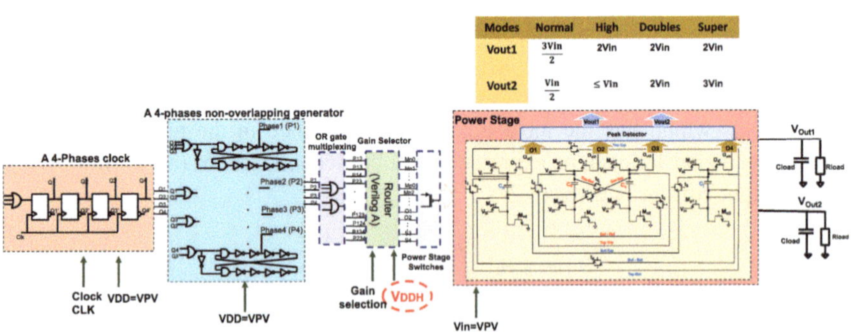

Fig. 8.11 Breakdown schematic of System diagram of proposed converter that operates in a non-overlapping four-phases clock

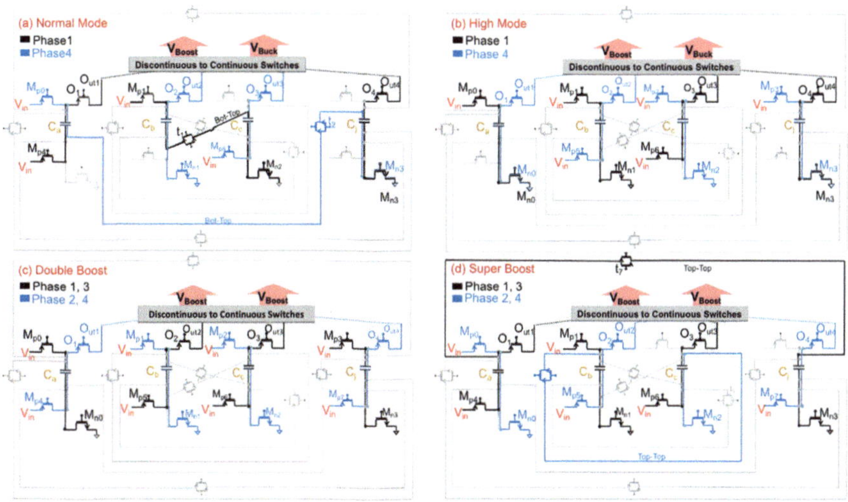

Fig. 8.12 The operation phases of the power stage network are at 4-different modes: (**a**) Normal Mode, (**b**) High Mode, (**c**) Double Boost and (**d**) Super Boost

The power circuit network comprises 20 switches to configure four integrated 1 nF MIM capacitors in series-parallel positions. However, only a few switches are required to configure the selection of mode, as shown in Fig. 8.12a–d. Charge/discharge actions of the capacitors are performed by the PMOS and NMOS transistors switches. The transmission gates are employed due to their bi-directional charge transfer, fast on-off transitions, and small turn-on resistance to minimise conduction power loss and avoid charge injection or clock feedthrough on both capacitors connected in the series arrangement. The circuit level configuration of the High mode is illustrated in Fig. 8.12b. Moreover, all the active switches to configure all four modes in 4-phases are described in Table 8.1. In the High mode, the two capacitors are charged in parallel by V_{in} through "M_{p0}, M_{n0}, M_{p1}, M_{n1}" to capacitors "C_a, C_b" for the phase1 and "M_{p2}, M_{n2}, M_{p3}, M_{n3}" are employed to charge capacitors "C_c, C_j" in the phase4. Meanwhile, in phase1 Net0, capacitor "C_c" is discharged in series with the input voltage to get $2V_{in}$ at "O_{ut3}" through "M_{p6}, O_3", and capacitor "Cj" is independently discharged at Net1 to the gain of V_{in} at "O_{ut4}" through "M_{n3}, O_4". Likewise, in phase 4, the Net0 output with the gain $2V_{in}$ is received at "O_{ut2}" via "M_{p5}, O_2", and Net1 output with the gain of V_{in} is achieved at "O_{ut1}" via "M_{n0}, O_1". As suggested in the topology, both discharged capacitors are swapped with two charged capacitors in the next phase. Thanks to the symmetry of the design, phase 2 and phase 3 can be described in a similar manner. Keeping the minimum length for all transistors, the width of PMOSs is 60 μm and 20 μm for NMOSs.

In all gain modes at 4-phases, the power switch protocols are designed to retain the optimum symmetric nature of the same on-resistances R_{on} value for consistency of outputs level. This is because different pathing for the same logic can lead to

Table 8.1 Transistor address for all modes

Phase	Net0						Net1			
Normal Mode										
P1		Mp1	t1	Mn2	Mp4	O1			Mn3	O4
P2		Mp3	t3	Mn0	Mp5	O2			Mn2	O3
P3		Mp2	t0	Mn1	Mp7	O4			Mn0	O1
P4		Mp0	t2	Mn3	Mp6	O3			Mn1	O2
High Mode										
P1	Mp0	Mn0	Mp1	Mn1	Mp6	O3			Mn3	O4
P2	Mp2	Mn2	Mp3	Mn3	Mp4	O1			Mn1	O2
P3	Mp0	Mn0	Mp1	Mn1	Mp7	O4			Mn2	O3
P4	Mp2	Mn2	Mp3	Mn3	Mp5	O2			Mn0	O1
Double Boost Mode										
P1		Mp0	Mn0		Mp5	O2	Mp3	Mn3	Mp6	O3
P2		Mp1	Mn1		Mp4	O1	Mp2	Mn2	Mp7	O4
P3		Mp0	Mn0		Mp5	O2	Mp3	Mn3	Mp6	O3
P4		Mp1	Mn1		Mp4	O1	Mp2	Mn2	Mp7	O4
Super Boost Mode										
P1	Mp1	Mn1	t7	Mn3	Mp4	O1			Mp6	O3
P2	Mp0	Mn0	t4	Mn2	Mp5	O2			Mp7	O4
P3	Mp0	Mn0	t4	Mn2	Mp5	O2			Mp7	O4
P4	Mp1	Mn1	t7	Mn3	Mp4	O1			Mp6	O3

slight variation in ΔV at internal outputs of power stage "O_{ut1}-O_{ut4}". For instance, Super Boost phase 1,4 "O_{ut1}" and phase 2,3 "O_{ut1}" configurations provide the same output value and share the same R_{on} values. Note that transistors "t_5, t_0, t_6, t_3" are reserved for future purposes. The voltage regulation can be further improved by introducing an adaptive pulsing unit in between the gain selector and power stage network since fast switching energy delivery for low and high loads can vary. Therefore, 5.8823 kHz for Normal, 666.66 kHz for High, 1.1 MHz for Double Boost and 2 MHz for Super Boost are chosen, respectively, as operation frequencies.

8.9 Output Results

For a 1 V input at a temperature of 27·C, the loads are varied from 50–100 kΩ. We show that a voltage conversion ratio of 98% can be achieved. At best, the average boost output mode produces 1.49 V (at 29.45 μA), and the buck is 0.467 V (at 9.36 μA) in Normal mode. Likewise, for the High mode, the average boost output produces 1.905 V (at 39.2 μA) and 0.98 V (at 19.6 μA) for the buck output.

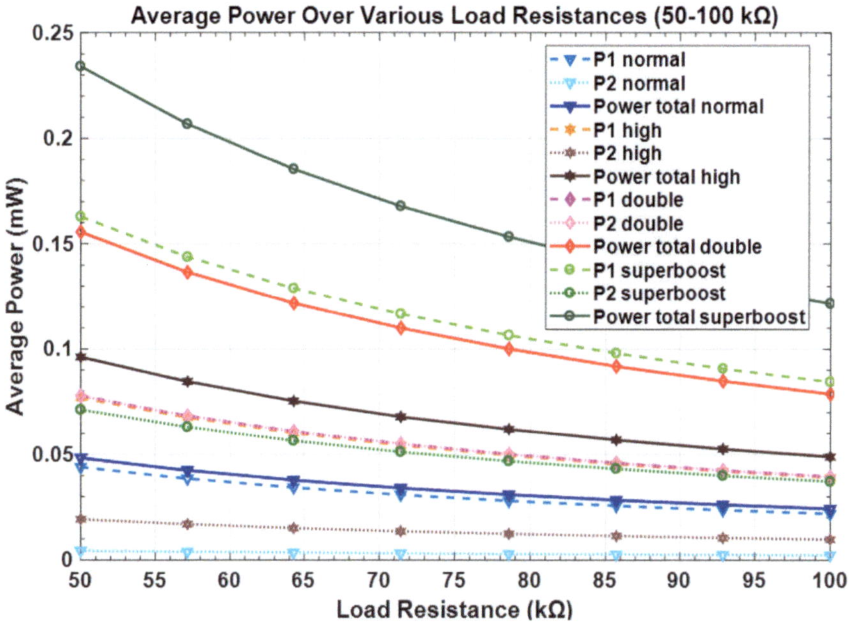

Fig. 8.13 Average output power for varying loads tested with 50–100 kΩ

Similarly, Double Boost produces 1.97 V (at 39.4 μA), and Super Boost yields 2.85 V (at 57.1 μA) and 1.89 V (at 37.8 μA) average outputs. The output ripples are recorded between 14–59 mV. The line transient indicates that the converter takes approximately 1–3.3 μs to get the first signal and 2.65–15 μs load transient time to reach saturation, depending on the selection of modes and the loads.

The output voltages and the current values of all gain modes tested with various resistance loads 50–100 kΩ are recorded. These values are then plotted in Fig. 8.13. When the load increase, the voltage at the output increases and the current dropped as expected from the Ohms law characteristic. Each mode has a different gradient change due to its different internal resistance Ron of the current paths. The diagram illustrates that although out1 of High, out1,2 of Double Boost and out2 of Super Boost modes, which altogether share the same gain of (2Vin), are configured differently with different switches, out1 of High and out1,2 of Double Boost modes provide the same I-V characteristic, since they overlapped, and closely match with Super's out2 line. The output current from 5.19–50.7 μA can be expected from the converter.

The average power output ranges from 0.0023–0.23 mW. Therefore, this power system is also suitable for implantable systems. Due to target output values being the same (×2 gain) for P1 double, P2 double and P1 High, it is observed that there are some overlapping performance lines, as shown in Fig. 8.14. These overlapping lines suggest that the proposed converter maintains the stable performances for the different configurations and the different switches of the same output. Assuming

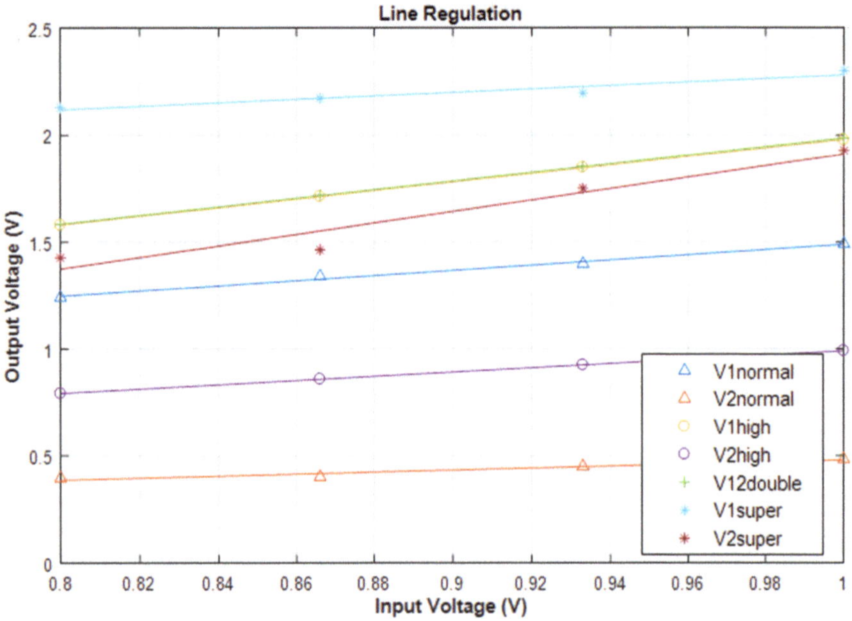

Fig. 8.14 Line regulation validated with the charging input of 1–0.8 V at 100 kΩ

input integrated battery degrades over time or supplied PV cell power degrades at night-time, line regulation with varying inputs of 1–0.8 V, simulated with a fixed load of 100 kΩ and the results shown in Fig. 8.9. Although the output voltages reduce from targeted values due to input changes, the linear line indicates that the converter still maintains correct gain operation relative to a given input.

The conduction power losses are calculated by inserting simulated result values into Eq. (8.2); which has resulted in 0.759 μW and 0.077 μW, respectively, for $P_{conduct\_Norm(boost,\ buck)}$, and 0.9239 μW and 0.231 μW for $P_{conduct\_High(boost,\ buck)}$. Similarly, 1.216 μW for $P_{conduct\_DoubleBoost}$ and 1.14 μW and 0.5013 μW for $P_{conduct(out1,\ out2)SuperBoost}$, respectively. Also, the switching power losses are calculated using Eq. (8.3) produces 4.3411×10^{-19} W for P_{sw_Norm}, 6.714×10^{-19} W for P_{sw_High}, 1.25×10^{-18} W for $P_{sw\_DoubleBoost}$, and 4.89×10^{-18} W for $P_{sw\_SuperBoost}$. Evaluated redistribution loss of all four modes in Eq. (8.4), produces 2.413 μW, 4.317 μW, 2.094 μW, and 2.929 μW, respectively. The simulated input power for all modes is 485.7 μW, 104 μW, 107.1 μW and 316.6 μW, respectively. Likewise, simulated power consumption at digital modules is 80.82 μW, 27.99 μW, 0.184 mW and 0.6189 mW, respectively. It is observed that digital modules consume huge power from high-frequency operations for Super and Double modes. The converter's efficiency is simulated from output power divided by input power for all four modes, yielding 85.257%, 72.884%, 53.6% and 47.85%, respectively.

The recent state-of-the-art single-input-multi-output converters, which were previously discussed, are summarised in the Table 8.2. In addition to [16], which

Table 8.2 Comparison of proposed work with state-of-the-art

Specification	[21]	[16]	[19]	[22]	[18]	This work
Technology	40 nm	45 nm	90 nm	350 nm	130 nm	180 nm
Input (V)	3.7	1	1.2	1.1–1.8	2.5	1
Output (V)	1.8, 0.8	0.66, 0.33, 1	0.76, 0.32	2, 3	4.81, −2.31	0.467–2.85
I_{load} (mA)	0.4,-1	–	0.4, 0.9	12	2, 2	0.04,0.06
VCR (x)	1/2,1/4	2/3, 1/3, 1	1/2, 2/3	2, 3 or 2, 2	2, −1	3/2,1/2 or 2,1 or 2,2 or 2,3
No. of outputs	2	3	2	2	2	2
P_{out_max} (mW)	1.3	1.3	1	60	9.62	0.0023–0.23
Ripple (mV)	60	9.6	19.5–40.6	56	95, 85	14–29
Efficiency (%)	70	90	68	90	92	85.26
Int. passive device	2.24n × (≥ 2)	3.7n ×(≥ 4)	5n × 26n × 2	4.7 μ × 2	1 μ × 2	1 nF × 4

produces three simultaneous outputs, the rest of the SC converters presented in the above table are SIDO designs. Depending on the application, required output power, and chip area availability, most methods use Nano Farad size capacitors. They typically employ four flying capacitors for dual output designs. Contrary to state-of-the-art, only two designs- this proposed work and [22] can reconfigure into different gain values. The rest of the SIMO converter designs have relatively fixed gain ratios. However, the topology of [22] has some idle time for every 180-degree phase shift due to it returning to the charging phase. Our proposed topology can simultaneously output the different VCRs simultaneously. Moreover, this proposed work has the highest reconfigurability since it can configure four various DC gain pairs.

Furthermore, only the [22] [18] designs are accessible for step-up conversion, while other SIDO converters only enable simultaneous down conversions. Although this proposed design has the lowest maximum power output compared, it is to note that the 1-nF charge pump capacitor is utilised. It is still the open-loop, and the maximum efficiency this converter can produce is comparable to most existing technologies. The efficiency value of this converter is described in the Table 8.2. is in consideration of digital circuitry and input power losses. If only charge pump power losses are considered, it can go as high as 98% efficiency.

References

1. M. Zamani, Y. Rezaeiyan, O. Shoaei, W.A. Serdijn, A 1.55 μW bio-impedance measurement system for implantable cardiac pacemakers in 0.18 μm CMOS. IEEE Trans. Biomed. Circuits Syst. **12**(1), 211–221 (2018)
2. Y. Rezaeiyan, M. Zamani, O. Shoaei, W.A. Serdijn, Mixed-signal IC with pulse width modulation wireless telemetry for implantable cardiac pacemakers in 0.18-μm CMOS. IEEE Trans. Biomed. Circuits Syst. **12**(3), 589–600 (2018)
3. A. Tanwear et al., Spintronic Eyeblink gesture sensor with wearable Interface system. IEEE Trans. Biomed. Circuits Syst. (2022)
4. H.A. Huynh, S. Kim, Design of on-chip power noise sensing circuit and its applications, in *IEEE Electrical Design of Advanced Packaging and Systems Symposium (EDAPS)*, (IEEE, 2015), pp. 209–212
5. A. Mosalmani, M. Khoee, O. Shoaei, A 9-bit 70-MS/s two-stage SAR ADC with passive residue transfer. IEEE Trans. Very Large Scale Integr. (VLSI) Syst. **28**(5), 1185–1194 (2020)
6. R. Li, H. Fariborzi, Design and demonstration of MEM relay-based arithmetic and sequential circuit blocks. IEEE Trans. Electron Dev. **68**(12), 6415–6421 (2021)
7. S. Shreya, A. Jain, B.K. Kaushik, Computing-in-memory architecture using energy-efficient multilevel voltage-controlled spin-orbit torque device. IEEE Trans. Electron Dev. **67**(5), 1972–1979 (2020)
8. N. Butzen, M. Steyaert, 12.2 A 94.6%-efficiency fully integrated switched-capacitor DC-DC converter in baseline 40nm CMOS using scalable parasitic charge redistribution, in *2016 IEEE international solid-state circuits conference (ISSCC)*, (IEEE, 2016), pp. 220–221
9. A. Mahmoud, M. Alhawari, B. Mohammad, H. Saleh, M. Ismail, A charge pump based power management unit with 66%-efficiency in 65 nm CMOS, in *IEEE International Symposium on Circuits and Systems (ISCAS)*, (IEEE, 2018), pp. 1–4

10. X. Yu, K. Moez, I.-C. Wey, M. Sawan, J. Chen, A fully integrated multistage cross-coupled voltage multiplier with no reversion power loss in a standard CMOS process. IEEE Trans Circuits Syst II Express Briefs **64**(7), 737–741 (2016)
11. T. Kimura, H. Ochi, A– 0.5 V-input voltage booster circuit for on-chip solar cells in 0.18 μm CMOS technology, in *15th International Symposium on Communications and Information Technologies (ISCIT)*, (IEEE, 2015), pp. 193–196
12. J. Jiang, W.-H. Ki, Y. Lu, Digital 2–/3-phase switched-capacitor converter with ripple reduction and efficiency improvement. IEEE J. Solid State Circuits **52**(7), 1836–1848 (2017)
13. X. Zhang, H. Lee, An efficiency-enhanced auto-reconfigurable x2/x3 SC charge pump for transcutaneous power transmission. IEEE J. Solid State Circuits **45**(9), 1906–1922 (2010)
14. H. Lee, Z. Hua, X. Zhang, A reconfigurable x2\x2.5\x3\x4 times SC DC–DC regulator with fixed on-time control for transcutaneous power transmission. IEEE Trans. Very Large Scale Integr (VLSI) Syst. **23**(4), 712–722 (2014)
15. W. Jung et al., 8.5 A 60%-efficiency 20nW-500μW tri-output fully integrated power management unit with environmental adaptation and load-proportional biasing for IoT systems, in *IEEE International Solid-State Circuits Conference (ISSCC)*, (IEEE, 2016), pp. 154–155
16. Y. Yang, Y. Zhao, G. Dong, A multi-output on-chip switched-capacitor DC–DC converter with unequal flying capacitors for different power modes. J. Circuits, Syst. Comput. **24**(04), 1550051 (2015)
17. D. Kilani, M. Alhawari, B. Mohammad, H. Saleh, M. Ismail, An efficient switched-capacitor DC-DC buck converter for self-powered wearable electronics. IEEE Trans. Circuits Syst. I Regul. Pap **63**(10), 1557–1566 (2016)
18. Y.-C. Lin, Y.-K. Luo, K.-H. Chen, W.-C. Hsu, Liquid crystal display (LCD) supplied by highly integrated dual-side dual-output switched-capacitor DC-DC converter with only two flying capacitors. IEEE Trans. Circuits Syst I Regul. Pap. **59**(2), 439–446 (2011)
19. N. De Clercq, T. Van Breussegem, W. Dehaene, M. Steyaert, Dual-output capacitive DC-DC converter with power distribution regulator in 90 nm CMOS, in *Proceedings of the ESSCIRC (ESSCIRC)*, (IEEE, 2012), pp. 169–172
20. X. Zhang et al., A fully integrated battery-powered system-on-chip in 40-nm CMOS for closed-loop control of insect-scale pico-aerial vehicle. IEEE J. Solid State Circuits **52**(9), 2374–2387 (2017)
21. T. Tong, S.K. Lee, X. Zhang, D. Brooks, G.-Y. Wei, A fully integrated reconfigurable switched-capacitor DC-DC converter with four stacked output channels for voltage stacking applications. IEEE J. Solid State Circuits **51**(9), 2142–2152 (2016)
22. Z. Hua, H. Lee, A reconfigurable dual-output switched-capacitor DC-DC regulator with sub-harmonic adaptive-on-time control for low-power applications. IEEE J. Solid State Circuits **50**(3), 724–736 (2015)
23. I. Chowdhury, D. Ma, Design of reconfigurable and robust integrated SC power converter for self-powered energy-efficient devices. IEEE Trans. Ind. Electron. **56**(10), 4018–4028 (2009)
24. Y. Wang, P. Luo, X. Zheng, B. Zhang, A 0.3 V–1.2 V ultra-low input voltage, reconfigurable switched-capacitor DC-DC converter for energy harvesting system, in *13th IEEE International Conference on Solid-State and Integrated Circuit Technology (ICSICT)*, (IEEE, 2016), pp. 1333–1335
25. K.O. Htet, H. Fan, H. Heidari, Switched capacitor dc-dc converter for miniaturised wearable systems, in *2018 IEEE international symposium on circuits and systems (ISCAS)*, (IEEE, 2018), pp. 1–5
26. D. Ma, R. Bondade, Power loss in switched-capacitor power converters: Causes and analysis, in *Reconfigurable Switched-Capacitor Power*, (Springer, Converters, 2013), pp. 59–69

Chapter 9
Power Management Integrated Circuits for Implantable Devices

Chuang Wang

9.1 Introduction

Implantable devices are inserted into the human body for various purposes such as monitoring diverse health conditions and/or assisting physical activities [1, 2]. For implantable devices, the physical volume and energy efficiency are two important factors. First, the physical volume is directly proportional to the burden placed on the human body. Therefore, any additional external passive components should be eliminated in the design of an implantable device. Second, because the implantable devices operate in energy-limited environments, the given energy must be optimally used to minimize power loss. For now, a few types of energy sources, such as batteries, harvested energies, or a combination of them [3–5] are used to power the implantable devices. Though various harvested energies are promising in the future, the batteries are currently reliable and effective. A cion or Li-ion battery voltage of 1.5–5 V usually provides the energy and acts as the energy source. The implantable device is typically powered by sub-1 V supply rail to improve the energy efficiency and prolong its life [6]. Thus, a step-down DC-DC converter with a relatively low voltage conversion ratio (VCR) is highly desirable to act as the power supply of the implantable device, as shown in Fig. 9.1.

To achieve the required low VCR, the topology of the DC-DC converter is ought to be appropriately chosen and designed according to the individual implantable devices. Figure 9.2 exhibits three potential solutions. Figure 9.2a employs the conventional single-step solution, namely two-level buck converter [7, 8], switched-capacitor (SC) converter [9], or linear regulator [10, 11]. But, the two-level buck converter has several problems when the VCR is low. First, the voltage swing

C. Wang (✉)
Beijing University of Posts and Telecommunications, Beijing, China

Huawei Technologies Co. Ltd., Beijing, China

R. Das, H. Heidari (eds.), *Wireless Power Technologies for Biomedical Devices*, Analog Circuits and Signal Processing, https://doi.org/10.1007/978-3-031-52834-7_9

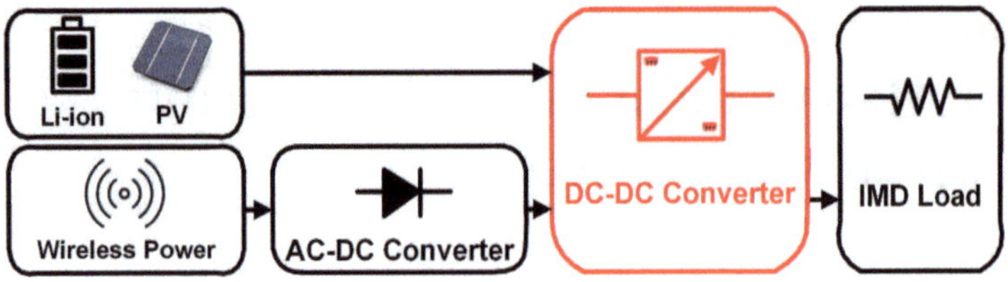

Fig. 9.1 Supply architecture for an implantable device

Fig. 9.2 Solutions for bridging the voltage gap between the battery voltage and the load supply

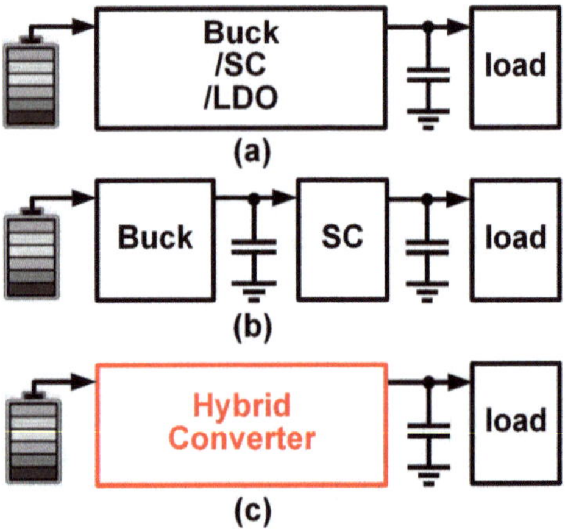

and thus the voltage stress on the power inductor is as large as the input voltage, resulting in a large power loss and inductor current ripple. Second, to decrease the inductor current ripple, the power inductor requires a large inductance and hence a large volume. Third, the VCR mathmatically equals the duty cycle D, which would be quite small for the low VCR, subsequently increasing the controller's design difficulty. Besides, the SC converter can enhance high power density, because the SC converter eliminates the current limiting component (inductor) and a capacitor in advanced technology has a higher energy density when compared with an inductor. However, an SC converter can only obtain high efficiency at its naturally fixed VCRs [12, 13], and its efficiency drops proportionally to its output voltage. Moreover, a linear regulator has extremely low efficiency when achieving a low VCR.

Figure 9.2b depicts the two-step solution by cascading one buck-converter and one SC converter to convert the battery voltage to the supply rail [14]. Despite this would be more flexible for multiple voltage domains in a system, two stages will limit the overall efficiency and increases the solution complexity. Moreover, the relatively large (compared with the flying capacitors) filtering capacitors have to be placed between the two stages [15].

To solve the previously mentioned issues, Fig. 9.2c gives another single-step conversion candidate i.e. the hybrid converter. For now, three kinds of hybrid converters, namely resonant switched-capacitor (ReSC) converter, multi-level buck converter, and multi-path buck converter are popular topologies. They contribute the good performances in the efficiency and power density by reducing the current spikes, voltage swings, and inductor currents, respectively. Therefore, this chapter starts by introducing a three-state (3Φ) ReSC converter. Next, a two-phase three-level (2P3L) buck converter with cross-connected flying capacitors (X-C_{FLY}) will be discussed in detail. Finally, a tri-path buck converter will be studied. Major technical and design details for this chapter have been taken from Refs. [16–20].

9.2 Three-State Resonant Switched-Capacitor Converter

To realize the low VCR of 1/4, an SC converter is reported by employing a three-state (3Φ) operation and only two integrated flying capacitors [21]. But, the efficiency is limited by two factors: (1) the parasitic capacitance loss resulting from the on-chip capacitors, and (2) the capacitor hard charging loss (current spike) existing in each SC converter. A resonant switched-capacitor (ReSC) converter is a candidate for implementing capacitor soft charging i.e. decrease the hard-charging current spikes by introducing an additional small resonant inductor [22–24]. However, the reported ReSC converters with two flying capacitors only achieve the VCRs of 1/2 or 1/3, as these converters operate in two phases [22, 23]. Moreover, the startup issue in previous integrated ReSC converters was ignored. Thereby, this subchapter presents a highly-integrated 3Φ-ReSC converter targeting a VCR of 1/4 [16] as shown in Fig. 9.3. It consists of seven power switches $S_1 - S_7$, two flying capacitors C_{F1} and C_{F2}, one resonant inductor L_R, and one load capacitor C_L. Theoretically, the resonant inductor L_R can be placed in series with C_{F2} or on the output path [17], and these L_R placements give the same soft charging effects. Here, L_R is selected in series with the bottom plate of C_{F2} to reduce C_{F2} parasitic capacitor loss.

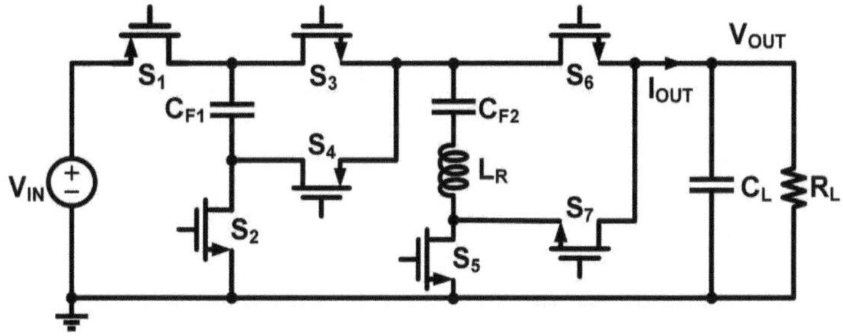

Fig. 9.3 3Φ-ReSC converter

9.2.1 Architecture and Operation Principle

Figure 9.4 presents the operation principle and circuit implementation of the 3Φ-ReSC converter. It has three phases, namely the charging phase Φ_1 and the discharging phases Φ_2 and Φ_3. In Φ_1, C_{F1}, C_{F2}, and L_R are in series between V_{IN} and V_{OUT}. In Φ_2, C_{F1} still connects in series with C_{F2} and L_R but reverses its polarity. In Φ_3, C_{F1} is floating, and C_{F2} reverses its polarity and supplies V_{OUT} through L_R. To deduce the VCR, the charges delivered to the load in each phase are defined as q_1, q_2, and q_3, respectively, and thus the total charge Q is $Q = q_1 + q_2 + q_3$. C_{F1} charges in Φ_1, and discharges in Φ_2, and thereby $q_1 = q_2$. C_{F2} charges in both Φ_1 and Φ_2, and discharges in Φ_3, and hence $q_3 = q_1 + q_2$. So, $q_1 = Q/4$. Further, by averaging q_1 and Q during one cycle, $I_{IN} = I_{OUT}/4$,

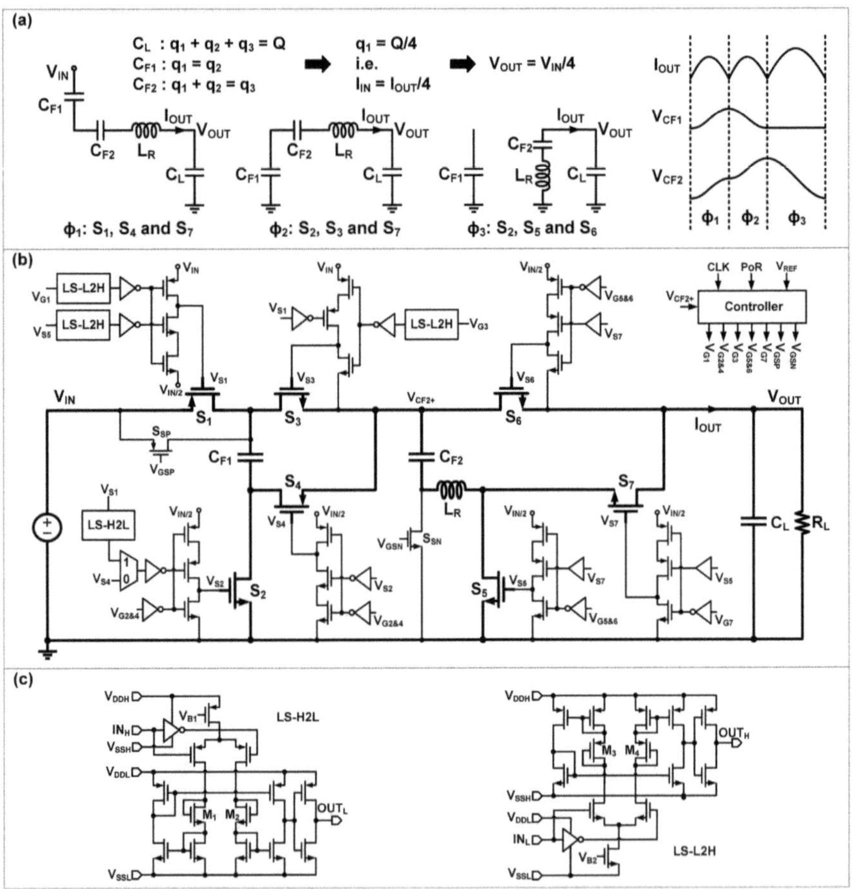

Fig. 9.4 Operation principle, and circuit implementation including the biased voltage level shifters of the 3Φ-ReSC converter

thus $V_{OUT} = ¼·V_{IN}$, i.e. the natural VCR of the 3Φ-ReSC converter is 1/4. By using the similar method and considering that the residual DC voltage on the resonant inductor V_{LR} is zero in full resonant operation, the steady state average voltage across each capacitor can be approximated to DC values as $V_{CF1} = ½ V_{IN}$, $V_{CF2} = ¼ V_{IN}$. Thereby, the 3Φ-ReSC converter has

$$V_{CF1} = \frac{1}{2}V_{IN}, V_{CF2} = \frac{1}{4}V_{IN}, V_{OUT} = \frac{1}{4}V_{IN}. \tag{9.1}$$

According to the operation states and Eq. (9.1), the ideal voltage stresses of the flying capacitor C_{F1}, and the switches S_1, S_2, S_3, and S_4 are $V_{IN}/2$, and the ideal voltage stresses of the flying capacitor C_{F2}, and the switches S_5, S_6, and S_7 are $V_{IN}/4$. For a specific load, the voltage stress on each component would have a fluctuation, equal to the flying capacitor voltage ripple.

Figure 9.4 also shows the detailed circuit implementation. The non-overlapping gate drivers are achieved with three-transistor inverter stages to avoid the shoot-through current. Besides, the biased dynamic voltage level shifters LS-L2H and LS-H2L [16] set most transistors as 1.2-V core devices to quickly shift the voltage swing of the driver signals and use the diode-connected transistors M_1, M_2, M_3, and M_4 to avoid device breakdown.

9.2.2 Resonant Inductor Current

The instant resonant inductor current, i.e., its shape is important for understanding the converter. Here, the infinite resonance quality factor i.e. the ignored parasitic resistance is assumed to derive instant resonant inductor current.

During Φ_1, by using KVL and KCL, the time-domain equations are got as,

$$V_{IN} = v_{CF11} + v_{CF21} + L_R\frac{di_{LR1}}{dt} + V_O, \tag{9.2}$$

$$i_{CF11} = C_{F1}\frac{dv_{CF11}}{dt} = i_{LR1}, \tag{9.3}$$

$$i_{CF21} = C_{F2}\frac{dv_{C21}}{dt} = i_{LR1}. \tag{9.4}$$

where i_{CF11}, v_{CF11}, i_{CF21}, and v_{CF21} are the instant currents and voltages of C_{F1} and C_{F2}, respectively; i_{LR1} is the instant inductor current in Φ_1. Differentiating Eq. (9.2) to time and combining with Eqs. (9.3) and (9.4) give

$$\frac{i_{LR1}}{C_{F1}} + \frac{i_{LR1}}{C_{F2}} + L_R\frac{d^2i_{LR1}}{dt^2} = 0. \tag{9.5}$$

By solving Eq. (9.5), a solution is got as

$$i_{LR1} = E \cdot \sin\left(\frac{1}{\sqrt{L_R\left(C_{F1}\middle\|C_{F2}\right)}}t\right), \tag{9.6}$$

where E is a constant relying on the input voltage and output current. By using the similar deduction for Φ_2 and Φ_3, the instant expressions are got as

$$i_{LR2} = F \cdot \sin\left(\frac{1}{\sqrt{L_R\left(C_{F1}\middle\|C_{F2}\right)}}t\right), \tag{9.7}$$

$$i_{LR3} = G \cdot \sin\left(\frac{1}{\sqrt{L_R C_{F2}}}t\right). \tag{9.8}$$

where F and G are the constants depending on the input voltage and output current, respectively.

According to Eqs. (9.6)–(9.8), the shape of the resonant inductor current will act as in Fig. 9.4, if the durations of each phase are correctly guaranteed.

9.2.3 Parasitic Loss Reduction

In CMOS technology, the on-chip flying capacitors have a considerly large bottom plate parasitic capacitor C_P, and hence have a parasitic loss (proportional to $C_P V_{SP}{}^2$, where V_{SP} is the voltage swing of C_P during normal operation). Despite vertically stacking the available on-chip capacitors for example the MIM+MOM+MOS capacitors reduces the ratio of C_P/C_F, the voltage swing of C_P is much larger than that of the flying capacitor. In the 3Φ-ReSC converter, the voltage swing of C_{F1P} is 2× of that of C_{F2P}, causing 4× larger parasitic loss if $C_{F1P} = C_{F2P}$. In a 65 nm process, C_{F1} may have >1% bottom plate parasitic C_{F1P} [16], causing 3%–4.5% efficiency reduction in simulation. In brief, C_{F1P} would cause around 15% of the overall power loss. Then, a tiny (01005-package) off-chip capacitor is chosen as C_{F1} to remove C_{F1P} and stacked on top of the die. For the on-chip C_{F2}, C_{F2P} reduces its parasitic loss because by the parasitic diodes of S_5 and S_7 conduct the inductor residual currents to soft charge/discharge C_{F2P} during the deadtime at the end of Φ_2 and Φ_3, respectively.

9.2.4 Pre-charge Startup

Figure 9.5 displays the pre-charge startup that eliminates the overvoltage stress on the switches. During the startup process, the startup switches S_{SP} and S_{SN} are on and the switches S_1, S_3, S_5, S_6, and S_7 keep off. S_2 and S_4 alternatively turn on with a duty cycle D_{S4}. When S_2 is on, I_{PC1} charges C_{F1} only. When S_4 is on, I_{PC2} charges both C_{F1} and C_{F2}. V_{CF1} should have a 2× slew rate than V_{CF2} by designing D_{S4}, leading to V_{CF1} could be close to $V_{IN}/2$ when V_{CF2} reaches $V_{IN}/4$. Hence, the charges on C_{F1} and C_{F2} sactifiy $C_{F1}\cdot V_{CF1} = I_{PC1}\cdot(1 - D_{S4})\cdot T_{PC} + I_{PC2}\cdot D_{S4}\cdot T_{PC}$ and $C_{F2}\cdot V_{CF2} = I_{PC2}\cdot D_{S4}\cdot T_{PC}$, respectively. So, $V_{CF1} = V_{IN}/2$, $V_{CF2} = V_{IN}/4$, and considering I_{PC1} and I_{PC2} as RC charging networks requires D_{S4} to be 0.67. To terminate the startup process, a dynamic comparator driven by the rising edge of V_{G2} compares V_{CF2+} with a reference voltage $V_{REF} = V_{IN}/4$. Figure 9.5b illustrates the gate signals during the startup process. Figure 9.5c reveals the dynamic comparator, which has two analog inputs V_{CF2+} and V_{REF}, a digital output OUT, and a clock input V_{G2}. The comparison is only executed at the rising edge of the clock. If $V_{CF2+} < V_{REF}$, OUT = 0, otherwise, OUT = V_{G2}. Once the startup

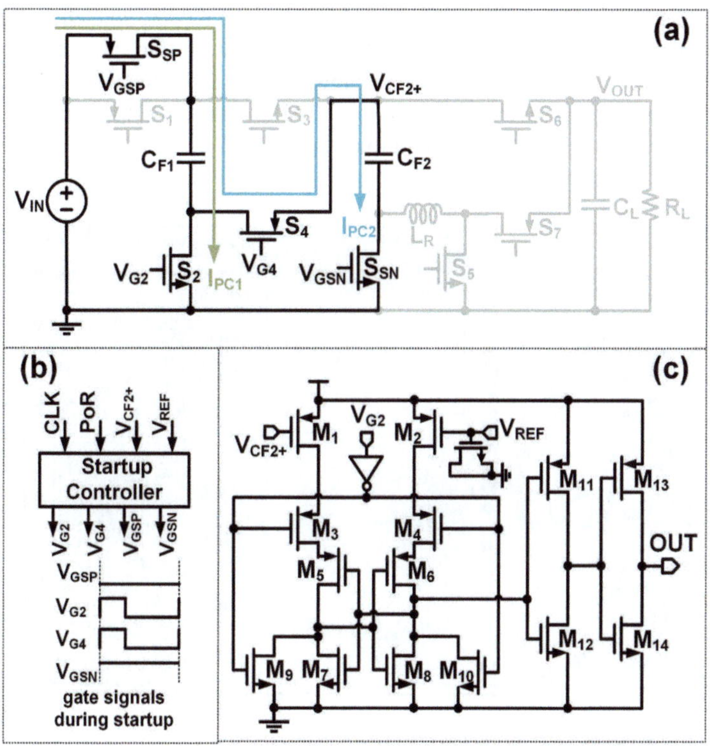

Fig. 9.5 (**a**) Operation principle, (**b**) gate signals, and (**c**) dynamic comparator for startup prc-charge

end-point is determined (the signal OUT has a rising edge), V_{GSP} and V_{GSN} are turned over for enabling normal operation. The dynamic comparator consumes no static current and uses the filter MOSFET capacitor to mitigate the kickback effect.

9.2.5 Measurement Results

The 3Φ-ReSC converter is fabricated in standard 65-nm CMOS and occupies a silicon area of 1.39 mm^2. The on-chip C_{F2} is 5.6 nF, and C_{F1} in a 01005 package is 5.6 nF placed on top of the die as shown in Fig. 9.6. Therefore, the die-attached capacitor and silicon IC fit at a height of <0.7 mm. Besides, the 0403 commercial inductor with a compact volume of 1.19 × 0.86 × 0.71 mm^3 together with the bonding wires and PCB traces serves as L_R. Thereby, a future system-in-package solution of this 3Φ-ReSC converter is expected to be as small as 1.2 × 2.1 × 0.71 = 1.79 mm^3, suitable for many implantable devices.

Figure 9.7 displays the measured waveforms in startup process and normal operaion. As shown in Fig. 9.7a, when V_{IN} increases from 0 to 2.4 V within only 1.1 ms, the 3Φ-ReSC converter pre-charges C_{F1} and C_{F2} to around 1.4 V and 0.6 V, respectively. Then, this converter enters the normal operation, charging V_{OUT} to $V_{IN}/4$ rapidly, and making V_{CF1} and V_{CF2} fluctuate at the individual nominal values $V_{IN/2}$ and $V_{IN/4}$, respectively. As shown in Fig. 9.7b, the resonant operation is achieved at a switching frequency F_{SW} = 10 MHz, and the output ripple is 25 mV for a load current of 30 mA with an output capacitor of 22 nF.

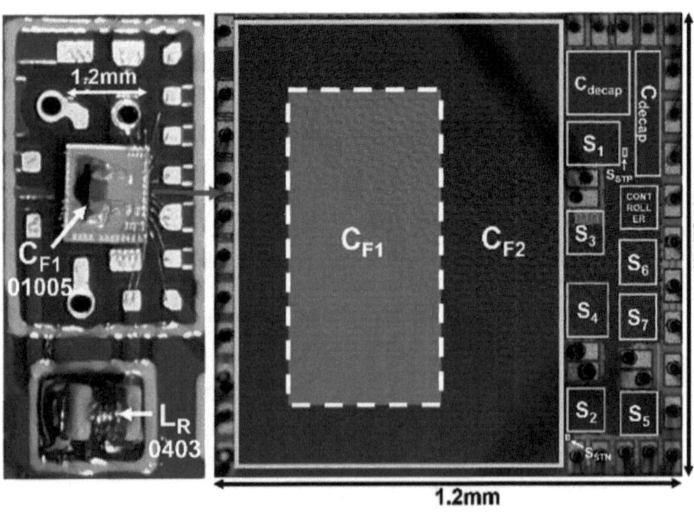

Fig. 9.6 Chip micrograph and package solution of the 3Φ-ReSC converter, with an off-chip C_{F1} placed on the top of the die

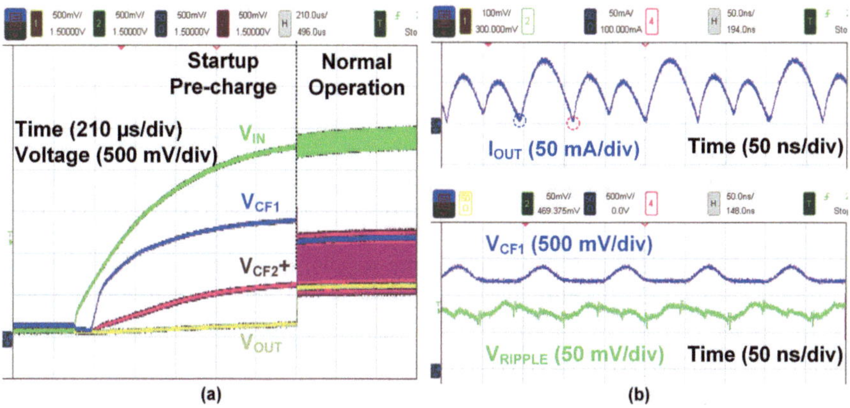

Fig. 9.7 Measured waveforms of in the (**a**) startup process and (**b**) normal operaion

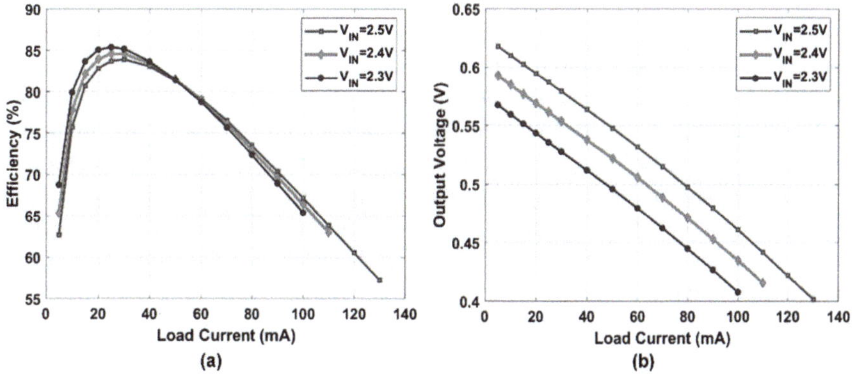

Fig. 9.8 Measured (**a**) efficiency, (**b**) output voltage versus load current

Figure 9.8 summarizes the measured power conversion efficiency and output voltage versus the load current. The 3Φ-ReSC converter in the full resonant operation (unregulated output voltage) achieves a peak efficiency of 85.5%, and a maximum output current of 130 mA. Accordingly, the output voltage range is from 0.4 V to 0.625 V.

9.2.6 Summary

This subchapter presents a highly-integrated 3-state (3Φ) resonant switched-capacitor (ReSC) converter with parasitic capacitor soft charging, significantly reducing the current stress. Both the 3Φ resonant operation and the parasitic loss reduction schemes improved the efficiency. Also, this 3Φ-ReSC converter implements a fast flying capacitor pre-charge scheme to guarantee a safe startup.

Finally, the fabricated 3Φ-ReSC converter in a three-dimensional mm-sized solution achieved a peak efficiency of 85.5%, a maximum output current of 130 mA, and a short startup time of 1.1 ms.

9.3 Two-Phase Three-Level Buck Converter with X-C_{FLY}

Three-level (3 L) buck converter has become popular due to its high efficiency and low ripple. Refs. [25, 26] introduces a single-phase 3 L buck converter, but it generally delivers a relatively limited current. Thus, two-phase 3 L (2P3L) [27], 4P3L [28] buck converters, etc., are used to deliver more current. But, they might have undesirable unbalanced inductor currents even with the current-balancing control [29]. In the worst case, the partial phases would become the load of the others. So, some feedforward [30, 31] or feedback current-balancing controls [32, 33] appeared for multi-phase buck converters to ease this problem. Unluckily, the feedforward or feedback method highly relies on the accuracy of high-frequency reconstruction of each phase current and output current or the accuracy of high-frequency sensing of each phase current. Because either the current reconstruction or current sensing at high frequency is challenging, the additional current-balancing controls would significantly increase the control difficulty. Therefore, this subchapter presents a 2P3L buck converter with cross-connected flying capacitors (X-C_{FLY}) as shown in Fig. 9.9. It suppresses the unbalanced inductor currents via alternately connecting the flying capacitors with the inductors of each phase, and also utilizes no extra power component.

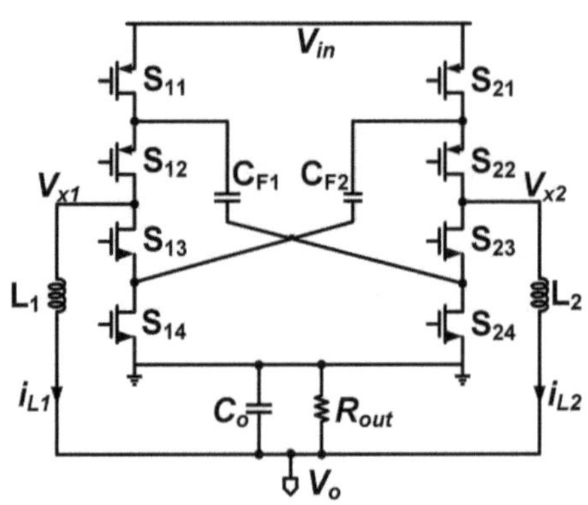

Fig. 9.9 2P3L buck converter with X-C_{FLY}

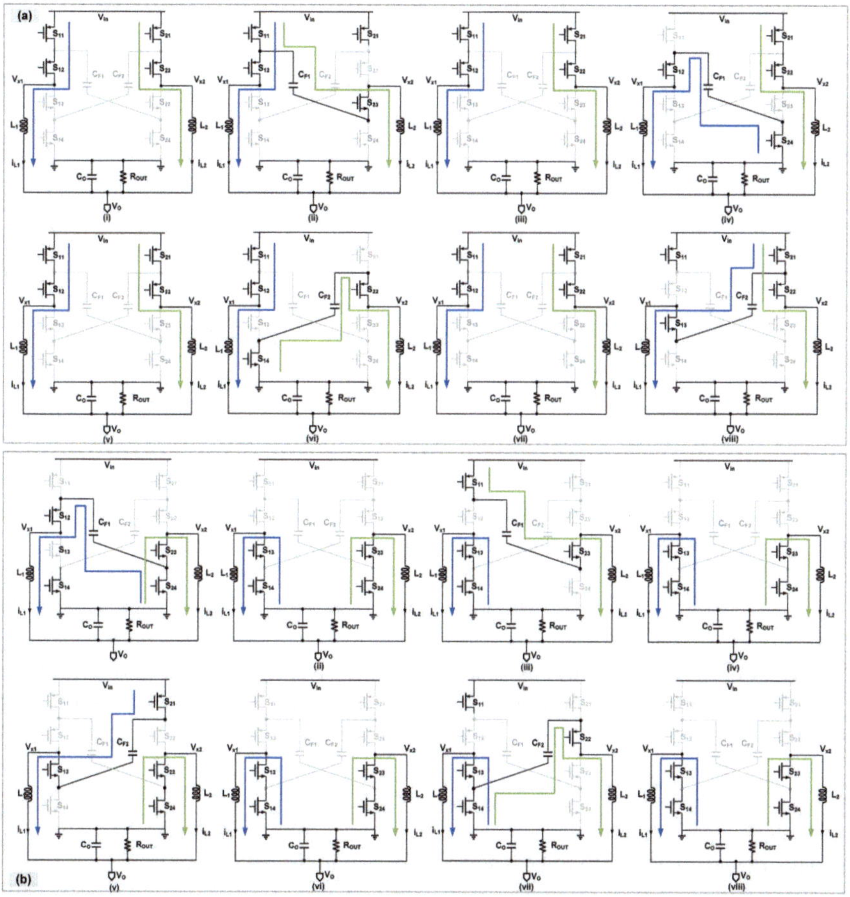

Fig. 9.10 Operating states of the 2P3L buck converter with X-C$_{FLY}$: V_{x1} and V_{x2} fluctuate (**a**) between $V_{in}/2$ and V_{in}, (**b**) between 0 and $V_{in}/2$

9.3.1 Operation Principle

Figure 9.10 illustrates the operation principle of the 2P3L buck converter with X-C$_{FLY}$. For the sake of readability, the case of Fig. 9.10a is selected for the detailed analysis. By employing four interleaved pulse width modulation (PWM) signals D_1, D_2, D_{S1}, and D_{S2} (the durations of zero voltage on V_{x1} or V_{x2}) to control the power switches S_{11} & S_{24}, S_{21} & S_{14}, S_{12} & S_{13}, S_{22} & S_{23}, respectively, the operation states of each cycle can be guaranteed (same figure). Thus, the switching nodes V_{x1} and V_{x2} follow $V_{in} \rightarrow V_{CF1} \rightarrow V_{in} \rightarrow V_{in} - V_{CF2}$ and $V_{in} \rightarrow V_{CF2} \rightarrow V_{in} \rightarrow V_{in} - V_{CF1}$, respectively. The switching nodes V_{x1} and V_{x2} bring the interaction between two phases and have similar amplitudes to those of the conventional 2P3L buck converter. Noticing that, D_1, and D_{S1} swap for the case in Fig. 9.10b.

Next, by applying the state-space average method [34], i.e. the voltage-second balance law to the two inductors L_1 and L_2, there are

$$(V_{in} - V_o)(D_1 - 0.5) + (V_{CF1} - V_o)(1 - D_1) + (V_{in} - V_o)(D_{s1} - 0.5) + (V_{in} - V_{CF2} - V_o)(1 - D_{s1}) = 0, \quad (9.9)$$

$$(V_{in} - V_o)(D_2 - 0.5) + (V_{CF2} - V_o)(1 - D_2) + (V_{in} - V_o)(D_{s2} - 0.5) + (V_{in} - V_{CF1} - V_o)(1 - D_{s2}) = 0. \quad (9.10)$$

Here, a good matching is assumed to make the mismatches between the two phases negligible. Hence, by setting the four duty cycles $D_1 = D_2 = D_{S1} = D_{S2} = D$, and solving Eqs. (9.9) and (9.10), the output voltage V_O is deduced as

$$V_O = D \cdot V_{in}. \quad (9.11)$$

The VCR is the same as that of the conventional single–/multi-phase two–/three-level buck converters. Then, bringing Eq. (9.11) into Eq. (9.10) will lead to

$$V_{CF1} = V_{CF2}. \quad (9.12)$$

This reveals that two flying capacitor voltages are ideally equal.

9.3.2 Inductor Current Balance

The 2P3L buck converter with X-C_{FLY} automatically balances the inductor current into one-twice of the output current. Using the case in Fig. 9.10b for the detailed analysis, at state (i), the flying capacitor C_{F1} is discharged through the first phase inductor current i_{L1}, but is charged by the second phase inductor current i_{L2} during state (iii). Similarly, the flying capacitor C_{F2} is charged by i_{L1} during state (v), and is discharged by i_{L2} during state (vii). For example, if initially$i_{L1} > i_{L2}$, the unequal charged and discharged energy will decrease and increase the flying capacitor voltage V_{CF1} and V_{CF2}, respectively. Hence, in the next cycle, the switching node V_{x1} (V_{CF1} in state (i) or $V_{in} - V_{CF2}$ in state (v)) decreases, while the switching node V_{x2} (V_{CF2} in state (vii) or $V_{in} - V_{CF1}$ in state (iii)) increases. This accordingly forces i_{L1} to decrease and i_{L2} to increase. As a result, the two inductor currents i_{L1} and i_{L2} will balance automatically. In brief, the constant capacitor average voltage in the steady state provides an inherent feedback loop to balance two inductor currents [35].

Next, the inductor current mismatch ΔI_L is quantitatively calculated by a large signal model. Considering again Fig. 9.10b for analysis, by applying KCL on the top plates of each flying capacitor and KVL in one period, there are

$$C_{F1}\frac{dV_{CF1}}{dt} = D_i \cdot I_{L1} - D_{iii} \cdot I_{L2}, \tag{9.13}$$

$$C_{F2}\frac{dV_{CF2}}{dt} = -D_v \cdot I_{L1} + D_{vii} \cdot I_{L2}, \tag{9.14}$$

$$L_1\frac{dI_{L1}}{dt} = D_i \cdot V_{CF1} - V_O - R_{L1} \cdot I_{L1} + D_v \cdot (V_{in} - V_{CF2}) - V_O - R_{L1} \cdot I_{L1}, \tag{9.15}$$

$$L_2\frac{dI_{L2}}{dt} = D_{iii} \cdot (V_{in} - V_{CF1}) - V_O - R_{L2} \cdot I_{L2} + D_{vii} \cdot V_{CF2} - V_O - R_{L2} \cdot I_{L2}. \tag{9.16}$$

where I_{L1} and I_{L2} are the average inductor currents for two-phase inductors L_1 and L_2, respectively; C_{F1} and C_{F2} are the flying capacitances; R_{L1} and R_{L2} represent the inductor DCRs and other resistances; V_{in} is the dc input voltage, and V_O is the average output voltage; V_{CF1} and V_{CF2} are the average flying capacitor voltages; D_i, D_{iii}, D_v, and D_{vii} are the duty ratios for the states (i), (iii), (v), and (vii), respectively. For simplicity, by assuming $L_1 = L_2 = L$, $R_{L1} = R_{L2} = R$, $C_{F1} = C_{F2} = C_F$, $D_i = D_{iii} = D_v = D_{vii} = D_x$, and subtracting Eq. (9.15) from Eq. (9.16), an equation on the unbalanced current ΔI_L, defined as the difference of two average inductor currents ($\Delta I_L = |I_{L1} - I_{L2}|$), is obtained as

$$L\frac{d\Delta I_L}{dt} = 2D_x\,(V_{CF1} - V_{CF2}) - R \cdot \Delta I_L. \tag{9.17}$$

Differentiating Eq. (9.17) to time and combining with Eqs. (9.13) and (9.14), there is

$$\frac{d^2\Delta I_L}{dt^2} + \frac{R}{L}\frac{d\Delta I_L}{dt} + \frac{2{D_x}^2}{LC_F}\Delta I_L = 0. \tag{9.18}$$

Equation (9.18) is a second-order differential equation of ΔI_L and is in the form of a damped harmonic oscillator [36]. The damping factor is

$$\zeta = \frac{R}{2D_x}\sqrt{\frac{C_F}{2L}}. \tag{9.19}$$

Because $R \neq 0$, and $C_F \neq 0$, there is $\zeta \neq 0$, i.e., ΔI_L will damp to zero eventually.

Qualitatively and quantitatively, the aforesaid analyses prove that the 2P3L buck converter with X-C_{FLY} ideally balances the inductor currents in the specific duty cycle range, and potentially eliminates the challenging high-frequency current reconstruction or current sensing.

9.3.3 Circuit Implementation

Figure 9.11 shows the circuit implementation of the 2P3L buck converter with X-C_{FLY}, including the power stage and feedback control loop.

According to the current transfer function deduced by [19], the corner frequency is got as

$$f_{OL} = \frac{1}{2\pi\sqrt{\left(\frac{L_k}{n}\right)C_o}}. \tag{9.20}$$

So, for $n = 2$ and $C_O = 1$ nF, L_1 and L_2 should be as 500 nH targeting the corner frequency f_{OL} around 10 MHz. Meanwhile, a switching frequency is set as 40 MHz. Because the voltage stresses of S_{12}, S_{13}, S_{22}, and S_{23} are equal to the input voltage V_{in} (1.8 V) and the voltage stresses of S_{11} and S_{21} in the start-up phase are equal to V_{in}, the power stage employs 1.8-V I/O devices rather than 1.2-V core devices in a 65 nm CMOS technology as the power transistors. Besides, two power switches S_{13} and S_{23} are placed in deep N-wells with a voltage potential of 1.8 V, for removing the source-body voltage stress.

To explore the current balancing feature of the X-connected topology, the output control loop uses a voltage mode compensator with no inductor current information. The feedback resistors R_{FB1} and R_{FB2} sense the output voltage with a feedback coefficient of 0.5. Then, the integrator, composed of an error amplifier, two compensation capacitors C_{C1} and C_{C2}, and one compensation resistor R_C, generates the control voltage V_{CTRL}. Then, the V_{CTRL} is compared to four ramp

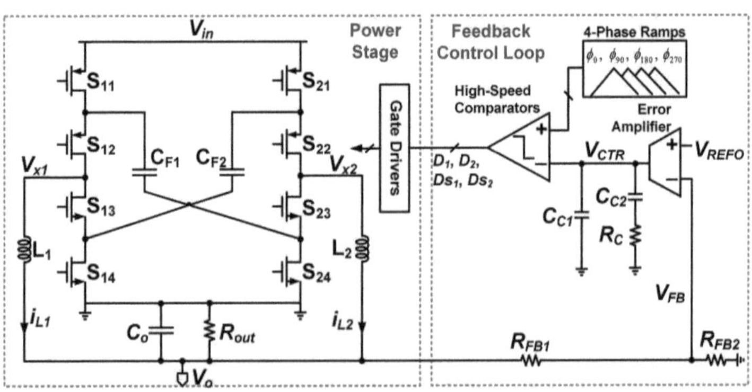

Fig. 9.11 Circuit implementation of the proposed converter

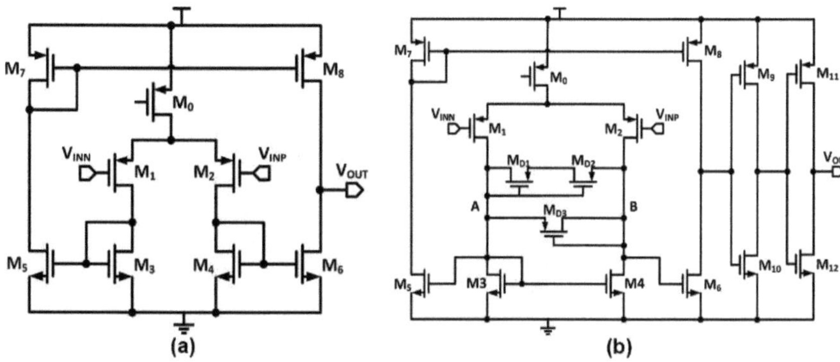

Fig. 9.12 (**a**) Error amplifier, (**b**) high-speed comparator

signals by the high-speed comparators to generate four interleaving PWM signals, which accordingly control the corresponding power switches.

The error amplifier as shown in Fig. 9.12a is obtained through a conventional current mirror operational transconductance amplifier [18, 19]. The error amplifier, by setting the channel lengths of $M_3 - M_8$ and the tail current to 2 μm and 4 μA, respectively, achieves a high gain of close to 50 dB and provides an appropriate output impendence of around 500 kΩ as the first pole. The high-speed comparator as shown in Fig. 9.12b is achieved by a two-stage push-pull comparator [18, 19]. Three body-drain connected transistors M_{D1}, M_{D2} and M_{D3} act as the diodes to limit the voltage range of node B. When $V_{INP} > V_{INN}$, the voltage at node A goes up and the voltage at node B drops. Once the voltage difference between nodes A and B is large enough, the parasitic diode M_{D3} turns on, and hence node B is clamped to a fixed value. Similarly, when $V_{INP} < V_{INN}$, the voltage at node A drops, and the voltage at node B goes up. Once the voltage difference between nodes B and A is larger than the built-in potential of the two series parasitic diodes in M_{D1} and M_{D2}, they turn on and node B is clamped to another fixed value. Hence, the comparison speed is improved by the limitation of the voltage range of node B (Fig. 9.12).

9.3.4 Measurement Results

As exhibited in Fig. 9.13, the 2P3L buck converter with X-C_{FLY} fabricated in a 65 nm CMOS with I/O devices has a chip area close to 2.95 mm^2. The common-centroid layout is employed to eliminate the mismatch between the two phases, and the N-cap and MIM capacitors are stacked to save the silicon area. Furthermore, the output capacitor with a capacitance of 1 nF and an ESR of 10 mΩ, and the inductors with an inductance of 500 nH and a DCR of 1.15 mΩ are chosen.

Figure 9.14 shows the output voltage and inductor currents for the output voltage of 1.35 V and 0.4 V, with a switching frequency of 40 MHz. For the output voltage

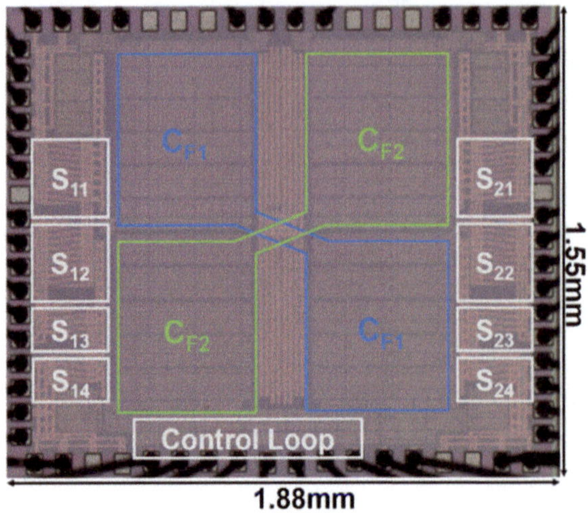

Fig. 9.13 Chip photograph of the 2P3L buck converter with X-C_{FLY}

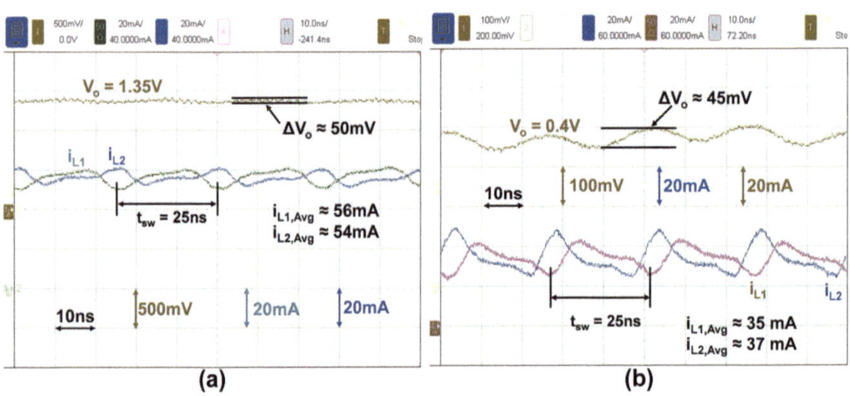

Fig. 9.14 Output voltage and inductor currents for (**a**) 1.35 V, and (**b**) 0.4 V output voltage

of 1.35 V with a load current of 110 mA, the output ripple is close to 50 mV with the inductor currents balanced around 55 mA. For the output voltage of 0.4 V with a load current around 72 mA, the output ripple is close to 45 mV with the inductor currents balanced around 36 mA. To evaluate the balance capacity, the unbalanced current ΔI_L is estimated. For both cases, the unbalanced average currents are less than 5 mA, which might be caused by the mismatches resulting from the process variations of on-chip and PCB power traces, and the bond-wire inductors.

Figure 9.15 displays the measured power conversion efficiency versus the output voltage and load current. The 2P3L buck converter with X-C_{FLY} achieves a peak efficiency of slightly less than 86%. Besides, this converter provides much higher efficiency than an ideal LDO at 0.4 V output voltage when the load current is larger than 35 mA.

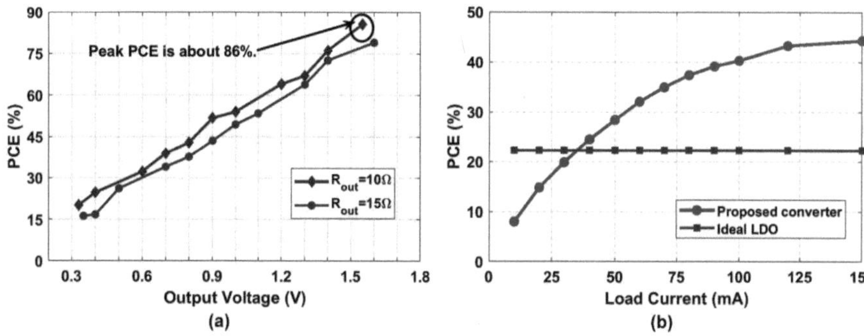

Fig. 9.15 Efficiency (**a**) with the fixed load resistances versus the output voltage, (**b**) versus the load current at $V_O = 0.4$ V

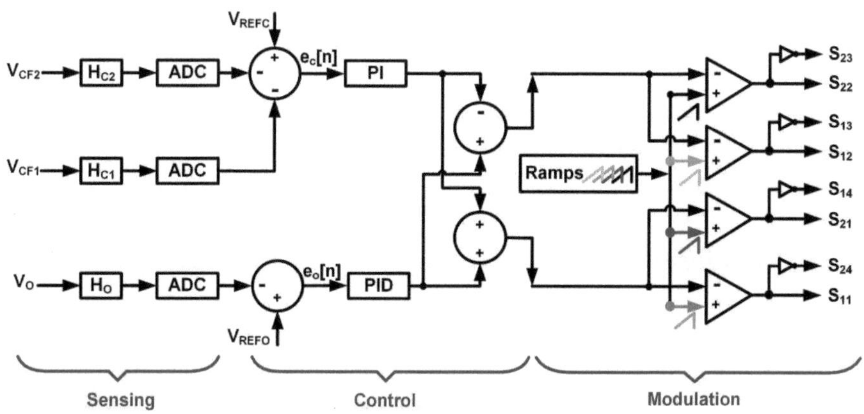

Fig. 9.16 Digital control strategy with flying capacitor voltage regulation

The potential of the 2P3L buck converter with X-C_{FLY} might be not fully gained because the controller of the aforesaid integrated prototype lacks the flying capacitor voltage regulation. An extra inner loop could be inserted at the outer output voltage loop to regulate the flying capacitors. For illustration, another prototype is constructed by using the discrete devices, and a digital control strategy on DSP is designed as shown in Fig. 9.16. The strategy uses the outer loop to regulate the output voltage and employs the inner loop to control the flying capacitor voltages. Because the flying capacitor voltages need to drift for inherent current balancing [35], the strategy controls the relationship between two flying capacitor voltages rather than the individual values. In this prototype, $V_{in} = 5$ V, $V_O = 0.9$ V, $F_{SW} = 100$ KHz, $C_{F1} = C_{F2} = 11$ μF, $C_O = 2$ μF, $R_{out} = 1$ Ω. Figure 9.17 reveals the steady-state waveforms of the discrete prototype with two configurations (a) $L_1 = L_2 = 4.7$ μH, and (b) $L_1 = 4.7$ μH, $L_2 = 10$ μH. In both configurations, the flying capacitor voltages are half input voltage, and the average inductor currents

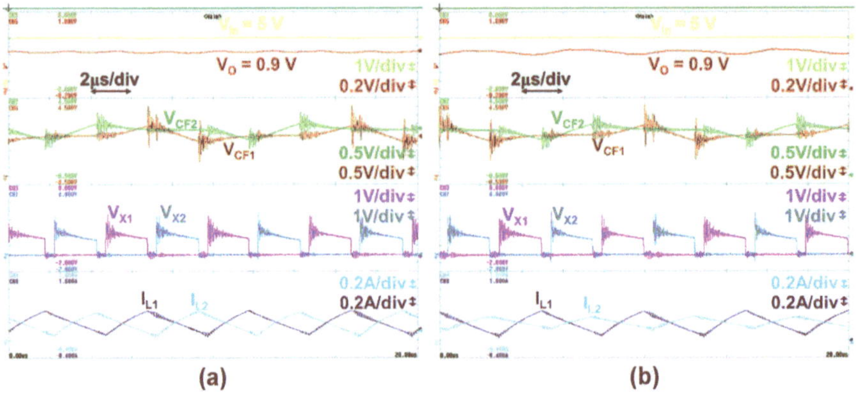

Fig. 9.17 Steady-state waveforms of the discrete prototype with two configurations (**a**) $L_1 = L_2 = 4.7\ \mu H$, and (**b**) $L_1 = 4.7\ \mu H$, $L_2 = 10\ \mu H$

are the same, achieving the current balancing. Further, the inductance mismatch only affects the inductor current ripple but rarely causes an unbalanced current.

9.3.5 Summary

This subchapter presented a 2P3L buck converter with X-C_{FLY} to suppress the unbalanced inductor currents via alternately connecting the flying capacitors with the inductors. After the inductor current balance is deduced in theory, two prototypes are built. Fabricated in a 65 nm CMOS technology, the integrated version achieved a low unbalanced current close to 5 mA and a low output ripple of 50 mV for a load current around 110 mA, and a high peak efficiency of 86%. On the other hand, the discrete version with a digital control almost has no unbalanced current for a load current of 0.9 A even under unequal inductances, and its flying capacitor voltages are equal to half the input voltage.

9.4 Tri-Path Buck Converter

To self-balance flying capacitor voltages and reduce the inductor current, [37] presented a dual-path buck converter with the VCR of $2D/(1 + D)$. But, the VCR is not low enough and the flying capacitor voltage is balanced as high as $V_{IN} - V_{OUT}$. Hence, [38] introduced another dual-path buck converter to balance the flying capacitor voltage at a small value V_{OUT} and to obtain a VCR of $D/(1 + D)$, while [39] showed an always dual-path buck converter to balance two flying capacitor voltages with small values $V_{IN} - 2V_{OUT}$ and V_{OUT}, respectively and to obtain a

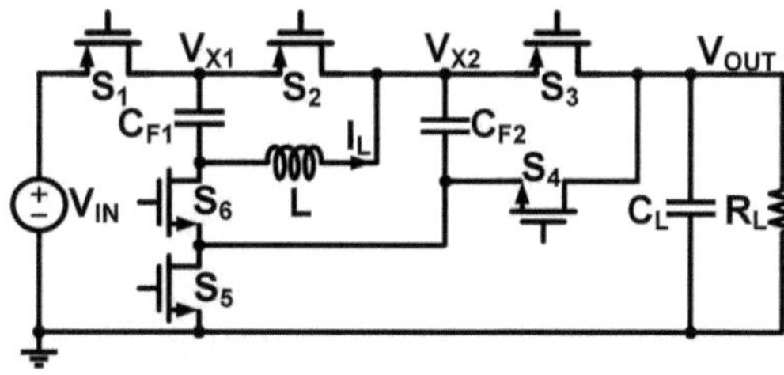

Fig. 9.18 Tri-path buck converter

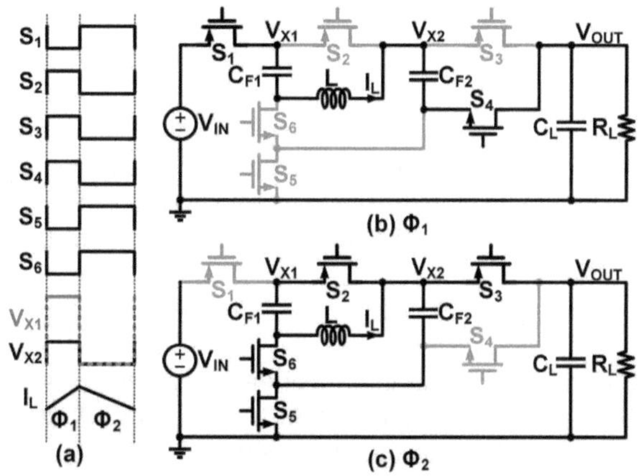

Fig. 9.19 (**a**) Gate signals, and operation states (**b**) Φ_1, (c) Φ_2 of the tri-path buck converter

VCR of 1/(3 − 2*D*). However, [38, 39] are open-loop designs, and they require other auxiliary dc-dc converters or external supplies for powering the gate drivers. Moreover, the VCR of 1/(3 − 2*D*) cannot fill the conversion requirement of from the battery voltage to the sub-1 V, while the VCRs of *D*/(1 + *D*), or 2*D*/(1 + *D*) are not low enough, and thus may require a small duty cycle, which limits the switching frequency going up with a smaller inductor volume. Thereby, this subchapter presents a highly-integrated tri-path buck converter with reduced inductor current, self-balanced flying capacitor voltage, and a lower VCR of *D*/(1 + 2*D*). Figure 9.18 reveals the tri-path buck converter composed of power switches $S_1 - S_6$, two flying capacitors C_{F1} and C_{F2}, one power inductor L, and one load capacitor C_{OUT}. The cascaded switches S_2 and S_3, S_5 and S_6 share the voltage stresses on the switching nodes, and hence tolerate higher voltage stresses.

9.4.1 Operation Principle

Figure 9.19 illustrates the two operation states ($\Phi_1 \rightarrow \Phi_2$) per cycle of the tri-path buck converter. In Φ_1, the flying capacitors, inductor, and output capacitor are connected in series V_{IN} and the ground. In Φ_2, the flying capacitors C_{F1} and C_{F2}, inductor L, and output capacitor C_{OUT} are in parallel.

By applying the state-space average method [34], i.e. the voltage-second balance law to the inductor implies, there is

$$D\left(V_{IN} - V_{CF1} - V_{CF2} - V_{OUT}\right) + (1 - D)\left(-V_{OUT}\right) = 0. \tag{9.21}$$

where D is the duty cycle. Moreover, the flying capacitors C_{F1} and C_{F2} re-distribute the charge with the output capacitor C_{OUT} during Φ_2, so the average flying capacitor voltages V_{CF1} and V_{CF2} are,

$$V_{CF1} = V_{CF2} = V_{OUT}. \tag{9.22}$$

V_{CF1} and V_{CF2} are ideally equal to the output voltage V_{OUT}. In other words, the tri-path buck converter automatically balances the flying capacitor voltages by using the charge-redistribution characteristics of the paralleled capacitors. Next, by combining Eqs. (9.21) and (9.22), the VCR of the tri-path buck converter is,

$$M = \frac{V_{OUT}}{V_{IN}} = \frac{D}{1+2D}. \tag{9.23}$$

Equation (9.23) illustrates that the tri-path buck converter can deliver any VCR of $D/(1 + 2D)$, ranging from 0 to 1/3. If a higher VCR is necessary, the tri-path buck can also operate as one dual-path buck [37, 38] with the VCR of $D/(1 + D)$ by keeping S_3 and S_5 always on and keeping S_4 always off.

9.4.2 Inductor Current Analysis

By applying the charge balance law to the two flying capacitors and the output capacitor in one periodic steady state, there is

$$\begin{aligned} I_{CF1} &= D\cdot I_{CF1,1} + (1 - D)\, I_{CF1,2} = 0 \\ I_{CF2} &= D\cdot I_{CF2,1} + (1 - D)\, I_{CF2,2} = 0 \\ I_{COUT} &= D\cdot I_{COUT,1} + (1 - D)\, I_{COUT,2} = 0 \end{aligned} \tag{9.24}$$

where I_{COUT}, $I_{COUT,1}$, and $I_{COUT,2}$ are the average currents of C_{OUT} in one cycle, Φ_1, and Φ_2, respectively; I_{CF1}, $I_{CF1,1}$ and $I_{CF1,2}$ are the average currents of C_{F1} in one cycle, Φ_1, and Φ_2, respectively; I_{CF2}, $I_{CF2,1}$, and $I_{CF2,2}$ are the average currents of C_{F2} in one cycle, Φ_1, and Φ_2, respectively. According to KCL, $I_{CF1,1} = I_{CF2,1} = I_L$, $I_{COUT,1} = I_L - I_{OUT}$, $I_{COUT,2} = I_L - I_{CF1,2} - I_{CF2,2} - I_{OUT}$. Then, the average inductor current and the flying capacitor currents in each state are solved as

$$\begin{aligned} I_L &= \frac{1}{1+2D} I_{OUT} = (1 - 2M)\, I_{OUT}, \\ I_{CF1,1} &= I_{CF2,1} = I_L, \\ I_{CF1,2} &= I_{CF2,2} = -\frac{D}{1-D} I_L = -\frac{M}{1-3M} I_L. \end{aligned} \tag{9.25}$$

Further, using the inductor current slope in Φ_1, the tri-path buck converter has an inductor current ripple as

$$\Delta I_L = \frac{V_{IN} - V_{CF1} - V_{CF2} - V_{OUT}}{L} DT_S. \tag{9.26}$$

where T_S is the switching cycle. Combining Eqs. (9.22), (9.23), and (9.26), the inductor ripple is rewritten as,

$$\Delta I_L = \frac{M(1-3M)}{1-2M} \frac{V_{IN}}{L} T_S. \tag{9.27}$$

Then, by using the same method, the average inductor currents of the multi-level (including the two-level) $I_{L,ML}$, and dual-path buck converters $I_{L,DP}$, as well as the inductor current ripples $\Delta I_{L,2L}$, $\Delta I_{L,NL}$, and $\Delta I_{L,DP}$ of the two-level, multi-level, and dual-path buck converters are deduced as

$$\begin{aligned} I_{L,ML} &= I_{OUT}, \\ I_{L,DP} &= \frac{D}{1+D} I_{OUT} = (1-M)\, I_{OUT}. \end{aligned} \tag{9.28}$$

$$\begin{aligned} \Delta I_{L,2L} &= M\,(1-M)\, \frac{V_{IN}}{L} T_S, \\ \Delta I_{L,NL} &= \left(\frac{fix[(N-1)M]}{N-1} - M\right)\left(M - \frac{fix[(N-1)M]}{N-1}\right) \frac{V_{IN}}{L} T_S, \\ \Delta I_{L,DP} &= \frac{M(1-2M)}{1-M} \frac{V_{IN}}{L} T_S. \end{aligned} \tag{9.29}$$

respectively, where *fix*(x) is the maximum integer less than x and N is the level number.

Figure 9.20 compares the average inductor current, the inductor current ripple, and root-mean-square (RMS) inductor current of the tri-path buck converter with those of the multi-level and dual-path buck converters at $V_{IN} = 3.3$ V, $T_S = 0.2$ μs, L = 1 μH, $I_{OUT} = 200$ mA. The tri-path buck converter significantly decreases the average inductor currents in comparison with the multi-level and dual-path converters. Taking a VCR of 1/4 as an example, the tri-path buck converter reduces the average inductor current by 50% or 25% when compared with the multi-level and dual-path converters, respectively. Also, the tri-path buck converter obtains smaller inductor current ripples when compared with the two-level and dual-path buck converters. Despite the tri-path buck converter having a relatively larger inductor current ripple than the multi-level buck converter, it has a much smaller RMS inductor current, i.e. smaller conduction loss on the inductor. With a VCR of 1/4 for example, the tri-path buck converter decreases the RMS inductor current by close to 50%, thus decreasing 75% loss caused by the inductor DCR.

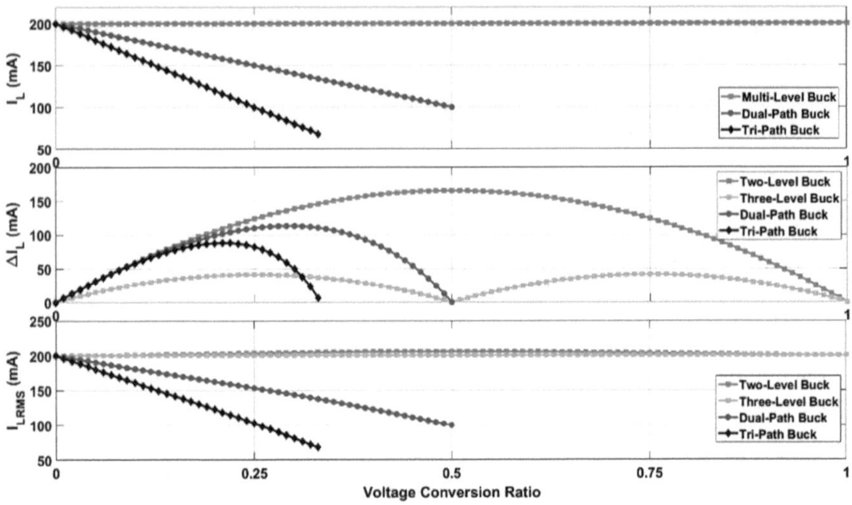

Fig. 9.20 Comparisons of (**a**) average inductor current, (**b**) inductor current ripple, and (**c**) RMS inductor current at $V_{IN} = 3.3$ V, $T_S = 0.2$ μs, L = 1 μH, and $I_{OUT} = 200$ mA

9.4.3 Small-Signal Model

For simplicity, using V_{OUT}/R to replace I_{OUT} and applying KVL and KCL, in Φ_1, there are

$$\begin{aligned} L\frac{dI_L}{dt} &= V_{IN} - V_{CF1} - V_{CF2} - V_{OUT} \\ C_{F1}\frac{dV_{CF1}}{dt} &= C_{F2}\frac{dV_{CF2}}{dt} = I_L \\ C_{OUT}\frac{dV_{OUT}}{dt} &= -\frac{V_{OUT}}{R} + I_L \end{aligned} \tag{9.30}$$

Similarly, in Φ_2, there are

$$\begin{aligned} L\frac{dI_L}{dt} &= -V_{OUT} \\ V_{CF1} &= V_{CF2} = V_{OUT} \\ C_{OUT}\frac{dV_{OUT}}{dt} &= -\frac{V_{OUT}}{R} + I_L - C_{F1}\frac{dV_{CF1}}{dt} - C_{F2}\frac{dV_{CF2}}{dt} \end{aligned} \tag{9.31}$$

Then, by ignoring the flying capacitor voltage ripple [19], Eqs. (9.30) and (9.31) are simplified as,

$$\begin{aligned} L\frac{dI_L}{dt} &= V_{IN} - 3V_{OUT} \\ C_{OUT}\frac{dV_{OUT}}{dt} &= -\frac{V_{OUT}}{R} + I_L \end{aligned} \tag{9.32}$$

$$\begin{aligned} L\frac{dI_L}{dt} &= -V_{OUT} \\ (C_{OUT} + C_{F1} + C_{F2})\frac{dV_{OUT}}{dt} &= -\frac{V_{OUT}}{R} + I_L \end{aligned} \tag{9.33}$$

respectively. Eqs. (32) and (33) can be rewritten as the state averaging equation, as

$$\frac{d}{dt}\begin{bmatrix} I_L \\ V_{OUT} \end{bmatrix} = \begin{bmatrix} 0 & \frac{-2D-1}{L} \\ \frac{D}{C_{OUT}} + \frac{1-D}{C_{OUT}+C_{F1}+C_{F2}} & -\frac{D}{RC_{OUT}} - \frac{1-D}{R(C_{OUT}+C_{F1}+C_{F2})} \end{bmatrix} \begin{bmatrix} I_L \\ V_{OUT} \end{bmatrix} + \begin{bmatrix} \frac{D}{L} \\ 0 \end{bmatrix} V_{IN}. \tag{9.34}$$

Next, by applying perturbation and linearization to Eq. (9.34), the open-loop control-to-output transfer function G_{vd} is got as

$$G_{vd} = \frac{\hat{v}_{out}}{\hat{d}} = \frac{\frac{V_{IN}}{L}\left(\frac{D}{C_{OUT}} + \frac{1-D}{C_{OUT}+C_{F1}+C_{F2}}\right)}{s^2 + \frac{1}{R}\left(\frac{D}{C_{OUT}} + \frac{1-D}{C_{OUT}+C_{F1}+C_{F2}}\right)s + \frac{1+2D}{L}\left(\frac{D}{C_{OUT}} + \frac{1-D}{C_{OUT}+C_{F1}+C_{F2}}\right)}. \tag{9.35}$$

Similar to the conventional two-level buck converter, the tri-path buck converter also has a pair of complex poles. So, a common Type-III compensation method can be used as well for output regulation.

9.4.4 Circuit Implementation

The tri-path buck converter mainly consists of the power stage including the gate drivers, and the PWM controller.

Figure 9.21 displays the power stage, and the non-overlapping gate drivers for the P-type and N-type power switches. With the voltage stress of the switches S_3, S_4, and S_5 as V_{OUT}, 1.2-V core transistors are utilized in these switches. Because the voltage stress of switches S_1 and S_2 is $V_{IN} - V_{OUT}$, while that of S_6 is $V_{IN} - 2V_{OUT}$, the 2.5-V I/O transistors are employed to implement them. To avoid the external supplies and other auxiliary dc-dc converters, the V_{DD} or V_{SS} rails of these gate drivers are acted by the existing switching terminals of the flying capacitors. In detail, the V_{DD} and V_{SS} of the driver of the switch S_2 come from the positive V_{CF1P} and negative V_{CF1N} terminals of C_{F1}; the drivers of the switches S_3, S_4, and S_6 are supplied by the positive V_{CF2P} and negative V_{CF2N} terminals of C_{F2}; the driver of the switch S_1 is feed by V_{IN} and the positive terminal V_{CF2P} of C_{F2} supply; the V_{DD} for the driver of the switch S_5 is V_{CF2P}. Besides, non-overlapping logic is also designed to avoid voltage overstress during transitions and the shoot-through current losses for reliability and efficiency. For instance, the gate signal V_{G1} for S_1 along with the pulse width signal D generates the gate signal V_{G2} for S_2. For quickly shifting the driver signals, a dynamic voltage high-to-low level shifter (LS) is got by removing the bias transistor of the biased level shifter in Fig. 9.4.

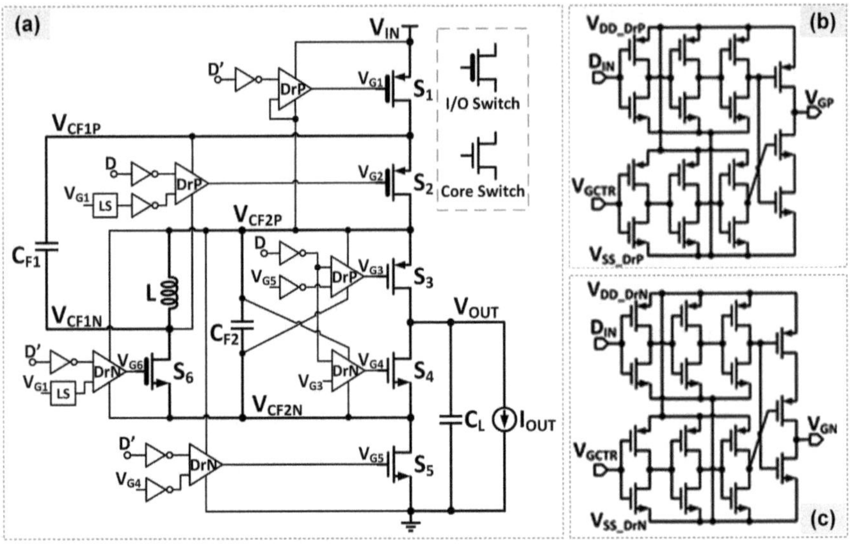

Fig. 9.21 (**a**) Power stage, (**b**) PMOS driver DrP, and (**c**) NMOS driver DrN

Fig. 9.22 PWM controller using Type-III compensator

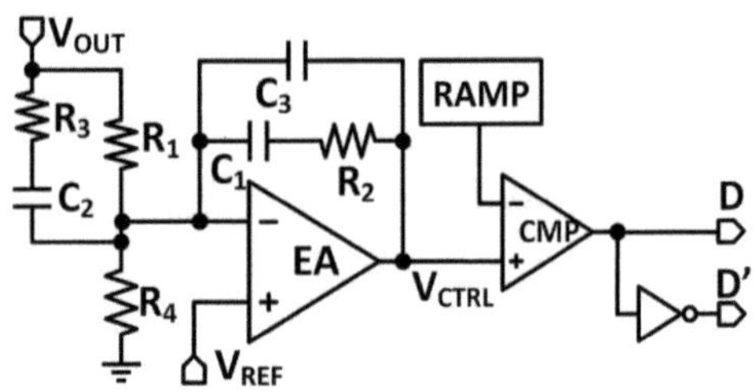

Figure 9.22 shows the PWM controller with a voltage mode Type-III compensator [25, 40]. The feedback resistors R_1, and R_4 sense V_{OUT} with a feedback coefficient of 0.5. Then, the integrator composed of the error amplifier (EA), the compensation resistors, R_2, and R_3, and the compensation capacitors C_1, C_2, and C_3 generate the control voltage V_{CTRL}. Thereby, the transfer function of the Type-III compensator can be obtained as,

$$G_C = \frac{V_{CTRL}}{V_{OUT}} = \frac{\left(1+\frac{s}{z_1}\right)\left(1+\frac{s}{z_2}\right)}{\frac{s}{p_0}\left(1+\frac{s}{p_1}\right)\left(1+\frac{s}{p_2}\right)} \tag{9.36}$$

Two zeros $-z_1 = (2\pi(R_1 + R_3)C_2)^{-1}$ and $- z_2 = (2\pi R_2 C_1)^{-1}$ are located at the frequencies lower than the complex pole pair to boost the phase, and two high-frequency poles $-p_1 = (2\pi R_3 C_2)^{-1}$ and $- p_2 = (2\pi R_2 C_3)^{-1}$ are located near the switching frequency to suppress the switching noise. Then, a high loop-gain bandwidth located between $-z_2$ and $- p_1$ is got.

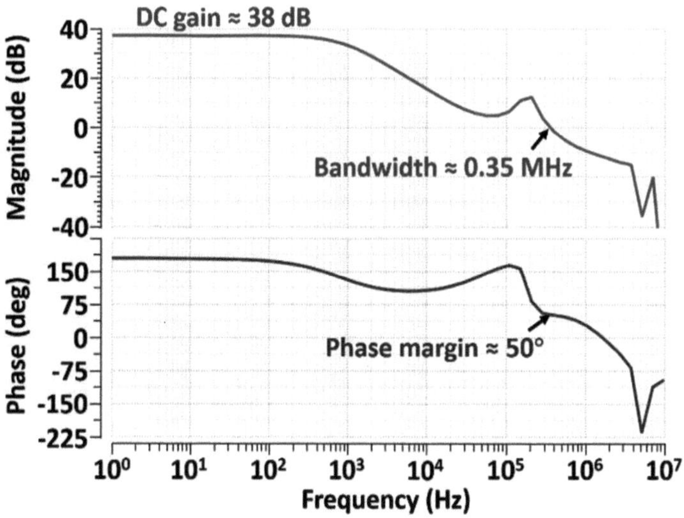

Fig. 9.23 Simulated bode plot of loop gain

Once the control voltage V_{CTRL} is generated by the Type-III compensator, the high-speed comparator CMP compares it with the ramp signal and generates the PWM signal D and D′ to control the power switches. By considering the amplitude of the ramp is V_M, the ramp-relational gain is $G_{RAMP} = 1/V_M$. Thus, the entire loop gain is expressed as

$$G_{LOOP} = G_C \cdot G_{vd} \cdot G_{RAMP} \tag{9.37}$$

Figure 9.23 displays the bode plot of the entire loop gain co-simulated by periodic steady state (PSS) and periodic stability. From Fig. 9.23, the tri-path buck converter is stable, with a phase margin of 50°, a DC gain of 38 dB, and a bandwidth of 0.35 MHz.

9.4.5 Measurement Results

Figure 9.24 reveals the chip photo of the tri-path buck converter. Fabricated in a 65 nm CMOS technology, this converter occupies a silicon area of 0.83 mm^2. For compact volume, a Coilcraft shielded power inductor XPL2010 with an inductance of 1 μH and a DCR of 80 mΩ is employed. Each 470 nF flying capacitance is implemented by one 0402 ceramic capacitor, and the 1-μF output capacitance C_{OUT} is implemented by paralleling two 0805 220 nF ceramic capacitors, one 0805 100 nF ceramic capacitor, and one 0402 470 nF ceramic capacitor.

Fig. 9.24 Chip photo of the tri-path buck converter

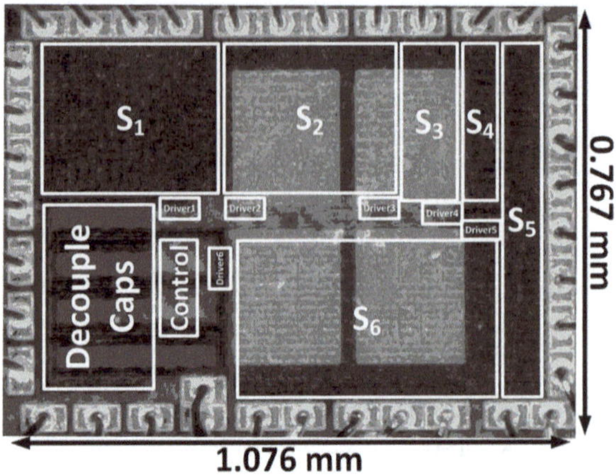

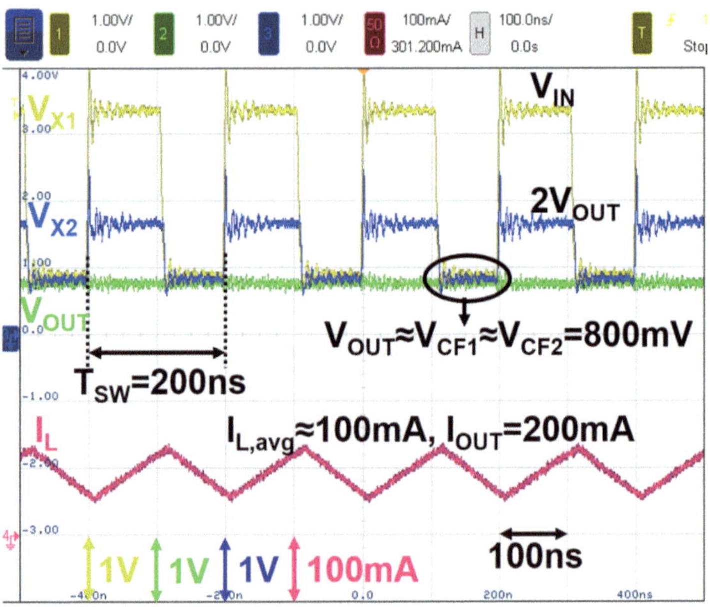

Fig. 9.25 Steady-state waveform with $V_{IN} = 3.3$ V, $V_{OUT} = 0.8$ V, and $I_{OUT} = 200$ mA [20]

Figure 9.25 displays the steady state waveform at the output voltage of 0.8 V and the output current I_{OUT} of 200 mA. For 200 mA I_{OUT}, the tri-path buck converter reduces the average inductor current by 50% to 100 mA. Also, the inductor current ripple is 70 mA, when the a switching frequency is 5 MHz. Besides, the V_{X1} and V_{X2} waveforms illustrate that the two flying capacitor voltages automatically balance as V_{OUT}.

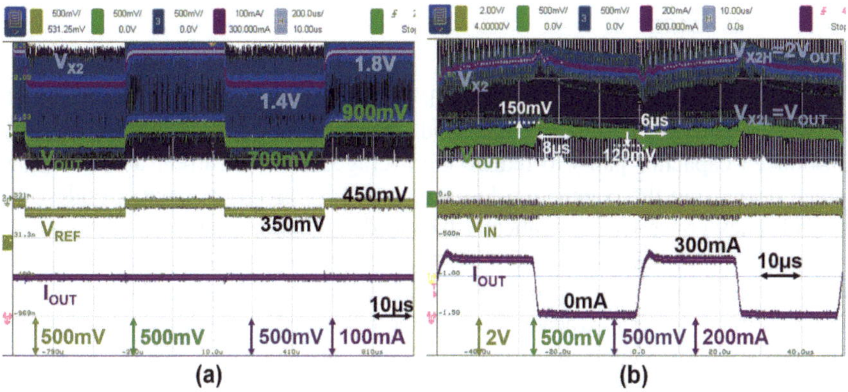

Fig. 9.26 (**a**) Reference tracking waveform, and (**b**) load transient response

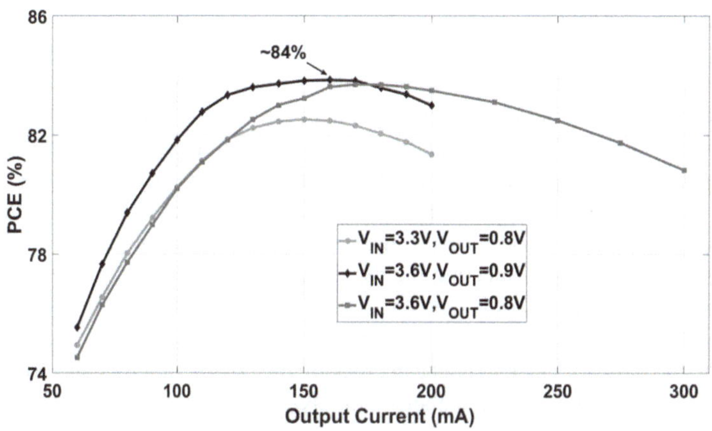

Fig. 9.27 Measured power conversion efficiency

Figure 9.26a reveals the reference tracking performance when the reference voltage steps between 350 mV and 450 mV. The tri-path buck converter takes 6 μs to regulate the output voltage between 700 mV and 900 mV. Figure 9.26b shows the load transient response when the load-current steps between 0 mA and 300 mA. For the up (down) transient, the tri-path buck converter has an undershoot of 120 mV (150 mV) and a recovery time of 6 μs (8 μs).

Figure 9.27 displays the measured power conversion efficiency versus the output current with different input and output voltages. The tri-path buck converter achieves a peak efficiency of 84% at V_{IN} = 3.6 V, V_{OUT} = 0.9 V, and I_{OUT} = 160 mA. Furthermore, the tri-path buck converter provides a maximum output current of 300 mA.

9.4.6 Summary

This subchapter presented a tri-path buck converter with reduced inductor current, self-balanced flying capacitor voltage, and a low VCR of $D/(1 + 2D)$. The converter introduced capacitor paths to reduce the average inductor current and its ripple, hence decreasing the conduction loss and core loss of the inductor. Besides, the tri-path buck converter automatically balanced the flying capacitors to the output voltage in the charge redistribution state. Moreover, it reuses the flying capacitor voltages for supplying the gate drivers, thus eliminating the external supply rails and additional power overheads. Finally, the fabricated prototype validated in 65-nm standard CMOS has a chip area of 0.83 mm^2, and got a peak efficiency of 84%, as well as provided a maximum output current of 300 mA.

9.5 Conclusions

This chapter described three different kinds of hybrid converters. First, we introduced a highly-integrated three-state resonant switched-capacitor converter with parasitic capacitor soft charging and fast flying capacitor pre-charge scheme. This resonant converter significantly improved the efficiency and guaranteed a safe startup. Second, we dwelled on a two-phase three-level buck converter with cross-connected flying capacitors. This converter suppressed the unbalanced inductor currents via alternately connecting the flying capacitors with the inductors. Finally, we presented a tri-path buck converter with reduced inductor current and self-balanced flying capacitor voltage. This tri-path converter introduced capacitor paths to reduce the inductor current, hence decreasing the conduction loss and core loss of the inductor, and automatically balanced the flying capacitors to output voltage.

References

1. N. Tran et al., A complete 256-electrode retinal prosthesis chip. IEEE J. Solid State Circuits **49**(3), 751–765 (2014)
2. K. Chen et al., A system verification platform for high-density epiretinal prostheses. IEEE Trans. Biomed. Circ. Syst. **7**(3), 326–337 (2013)
3. C. Wang, K. Zhao, Z. Li, Q. Xiong, Single-inductor dual-input triple-output buck-boost converter with clockless shortest power path control strategy for IoT nodes. IEEE Trans. Power Electron. **35**(2), 2044–2052 (2020)
4. X. Li, X. Meng, C. Tsui, W. Ki, Reconfigurable resonant regulating rectifier with primary equalization for extended coupling- and loading range in bio-implant wireless power transfer. IEEE Trans. Biomed. Circuits Syst. **9**(6), 875–884 (2015)
5. C. Wang, Z. Li, K. Zhao, Q. Guo, Efficient self-powered convertor with digitally controlled oscillator-based adaptive maximum power point tracking and RF kick-start for ultralow-voltage thermoelectric energy harvesting. IET Circuits Devices Syst. **10**(2), 147–155 (2016)

6. W. Jung et al., Dual-path three-level buck converter with loop-free auto-calibration for flying capacitor self-balancing. IEEE Trans. Power Electron. **36**(1), 51–55 (2020)
7. S.S. Amin, P.P. Mercier, A fully integrated Li-ion-compatible hybrid four-level DC–DC converter in 28-nm FDSOI. IEEE J. Solid State Circuits **54**(3), 720–732 (2019)
8. P. Wang, Y. Huang, T. Kuo, A reconfigurable transient optimizer applied to a four-phase buck converter for optimizing both DVS and load transient responses. IEEE Trans. Circuits Syst. II Express Briefs **67**(1), 52–56 (2020)
9. C. Ting, J. Lin, C.C. Chen, A quasi-V2 hysteretic buck converter with adaptive COT control for fast DVS and load-transient response in RF applications. IEEE Trans. Circuits Syst. II Express Briefs **67**(3), 531–535 (2020)
10. S. Kim et al., A 3nm GAAFET analog assisted digital LDO with high current density for dynamic voltage scaling mobile applications, in *2022 IEEE Symposium on VLSI Technology and Circuits*, (2022), pp. 190–191
11. X. Liu et al., A 0.76V Vin triode region 4A analog LDO with distributed gain enhancement and dynamic load-current tracking in Intel 4 CMOS featuring active feedforward ripple shaping and on-chip power noise analyzer, in *2022 IEEE International Solid-State Circuits Conference (ISSCC)*, (2022), pp. 478–480
12. V.W. Ng, S.R. Sanders, A high-efficiency wide-input-voltage range switched capacitor point-of-load DC-DC converter. IEEE Trans. Power Electron. **28**(9), 4335–4341 (2013)
13. Y. Jiang, M.-K. Law, P.-I. Mak, R.P. Martins, Arithmetic progression switched-capacitor DC-DC converter topology with soft VCR transitions and quasi-symmetric two-phase charge delivery. IEEE J. Solid State Circuits **57**(10), 2919–2933 (2022)
14. J. Jiang, Y. Lu, C. Huang, et al., 20.5 A 2−/3-phase fully integrated switched-capacitor DC-DC converter in bulk CMOS for energy-efficient digital circuits with 14% efficiency improvement, in *2015 IEEE International Solid-State Circuits Conference (ISSCC)*, (2015), pp. 1–3
15. Z. Ye, Y. Lei, R.C. Pilawa-Podgurski, The cascaded resonant converter: a hybrid switched-capacitor topology with high power density and efficiency. IEEE Trans. Power Electron. **35**(5), 4946–4958 (2020)
16. C. Wang, Y. Lu, R.P. Martins, A highly-integrated 3-phase 4:1 resonant switched-capacitor converter with parasitic loss reduction and fast pre-charge startup. IEEE Trans. Circuits Syst. II Express Briefs **68**(7), 2608–2612 (2021)
17. C. Wang, Y. Lu, N. Sun, R.P. Martins, A 3-phase resonant switched capacitor converter for data center 48-V rack power distribution. IEEE Trans. Circuits Syst. I Regul. Pap. **68**(6), 2717–2724 (2021)
18. C. Wang, Y. Lu, M. Huang, R.P. Martins, A two-phase three-level buck DC-DC converter with X-connected flying capacitors for current balancing. IEEE Solid-State Circuits Lett. **3**, 442–445 (2020)
19. C. Wang, Y. Lu, M. Huang, R.P. Martins, A two-phase three-level buck converter with cross-connected flying capacitors for inductor current balancing. IEEE Trans. Power Electron. **36**(12), 13855–13866 (2021)
20. C. Wang, Y. Lu, R.P. Martins, A highly integrated tri-path hybrid buck converter with reduced inductor current and self-balanced flying capacitor voltage. IEEE Trans. Circuits Syst. I Regul Pap **69**(9), 3841–3850 (2022)
21. J. Jiang, W.-H. Ki, Y. Lu, Digital 2−/3-phase switched-capacitor converter with ripple reduction and efficiency improvement. IEEE J. Solid State Circuits **52**(7), 1836–1848 (2017)
22. P.H. McLaughlin, Z. Xia, J.T. Stauth, 11.2 A fully integrated resonant switched-capacitor converter with 85.5% efficiency at 0.47W using on-chip dual-phase merged-LC resonator, in *2020 IEEE International Solid-State Circuits Conference (ISSCC)*, (Feb. 2020), pp. 192–194
23. P. Renz, M. Kaufmann, M. Lueders, B. Wicht, 8.6 A fully integrated 85%-peak-efficiency hybrid multi ratio resonant DC-DC converter with 3.0-to-4.5 V input and 500μA-to-120mA load range, in *2019 IEEE International Solid-State Circuits Conference (ISSCC)*, (2019), pp. 156–158
24. S. Webb, T. Liu, Y.-F. Liu, Zero inductor-voltage multilevel bus converter. IEEE Trans. Power Electron. **36**(10), 11565–11578 (2021)

25. X. Liu, C. Huang, P.K. Mok, A high-frequency three-level buck converter with real-time calibration and wide output range for fast-DVS. IEEE J. Solid State Circuits **53**(2), 582–595 (2018)
26. P. Kumar et al., A 0.4V∼1V 0.2A/mm2 70% efficient 500MHz fully integrated digitally controlled 3-level buck voltage regulator with on-die high density MIM capacitor in 22nm tri-gate CMOS, in *2015 IEEE Custom Integrated Circuits Conference (CICC)*, (2015), pp. 1–4
27. H.D. Gwon et al., 2-phase 3-level ETSM with mismatch-free duty cycles achieving 88.6% peak efficiency for a 20-MHz LTE RF power amplifier. IEEE Trans. Power Electron. **33**(4), 2815–2819 (2018)
28. W. Kim, D. Brooks, G.Y. Wei, A fully-integrated 3-level DC-DC converter for nanosecond-scale DVFS. IEEE J. Solid State Circuits **47**(1), 206–219 (2012)
29. P.Y. Wu, P.K. Mok, A two-phase switching hybrid supply modulator for RF power amplifiers with 9% efficiency improvement. IEEE J. Solid State Circuits **45**(12), 2543–2556 (2010)
30. J. Han, J.H. Song, Phase current-balance control using DC-link current sensor for multiphase converters with discontinuous current mode considered. IEEE Trans. Ind. Electron. **63**(7), 4020–4030 (2016)
31. Y. Cho et al., An active current reconstruction and balancing strategy with DC-link current sensing for a multi-phase coupled-inductor converter. IEEE Trans. Power Electron. **27**(4), 1697–1705 (2012)
32. M. Sun et al., A 6 a, 93% peak efficiency, 4-phase digitally synchronized hysteretic buck converter with ±1.5% frequency and ±3.6% current sharing error. IEEE J. Solid State Circuits **52**(11), 3081–3094 (2017)
33. Y. Su, K.Y.B. Cheng, W. Wu, High-efficiency multiphase DC-DC converters for powering processors with turbo mode based on configurable current sharing ratios and intelligent phase management, in *2017 IEEE Applied Power Electronics Conference and Exposition (APEC)*, (Mar. 2017), pp. 191–196
34. R.W. Erickson, D. Maksimovic, *Fundamentals of power electronics*, 2nd edn. (Springer, New York, 2007)
35. P.S. Shenoy, M. Amaro, J. Morroni, D. Freeman, Comparison of a buck converter and a series capacitor buck converter for high-frequency, high-conversion-ratio voltage regulators. IEEE Trans. Power Electron. **31**(10), 7006–7015 (2016)
36. P.S. Shenoy et al., Automatic current sharing mechanism in the series capacitor buck converter, in *IEEE Energy Conversion Congress and Exposition (ECCE)*, (2015), pp. 2003–2009
37. Y. Huh, S.-W. Hong, G.-H. Cho, A hybrid structure dual-path stepdown converter with 96.2% peak efficiency using 250-mΩ large-DCR inductor. IEEE J. Solid-State Circuits **54**(4), 959–967 (2019)
38. K. Hata, Y. Yamauchi, T. Sai, et al., 48V-to-12V dual-path hybrid DCDC converter, in *2020 IEEE Applied Power Electronics Conference and Exposition (APEC)*, (2020), pp. 2279–2284
39. K. Hata, Y. Jiang, M.K. Law, et al., Always-dual-path hybrid DC-DC converter achieving high efficiency at around 2:1 step-down ratio, in *2021 IEEE Applied Power Electronics Conference and Exposition (APEC)*, (2021), pp. 1302–1307
40. S. Lee, Demystifying type II and type III compensators using op-amp and OTA for DC-DC converters. Texas Instruments Application Report SLVA662, 2014

MIX
Papier aus verantwortungsvollen Quellen
Paper from responsible sources
FSC® C105338

If you have any concerns about our products,
you can contact us on
ProductSafety@springernature.com

In case Publisher is established outside the EU,
the EU authorized representative is:
Springer Nature Customer Service Center GmbH
Europaplatz 3, 69115 Heidelberg, Germany

Printed by Libri Plureos GmbH
in Hamburg, Germany